Getting a Dial Tone

The **Institute of Southeast Asian Studies (ISEAS)** was established as an autonomous organization in 1968. It is a regional centre dedicated to the study of socio-political, security, and economic trends and developments in Southeast Asia and its wider geostrategic and economic environment.

The Institute's research programmes are the Regional Economic Studies (RES, including ASEAN and APEC), Regional Strategic and Political Studies (RSPS), and Regional Social and Cultural Studies (RSCS).

ISEAS Publishing, an established academic press, has issued almost 2,000 books and journals. It is the largest scholarly publisher of research about Southeast Asia from within the region. ISEAS Publishing works with many other academic and trade publishers and distributors to disseminate important research and analyses from and about Southeast Asia to the rest of the world.

Getting a Dial Tone

Telecommunications Liberalisation in Malaysia and the Philippines

LORRAINE CARLOS SALAZAR

INSTITUTE OF SOUTHEAST ASIAN STUDIES
Singapore

First published in Singapore in 2007 by
Institute of Southeast Asian Studies
30 Heng Mui Keng Terrace
Pasir Panjang
Singapore 119614

E-mail: publish@iseas.edu.sg
Website: http://bookshop.iseas.edu.sg

ISEAS Library Cataloguing-in-Publication Data

Salazar, Lorraine Carlos.
Getting a dial tone : telecommunications liberalisation in Malaysia and the Philippines
1. Telecommunication—Government policy—Malaysia.
2. Telecommunication—Government policy—Philippines.
3. Privatisation—Malaysia.
4. Deregulation—Philippines.
I. Title.
HE8390.6 Z5S26 2007

ISBN 978-981-230-381-3 (soft cover)
ISBN 978-981-230-382-0 (hard cover)
ISBN 978-981-230-566-4 (PDF)

Cover Photo: Condo Diwa, an ex-Moro National Liberation Front combatant turned seaweed farmer, uses his mobile phone to obtain and compare market prices for seaweeds. Wider coverage and more affordable services are some of the benefits of telecommunication liberalisation. Photo courtesy of USAID's Growth with Equity in Mindanao (GEM) Project.

Typeset by International Typesetters Pte Ltd
Printed in Singapore by Mainland Press Pte Ltd

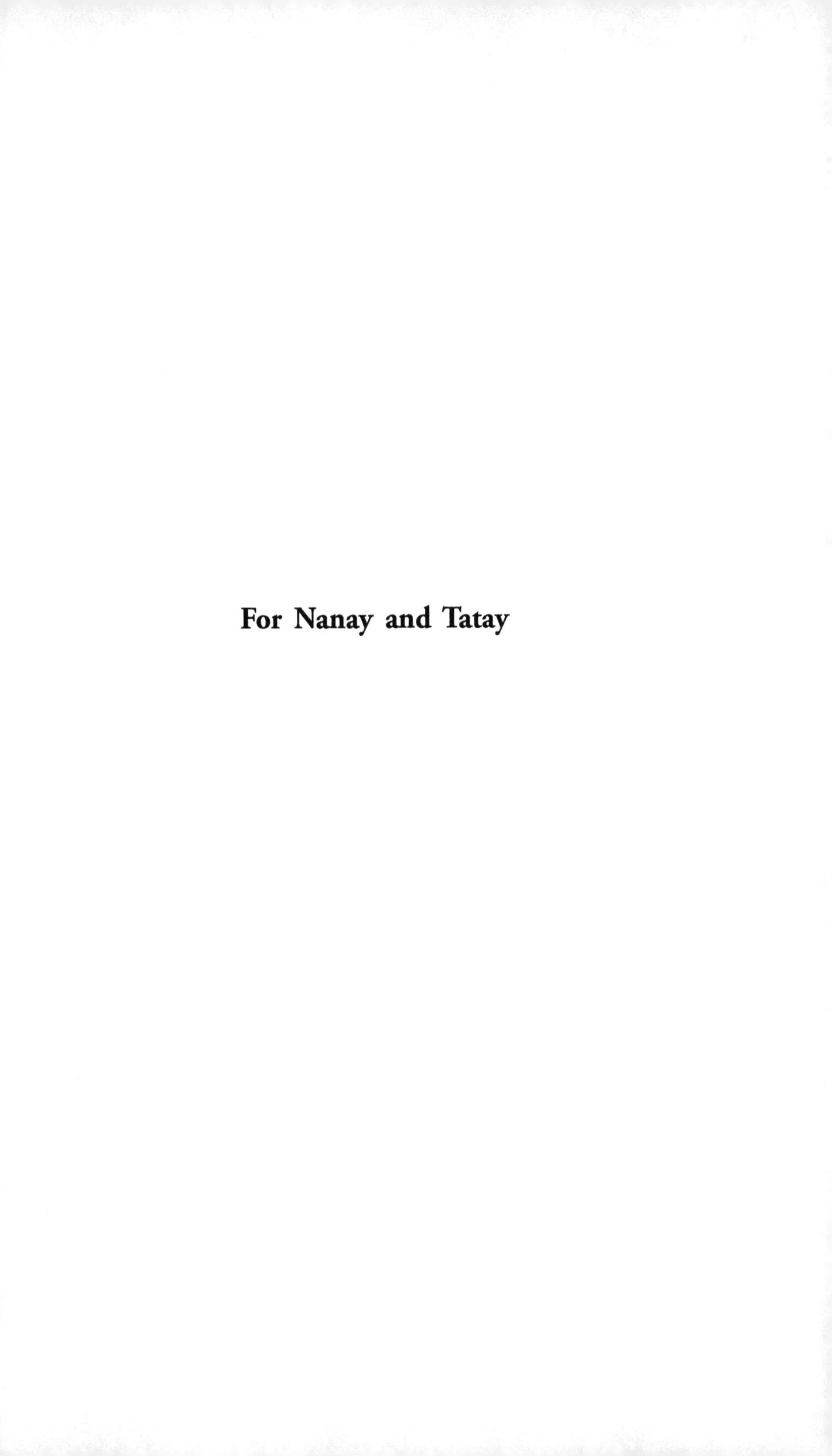

For Nanay and Tatay

Contents

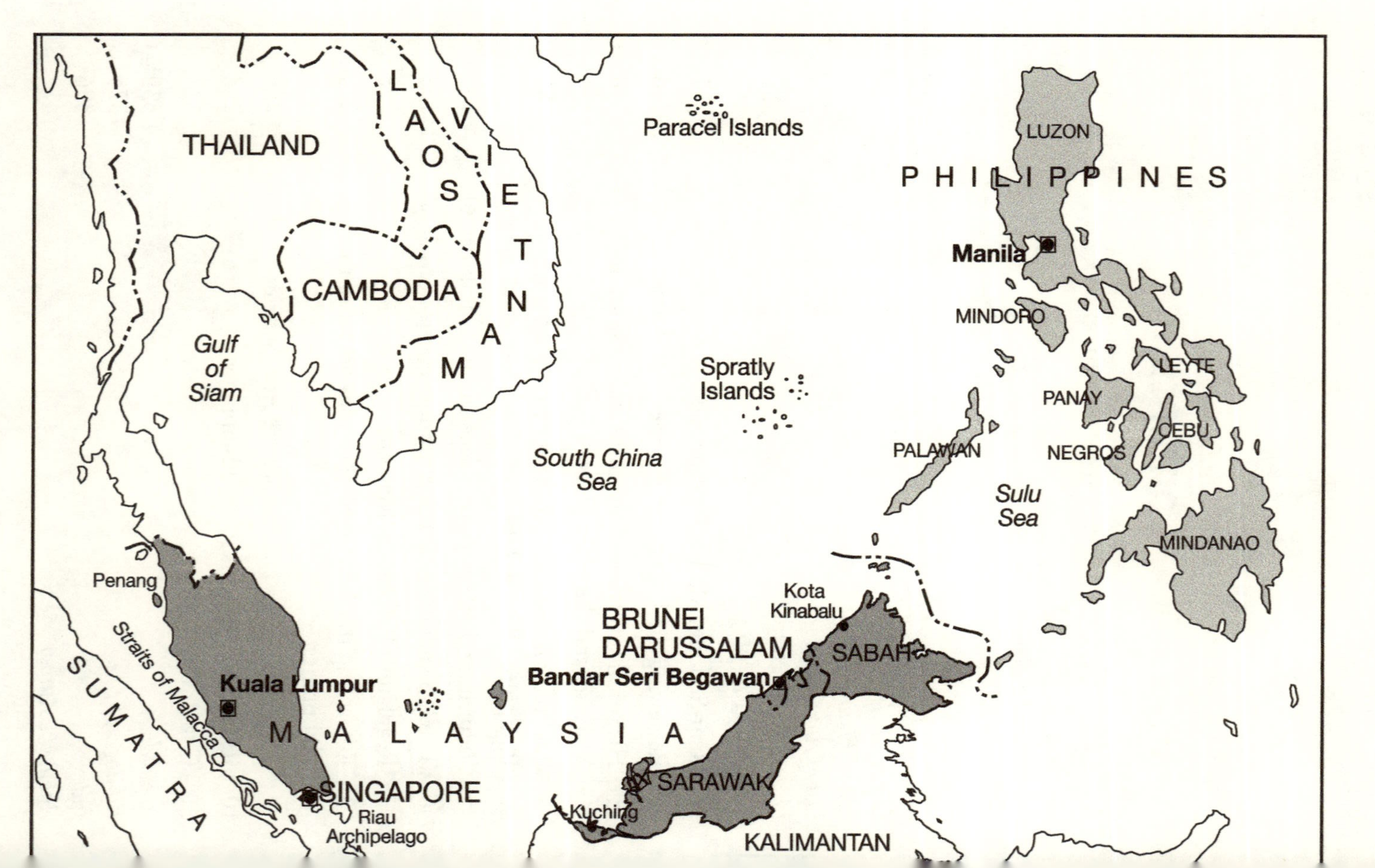
THAILAND
LAOS
VIETNAM
CAMBODIA
Gulf of Siam
Paracel Islands
Spratly Islands
South China Sea
PHILIPPINES
LUZON
Manila
MINDORO
PALAWAN
PANAY
NEGROS
CEBU
LEYTE
MINDANAO
Sulu Sea
Penang
Straits of Malacca
SUMATRA
Kuala Lumpur
MALAYSIA
SINGAPORE
Riau Archipelago
BRUNEI DARUSSALAM
Bandar Seri Begawan
Kota Kinabalu
SABAH
SARAWAK
Kuching
KALIMANTAN

List of Tables

List of Figures

List of Appendices

List of Abbreviations

ACA	Anti-Corruption Agency
ACCIM	Associated Chinese Chambers of Commerce and Industry of Malaysia
ADB	Asian Development Bank
AGILE	Accelerating Growth, Investment, and Liberalization with Equity
AMPS	Advanced Mobile Phone System
BayanTel	Bayan Telecommunications Holding Corporation
BellTel	Bell Telecommunications Corporation
BN	Barisan Nasional
BOT	build-operate-transfer
BTA	Basic Telecommunications Agreement
Butel	Bureau of Telecommunications
CA	Commonwealth Act
CAP	Consumers Association of Penang
CapWire	Capitol Wireless
CDMA	Code Division Multiple Access
Celcom	Cellular Communications Berhad
CEO	chief executive officer
CIC	Capital Issues Committee
CIDA	Canadian International Development Aid
CMA	Communications and Multimedia Act
CMC	Communications and Multimedia Commission
CMTS	Cellular Mobile Telephone System

CPCN	Certificate of Public Convenience and Necessity
CPP	Communist Party of the Philippines
CUEPACS	Congress of Unions of Employees in the Public and Civil Services
DAI	Development Alternatives Incorporated
DAP	Democratic Action Party
DC	Department Circular
Digitel	Digital Telecommunications Philippines Inc.
DNU	Department of National Unity
DOMSAT	Domestic Satellite Company
DOTC	Department of Transportation and Communications
EO	Executive Order
EOI	export-oriented industrialisation
EON	Edaran Otomobil Nasional
EPF	Employees' Provident Fund
EPU	Economic Planning Unit
ETACS	Extended Total Access Communication System
ETPI	Eastern Telecommunications Philippines Incorporated
Extelcom	Express Telecommunications, Incorporated
FELDA	Federal Land Development Authority
FIC	Foreign Investment Committee
GATS	General Agreement on Trade in Services
GFIA	General Framework for Interconnection and Access
Globe	Globe Telecom (formerly Globe Mackay Cable and Radio, Incorporated)
GR	General Register
GSM	Global System for Mobile Communications
GTE	General Telephone and Electronic Corporation
HICOM	Heavy Industries Corporation of Malaysia
IBRD	International Bank for Reconstruction and Development
ICA	Industrial Coordination Act
ICC	International Communication Corporation
ICT	Information and Communications Technology

ICU	Implementation and Coordination Unit
IDD	international direct dialling
IFC	International Finance Corporation
IGF	International Gateway Facility
IMF	International Monetary Fund
ISI	import substitution industrialisation
Islacom	Isla Communications Co. Inc.
IT	information technology
ITU	International Telecommunications Union
IWC	International Wireless Corporation
JICA	Japan International Cooperation Agency
JKH	Jadual Kadar Harga (Schedule of Standards and Rates)
JTM	Jabatan Telekom Malaysia (Telecommunications Department)
KLSE	Kuala Lumpur Stock Exchange
KWAP	Kumpulan Wang Amanah Pencen
Lakas ng Bansa –NUCD–UMCD	Strength of the Nation–National Union of Christian Democrats–United Muslim Christian Democrats
LDP	Laban ng Demokratikong Pilipino
LRIC	long-run incremental cost
LTAT	Lembaga Tabung Angkatan Tentera
MARA	Majlis Amanah Rakyat
MAS	Malaysian Airlines
MCA	Malaysian Chinese Association
MIC	Malaysian Indian Congress
METP	Ministry of Energy, Telecommunications, and Posts
MECM	Ministry of Energy, Communications, and Multimedia
MNLF	Moro National Liberation Front
MOF	Ministry of Finance
MORE Phones	Movement for Reliable and Efficient Phone System
MPH	Multi-Purpose Holdings
MRCB	Malaysian Resources Corporation
MSC	Multimedia SuperCorridor
MTPDP	Medium-Term Philippine Development Plan

MTSU	Malaysian Technical Services Union
NCC	National Consultative Council
NCR	National Capital Region
NDP	New Development Plan
NEB	National Electricity Board
NEDA	National Economic and Development Authority
NEP	New Economic Policy
NGO	non-governmental organization
NSC	National Security Council
NTC	National Telecommunications Commission
NTDP	National Telecommunications Development Plan
NTP	National Telecommunications Policy
NTT	Nippon Telephon and Telegraph Company
NUTE	National Union of Telecommunications Employees
OPP	Outline Prospective Plan
OSA	Official Secrets Act
PAP	Privatisation Action Plan
Paptelco	Philippine Association of Private Telephone Companies Incorporated
PAS	Parti Islam SeMalaysia
PCGG	Presidential Commission on Good Government
PCIJ	Philippine Center for Investigative Journalism
PCN	Personal Communications Network
PCS	Personal Communications System
Pernas	Perbadanan Nasional
PHI	Prime Holdings Incorporated
PHILCOM	Philippine Global Communications Company
Philcomsat	Philippine Communications Satellite Corporation
Piltel	Pilipino Telephone Company
PLDT	Philippine Long Distance Telephone Company
PLUS	Projek Lebuhraya Utara Selatan
PMP	Privatisation Masterplan
PNB	Permodalan Nasional Berhad (National Equity Corporation)
PSC	Public Services Commission
PSDN	Public Data Switched Network
PSTN	public switched telephone network

PTIC	Philippine Telecommunications Investment Corporation
PT&T	Philippine Telegraph & Telephone Corp.
RA	Republic Act
RAM	Rebolusyonaryong Alyansang Makabayan
Retelco	Republic Telephone Company
RIDA	Rural and Industrial Development Authority
RTDP	Regional Telephone Development Project
SAS	Service Area Scheme
SBC	Southwestern Bell Company
SC	Supreme Court
SEC	Securities and Exchange Commission (Philippines)
SEDC	State Economic and Development Corporation
SingTel	Singapore Telecommunications
SIP	Subscribers Investment Plan
Smart	Smart Telecommunications Company
SMS	short messaging service
SOE	state-owned enterprises
SSS	social security system
STM	Syarikat Telekom Malaysia
STW	Syarikat Telefon Wireless
Telekom/TM	Telekom Malaysia
TelicPhil	Telecommunications Infrastructure Philippines
Telof	Telecommunications Office
TRD	Telecommunications Regulatory Determination
TRI	Technology Resources Industries
UDA	Urban Development Authority
UEM	United Engineers Malaysia
UMNO	United Malays National Organisation
UMW	United Motor Works
USAID	United States Aid for International Development
USO	Universal Service Obligation
VANS	value-added network services
VAS	value-added services
WB	World Bank
WLL	wireless local loop
WTO	World Trade Organization

Acknowledgements

Writing this book has been a tremendously exciting and rewarding journey, which would not have been possible without the help, generosity, and kindness of so many people. Thus, I would like to acknowledge their contribution, though all errors and shortcomings are my responsibility alone.

First of all, the book was based on my Ph.D. dissertation at the Australian National University. I benefited greatly from having Ben Kerkvliet as a mentor. His capacity for critical, open-minded, and thorough thinking has been exemplary, as has his kindness and friendship. Harold Crouch, Hal Hill, Natasha Hamilton, Virginia Hooker, Terence Gomez, and Boying Lallana provided me with helpful advice, engaging comments, and encouragement from the start of my research up to the end. Before going to ANU, Dodong Nemenzo challenged me to pursue a Ph.D. rather than going to law school. His generosity and example of melding activism and scholarship have been a source of great inspiration for me.

Second, the Institute of Southeast Asian Studies provided me the space and opportunity to rewrite the dissertation into a book. For this, I am grateful to ISEAS Director Ambassador K. Kesavapany and Regional Economic Studies Programme Convenor Denis Hew. Also, I would like to thank the excellent and supportive staff of ISEAS Publications Unit led by Mrs Triena Ong for the completion of this book.

Third, an earlier version of Chapter 6 was published as "Privatisation, Patronage and Enterprise Development: Liberalising

Telecommunications in Malaysia", in *The State of Malaysia: Ethnicity, Equity and Reform*, edited by Edmund Terence Gomez (London: RoutledgeCurzon, 2004), pp. 194–228. I thank RoutledgeCurzon for their permission to reprint the chapter. I also would like to thank the United States Agency for International Development's (USAID) Growth with Equity in Mindanao (GEM) Project for giving me permission to use the photo of Condo Diwa, an ex-Moro National Liberation Front fighter turned seaweed farmer, for the book cover.

During fieldwork in Malaysia and the Philippines, the help and cooperation of a lot of people facilitated data gathering for this research. I am very grateful to all my interviewees who gave me time and agreed to share information, although some of them have requested anonymity. Without them, this book would not have been written. In Malaysia, I would like to specially thank: Leo Moggie, Nuraizah Abdul Hamid, Hod Parman, Munir Majid, Mohamad Said Mohamad Wira, Zamani Zakariah, Jaafar Ismail, and Vijay Kumar. In the Philippines, I would like to express my heartfelt gratitude to former President Fidel Ramos, Jose Almonte, Antonio Carpio, Anthony Abad, Simeon Kintanar, Josefina Lichauco, Helen Mendoza, Edgardo Cabarios, Philip Varilla, David Fernando, Antonio Samson, Rudy Salalima, Rodolfo Salazar, Mario Lamberte, Monette Serafica, and Jaime Faustino.

I am also thankful to the late Noordin Sopiee and the Institute of Strategic and International Studies-Malaysia (ISIS-Malaysia) and the late Ishak Shaari of IKMAS for giving me an institutional home and a research base while I was in Malaysia in 2000 and 2001. I benefited greatly from discussions with Terence Gomez, Khoo Khay Jin, Jomo K.S., Cassey Lee, Sumit Mandal, James Ongkili, Bridget Welsh, Saliha Hassan, Wan Faisal Wan Hamzah, Mohamad Jawhar, Zainal Aznam, Stephen Leong, Francis Loh, James Jesudason, Jennifer Jacobs, and B.K. Sidhu. Special thanks go to Lee Kam Hing for facilitating my use of the *Star* newspaper library. I am also grateful to Mr Kana of ISIS for organising my lodging, for Usha Rani and Shaleen for providing me a home in Kuala Lumpur, and for Pastor Vic Lumanglas and the Filipino congregation in Kuala Lumpur for their kindness and support. For their friendship and introduction to things Malaysian and otherwise, I thank Rehman Rashid, Angela Yap, Lee Hwok Aun,

Jacqui Tan, Denis Hew, Ahmad Shabery Chik, Neil Khor, Leong Choon Cheng, Patrick Pillai, Malik Hakim, and Noorul Ainur. In the Philippines, thanks are due to teachers and friends: Alex Magno, Boying Lallana, Temy Rivera, Ledi Carino, Luis Dery, Tesa Tadem, Patricia Pascual, Rudy and Kimi Quimbo, Shelah Lardizabal, Rhani Andam, Jaynee Saure, Zorah Andam, Chris Monterola, Cynthia Pagba, Mark Pierre Dimamay, Daphne Oh, Vanessa Velasco, Rey de Luna, Mark Uy, Third Fermin, Deo David, Toto Bacolcol, Germaine Santos, Emmylou Hallig, Chona Reyes, and Leeboy De Velez.

For their valuable documentary collections and accommodating staff, I would like to thank the libraries of the following institutions in Malaysia: Institute for Strategic and International Studies, University of Malaya, Universiti Kebangsaan Malaysia, Universiti Sains Malaysia, Ministry of Energy, Communications and Multimedia, Communications and Multimedia Commission, Securities Commission, Star Publications, New Straits Times, Telekom Malaysia, Maxis, and Celcom. In addition, the Singapore-Malaysia Collection at the National University of Singapore and the Institute for Southeast Asian Studies (ISEAS) library in Singapore proved to hold important materials on Malaysia and the Philippines. In the Philippines, I am grateful for access to the documentary collections of the following institutions: University of the Philippines Main Library, UP College of Law Library, National Centre for Public Administration and Governance Library, UP School of Economics Library, Philippine Centre for Investigative Journalism, Philippine Institute for Development Studies, Congressional and Senate Libraries, National Telecommunications Commission, Department of Transportation and Communications, Telecommunications Office, National Economic and Development Administration (NEDA), Securities and Exchange Commission, Congressional Planning and Budget Office, Tariff Commission, Accelerating Growth, Investment and Liberalisation with Equity (AGILE) office, Foundation for Economic Freedom, Philippine Long Distance and Telephone Company, Globe Communications, Bayantel, Smart Telecommunications, Philcom, RCPI, Philcomsat, and the Makati Business Club.

For financial and institutional support for the fieldwork research, I would like to thank the Department of Political Science and the

University of the Philippines for giving me study leave to go to ANU; the ANU and the Australian Department of Education, Training and Youth Affairs for granting me tuition and stipend scholarships, and the Faculty of Asian Studies for fieldwork funding.

Friends that I met in ANU and some who have moved on to other places since have made life fun, pleasant, and meaningful. For their friendship, animated conversations, and inspiration, I thank Jackie Siapno, Fernando de Araujo, Kaori Maekawa, Deb Johnson, Chalinee Hirano, Tomomi Ito, Julius Bautista, Kit Collier, Malcolm Cook, Mike Poole, Liew Chin-Tong, Nathan Quimpo, Neilson Mersat, Rommel Curaming, Emma Pennel, Tristan Stephens, Gevy Vega and Rob Chambers, Edgi Vega, Amy Chan, Yae Sano, and Kyoung-Hee Moon.

For being my parents in Canberra, I am truly grateful to Melinda and Ben Kerkvliet. I also would like to thank the active and engaging members of the Philippines Study Group and the Malaysian Study Group to whom I presented aspects of this study, and in return received constructive comments and delicious dinners! I also would like to thank my grandparents, my Lolo Enchong and Lola Ciony who, even if they have both recently passed away, have been a constant source of love, support, and humour. My Aunties Susan, Nita, Odette, Anna, and Carmen, Uncles Manuel, Manding, Boy, and Pat, Ate Jo and my cousins deserve to be thanked for giving me enjoyable times off work. In the Philippines, my siblings Cynthia, Karen, and Alvin have also been loving, supportive, and productive — and thanks to them, I now have three nephews and a niece!

In writing this book, my husband, Francis Hutchinson's help, diversions, love, encouragement, and support were crucial. Thanks to him, and the words of wisdom from my mother-in-law, Bea Cervantes Hutchinson, this book has been successfully completed.

My parents, Elvin and Perlita Salazar in Villasis, Pangasinan, deserve tons of appreciation and gratitude. They have continually amazed me with how, even as they live in and depend on a small rural economy, they were able to send all their four children to school. Their hard work, love, support, and example gave me motivation and strength to accomplish this task. It is to them that I dedicate this book.

Finally, for His goodness, sovereignty, and enduring love, I recognise and thank God. May all glory and honour be His alone.

1

RENT SEEKING, MARKET REFORMS, AND THE STATE

The puzzle of why the Philippines has failed to develop economically has long been of interest to scholars studying the country and the region. As Paul Hutchcroft has put it, why is the Philippines in a "developmental bog?"[1] The Philippines has widely been considered the exception to the rule or a laggard amidst the so-called economic successes of Southeast Asian newly industrialising countries (NICs). Some have attributed the cause of this problem to a weak state captured by a strong oligarchic class, while others have emphasised the historical impact of neo-colonialism and the Philippines' continued dependence on the United States. Still others emphasise the Philippines' poor macroeconomic policies and the country's inward-looking, protected economy.

Against the Philippines — an incessant aspirant to development — stands the contrasting example of Malaysia, a seemingly successful case of rapid economic development. Compared with the weak Philippines state, the Malaysian state is considered strong. This is because it has been consistently able to impose a development policy that seeks to eradicate poverty irrespective of race, and to restructure society by removing the identification of race with economic function.[2] Also, Malaysia has kept an open economy and a tight rein over its macroeconomic policies. While the post-colonial political and social systems in both countries are substantially different, patronage and rent-seeking stand out as clear similarities in Malaysia and the Philippines. Yet, the persistence of patronage and rent-seeking has apparently not prevented growth in Malaysia, whereas it is considered a major explanatory variable for the continued economic underdevelopment of the Philippines.

The 1997 financial crisis that hit Southeast Asian countries brought the economic impact of rent-seeking, cronyism, and corruption to the fore. A considerable number of scholars were unanimous in blaming cronyism and rent-seeking as the root causes of the crisis. Before 1997, cronyism was a term largely used by academics and journalists to describe the Philippines and Indonesia. In fact, cronyism was coined to describe the process whereby Ferdinand Marcos distributed largesse to his family members and close friends. The practice existed in the rest of Southeast Asia, including in Malaysia, but it was brushed aside because economies were booming. The 1997 financial crisis was precipitated by the interaction between short-term capital flows and highly leveraged companies, which were owned by politically well-connected people, and made a reassessment of the role of cronyism and rent-seeking in economic development timely. Nevertheless, attributing the crisis solely to domestic factors is myopic and ignores the widespread rent-seeking and corruption during the high growth years in Asian countries. Thus, it is critical to understand why patronage, corruption, cronyism, and rent-seeking have led to underdevelopment in some countries but not in others.[3]

For neoliberals, corruption and rent-seeking are costly, welfare-reducing activities that have to be eliminated for the economy to develop. Because rents are created in direct proportion to the state's involvement in the economy, the goal therefore is to liberalise and create a competitive economy with minimal state involvement. The adoption of market-oriented reforms through privatisation, liberalisation, and deregulation is the core policy advice of the neoliberal perspective encapsulated in the "Washington Consensus".

In the Philippines, where rent-seeking, cronyism, and patronage structures were identified as barriers to economic growth, the government adopted policies to open the market to competition and remove monopoly control over key sectors of the economy after the fall of President Marcos in 1986. Economic policy in the Philippines shifted towards market-oriented reforms in the belief that limited state involvement in the economy would solve the country's lagging development status. In Malaysia, privatisation and liberalisation policies were also adopted in the mid-1980s as a response to an economic slowdown, burgeoning public sector deficits, and inefficient state enterprises.

Ironically, market reforms that aim to remove state intervention create further occasions for rent-seeking because the policy changes are themselves government activities that have potentially large distributional and political consequences.[4] Ha-Joon Chang argues that while the goal of market liberalising reforms is to create a competitive, dynamic economy where the state does not intervene, even competitive markets are bound to create a lot of rents, and people are bound to spend resources to capture them.[5] Hector Schamis contends that the adoption of market liberalisation, in some cases, was successful because of support from rent-seekers or "distributional coalitions" that were aware of the market opportunities that liberalisation made available.[6] Thus, scholars now challenge the widely shared belief among policy makers that developmental success is related to the absence of rent-seeking. They argue that the effects of extensive corruption, clientelism, and other forms of rent-seeking differ across countries. The analytical task is to identify the determinants of these differences and the conditions that give rise to them.[7]

In developing countries, the issue of the state's capacity to introduce economic reform, in general, and market-oriented reforms, in particular, has warranted much scholarly attention. Such capabilities are often seen as lacking, given the state's weakness. Influential interest groups that benefit from the status quo are expected to resist change. The existing literature is generally pessimistic about the ability of weak, penetrated states to institute credible reforms that will break down the degree to which the economy has been captured by vested interests. Hence, it is important to examine the conditions that allow for the adoption and implementation of reform. Statist arguments emphasize the need for a "strong" and "autonomous" state that has the political will to adopt and implement policy changes towards a liberalised market.[8] More recent studies, however, emphasize the role of distributional coalitions that work with the policy makers for reform.[9]

Arguments

Despite very real differences, both Malaysia and the Philippines are similarly permeated by rent-seeking and patronage networks. Yet, one is considered a strong state while the other weak. Given the portrayal of the Philippine and Malaysian states in existing literature, one will expect the Philippine state, due to its weakness, lack of autonomy, and low capacity to be less successful in its attempt to introduce market reforms. On the other hand, the reverse is expected of the Malaysian state with its relative strength, autonomy, and capacity. However, as the subsequent chapters will show, the Philippines was more successful in introducing market reforms in its telecommunications sector than Malaysia. These findings do not neatly sit within conventional expectations and, therefore, beg explanation.

With the matters of rent-seeking, market reforms, and state capacity in mind, this book examines the case of telecommunications reform in Malaysia and the Philippines. The book seeks to answers the following questions. First, how does the strength or weakness of the Malaysian and Philippine states and the presence of influential rent-seeking and patronage networks affect the possibility and sustainability of policy reform? Second, who composes the crucial constituency for market-oriented reform? To what extent is reform affected by the nature of groups that support it? Third, what are the outcomes of market liberalisation? Does market liberalisation invariably remove rents? If not, are the effects of policy change in the face of rent-seeking always deleterious? What sorts of rents were created, who captures them, and how?

In answering these questions, the book offers three interrelated arguments that provide insights into the political economy dynamics of both countries. First, the study challenges conventional depictions of the Malaysian and Philippine states. Despite the weakness of the

Philippine state, reform occurred through a coalition for reform that outmanoeuvred vested interests. In Malaysia, although considered a strong state, patronage and rent-seeking played key roles in policy adoption and implementation.

Second, the nature of groups supporting reform shape policy implementation and its outcomes. In Malaysia, lobbying for liberalisation came from rent-seekers who were aware of potential benefits to be obtained. In the Philippines, by contrast, the coalition favouring liberalisation did not benefit directly from market entry. Third, market reforms do not invariably remove rents. While liberalisation removed monopoly rents in the Philippines, it created new ones in Malaysia. These arguments are summarised in Table 1-1.

This study provides the first major political economy analysis of telecommunications reform in Malaysia and the Philippines. Telecommunications reform was part and parcel of a bigger reform agenda involving the adoption of market-oriented policies in both countries. Telecommunications was chosen as an illustrative case of market reform because it is one of the first sectors to be liberalised. Thus, it is a rich source of empirical material for the questions that this book endeavours to answer. The study explores the reform process by investigating the type of reforms adopted, the reasons behind the policy shift, and the roles of major actors involved in the process, and by identifying the beneficiaries of reform. The outcomes of reform are then assessed in terms of efficiency improvements in the industry, on ownership patterns, and on the economic structure. Finally, the study assesses the new regulatory institutions that were created to monitor the liberalised industry.

While similar reform policies were adopted in each country, the reasons behind their adoption as well as their outcomes differed. The character of the state and business class and the interactions between the two can explain much of these divergences in the reform processes and outcomes. Furthermore, although technological changes, economic crises, and the impact of international actors and pressures are considered, the main focus is on the interactions of domestic actors, which are key to understanding what took place.[10] A political economy approach is utilised which views policy reform not so much

Table 1-1: Puzzles and Arguments

Puzzles	Malaysia	Philippines	Arguments
Type of state and its capacity to adopt reform	The state is strong and capable, and has the capacity to draw up and implement an ambitious economic development and restructuring program. Yet, during privatisation and liberalisation, the state under Mahathir purposively created rents and distributed them to a chosen few, who in turn, actively sought the state-created market opportunities. Patronage and rent-seeking are important factors in policy adoption and implementation despite the presence of a strong, capable state.	The state is weak, has no capacity to make consistent and coherent economic plans, and is overwhelmed by particularistic demands from influential elites. The problem is how to introduce reform in the face of powerful vested interests.	In Malaysia, there was a general policy framework for privatisation, but the way that the telecommunications industry was liberalised shows the active lobbying for market entry by rent-seekers who were politically well-connected. The role of rent-seeking and patronage raises questions about the portrayal of the Malaysian state as insulated. Reform in the Philippines was made possible by the work of a coalition for reform. The occurrence of reform in the face of intense opposition from vested interest raises doubt on the usual portrayal of the Philippine state as weak and incapable.

Role and type of coalition for reform	Those who lobbied for liberalisation were rent-seekers who advocated reform because they were aware of the benefits that they would gain from the change in rules.	The coalition for reform that supported the liberalisation of the telecommunications industry (and other key economic sectors) in the Philippines was composed of academics, professionals, and advisers to then President Ramos.	The rent-seekers in Malaysia cannot be called a "distributional coalition" because they did not lobby as a group. Rather, they sought market licenses, in competition with each other, through existing patronage networks based on racially organized political parties or through personal linkages to individual politicians. In the Philippines, the members of the reform coalition were not the direct beneficiaries of market entry.
Economic consequences of rent-seeking and rent-outcomes	Rents have not been a decisive hindrance to economic development.	Rents have been a major barrier to economic development.	To understand the differences in the economic impacts of rents, the types of rents created, how they are captured, and who are their recipients should be analysed. The types of rent outcomes are shaped by the characteristics and interactions of the state and business class in each country, and most importantly, who supported the state's move toward market reform.

as a function of its intended economic effects, but as the outcome of interactions among politicians, bureaucrats, and interest groups operating within a given institutional context.[11]

This book thus provides evidence that the characteristics and existing power relations between the state and societal actors determined the type of policy reform adopted, the policy beneficiaries, and the type of regulatory environment developed after liberalisation.

With its focus on rent-outcomes, the book also contributes to the post-1997 debate on how money politics, corruption, and rent-seeking can have mixed effects, depending on character of state and supporters of policy adopted. Finally, this book contributes to the literature on the state and its role in economic development. It provides evidence that a more nuanced and focused definition of state strength can be arrived at by examining the state's level of autonomy and capacity in a specific policy area. It argues that reform can be successfully adopted and implemented even if it is not expected of a weak, penetrated state like the Philippines. Meanwhile, a strong state like Malaysia can simultaneously be penetrated and adopt policies precisely because of rent-seeking activities by politically well-connected businessmen and not because policy elites independently crafted them.

The Case Studies

A country's telecommunications infrastructure is a component central to its capacity to prosper in these increasingly globalised times. In an era tagged as the "Information Age", telecommunications link people and facilitate information exchange via telephone, e-mail, or the Internet. As countries move towards economic and social systems where information is an important resource, the development of telecommunications capability becomes all the more crucial. In recognition of this, developing countries are reforming their telecommunications sectors in order to put into place the "missing link"[12] of basic telephony services. In 1997, the United Nations General Assembly actually adopted a resolution stating that access to communication is a basic human right.[13]

The level of modernisation of a country's telecommunications infrastructure is an accepted indicator of economic development. The development of the industry in turn has a multiplier effect on investment, and becomes a source of growth in itself. Its reach, availability, and cost are established measures of the economic status of a nation.[14]

Two major phenomena have wrought sweeping changes in the telecommunications sector over the past two decades, leading to the sector's reform worldwide. These are the rise of neoliberal reform and technological changes. The telecommunications sector was one of the first sectors targeted for neoliberal reform. Roger Noll argues that the economic crises during the 1970s and 1980s led to a re-evaluation of all economic policies, including that of the telecommunications sector.[15] The telecommunications sector and other state-owned infrastructural enterprises characterised by poor performance became objects of reform as the state began to cut subsidies, create a new tax base, and generate revenue from privatisation sales. The World Bank, which primarily funded infrastructure investments in the 1960s, shifted focus to sectoral reforms including telecommunications privatisation in the 1980s. From 1980 to 1998, Li, Quiang, and Xu found that 167 countries reformed their telecommunications sectors due to poor sector performance and World Bank loan conditionality.[16]

The second impetus for reform in telecommunications came from rapid technological development that made enhanced and multiple services through a variety of transmission media possible. The increasing convergence between telephony, information technology, and broadcasting has transformed these industries worldwide, triggering changes in market structures, government policies, and international regimes.[17] What was once seen as a "natural monopoly" due to the presence of economies of scale and scope was now seen to be best served by competition.[18] In particular, convergence, which refers to the blurring of boundaries between telecommunications, computing, and broadcasting, was made possible by developments in digital microelectronic technologies which allowed satellite, wireless microwave, cable TV networks, and telephone lines to be interconnected into one overall system performing similar functions — such as watching television (TV) shows on one's personal computer, making phone calls from the TV set, or downloading movies over phone lines. The internet[19] is probably today's best example of technological convergence, where people can search and download information, watch a movie or TV show, listen to a radio broadcast, make a phone call, or do shopping. While the future applications and the vision of "a networked, global information society", are replete with promises, Robin Mansell signals the absence of basic telecommunications infrastructure in most parts of the world. A report entitled "The Missing Link" draws attention to the fact that many people in developing countries do not even have access to basic telephone services.[20]

In trying to improve the provision of telecommunications services, the adoption of neoliberal reforms involving privatisation of nationalised corporations and the introduction of competition to the industry were seen as important policy tools.[21] In most cases, market opening greatly improved the availability of telephone services. However, regulations were needed to ensure the development of infrastructure in line with development needs, guarantee presence of competition, and the protection of the public interest.[22]

Malaysia

In Malaysia, the telecommunications industry was established and organised under a government department by the British colonial government in 1908. After independence in 1957, Malaysia was left with a modern telecommunications infrastructure, and one of the highest rates of teledensities in Southeast Asia. However, the Jabatan Telekom Malaysia (JTM) was reputedly inefficient and unable to meet the growing demand for telephones.[23] In 1980, the 4th Malaysia Plan, Malaysia's five-year development blueprint, aimed to expand telecommunications services from 400,000 lines to a target of 1.2 million lines in five years. Because of JTM's incapacity to meet increasing telephone demand and its low quality of service, the modernisation of the infrastructure was placed in the hands of private contractors. Such moves were in line with then new Prime

Minister Mahathir Mohamad's announcement of his government's adoption of privatisation. Official guidelines followed two years later. Privatisation has been generally defined as the transfer of enterprise ownership from the public to the private sector. In Malaysia's case, however, privatisation was more attuned to bolstering the New Economic Policy goals, particularly the creation of a *bumiputera* capitalist class.[24]

During 1987, JTM became the industry regulator when its original role was transferred to Syarikat Telekom Malaysia Berhad, a wholly government-owned company. Telekom Malaysia was given a 20-year monopoly for all basic telecommunications services. In 1990, Telekom Malaysia was listed on the Kuala Lumpur Stock Exchange and about 20 per cent of its shares were floated. As of 2000, the government continues to own 76 per cent of Telekom, giving it effective control over appointments to the Board as well as the corporate decision-making process.

Along with the policy of privatising Telekom Malaysia, the Malaysian government introduced competition in cellular telephony, radio paging, and Value-Added Network (VAN) services. Celcom, a cellular company originally owned by Telekom Malaysia, began operations in 1989, while others were allowed entry in 1994, thus prematurely terminating Telekom's 20-year monopoly. This resulted in a market with nine players who were licensed to provide fixed-line networks, international gateways, cellular networks, and Internet service. Teledensity improved from 6.5 in 1986 to 18.3 in 1996. While the government had a general policy framework for privatisation, the way in which the telecommunications industry was liberalised shows the active role of lobbying for market entry by rent-seekers who are politically well-connected. This study argues that telecommunications privatisation and liberalisation in Malaysia signals two things: first, the state decided to implement market-oriented reform primarily to create rents to award to a group of chosen few, ostensibly to create a bumiputra business class; second, this strategy corresponded with the interests of influential businessmen (both bumiputera and non-bumiputera) who were seeking entry to the industry. Thus, as was mentioned earlier, the role of rent-

Table 1-2: Teledensity In Malaysia and the Philippines, 1981–2001 (No. of telephones per 100 People)

Year	Malaysia	Philippines
1981	3.5	0.9
1986	6.5	1.0
1991	10.0	1.0
1996	18.3	2.5
1999	20.33	3.88
2001	19.58	4.24

Sources: Roger Noll, "Telecommunications Reform in Developing Countries", Working Paper 99–10 (Washington: AEI-Brookings Joint Center for Regulatory Studies, November 1999); International Telecommunications Union, *Asia-Pacific Telecommunications Indicators 2002* (Geneva: ITU, November 2002); World Development Indicators Online, http://www.worldbank.org/publications/wdi.

seeking and patronage in the reform process raises doubts about the often heavy emphasis on the strength and autonomy of the Malaysian state.

The Philippines

In the Philippines, the Philippine Long Distance Telephone Company (PLDT) was the first company to operate telephone services nationwide. Established in 1928 by the American General Telephone and Electric Company, PLDT changed hands in 1967, in anticipation of the end of the parity rights which granted Americans equal economic rights in the Philippines. With the intervention of then President Ferdinand Marcos, PLDT was sold to a company that was reportedly owned by a dummy on his behalf. Under martial law, PLDT consolidated its status and privilege as a monopoly. Local telephone services were virtually monopolized by PLDT, which controlled about 94 per cent of the national backbone transmission infrastructure. Although there were around 60 small provincial telecommunications companies, a Government Telephone System, and two international submarine cable companies, PLDT owned and controlled the domestic infrastructure where all calls passed through. One by one, PLDT bought the provincial companies, including the government telephone system. In addition, PLDT dominated international and cellular mobile telephone services, with minimal competition. Telephone density stood at 1.0 telephone for every 100 individuals for two decades until 1990. Service quality was very poor, and the government regulatory body, the National Telecommunications Commission (NTC), was unable to push PLDT to develop its network or improve its services.

After the 1986 "EDSA Revolution",[25] the government focused on the creation of a more open economy. However, it was only when Fidel Ramos became President in 1992 that the telecommunications industry was liberalised. Ramos argued that it was time to "dismantle the oligarchy whose rent-seeking elite have dominated the economy not through possession of skill at entrepreneurship or superior intelligence but through monopoly or access to political power".[26] Two executive orders (EOs) issued in 1993 initiated changes in the industry. The first, EO 59, provided for mandatory interconnection among all telecommunications carriers. The second, EO 109, laid down the "Universal Service Policy" that divided the country into 11 service areas. Nine new companies were allowed to operate cellular or international gateway facilities. Cellular phone companies were required to expand the national infrastructure by installing 400,000 lines in three years, while international carriers were required to install 300,000 in five years. To secure these improvements, the Public Telecommunications Act (Republic Act 7925) was enacted in 1995. RA 7925 designated the NTC to exercise regulatory power over all telecommunications services. Teledensity rose from 1.0 in 1991 to 9.08 in 1998. The cost of national and international calls decreased as much as 66 per cent per minute.[27] With nine international gateway facilities, five cellular mobile phone providers, and 14 paging companies at its height, consumers were given a much wider choice of services. This study explains how and why the Ramos government was able to bring about these remarkable changes. They occurred despite the usual assessment that the Philippine state is weak and incapable of introducing reforms because it is dominated by vested interests.

Preview of Chapters Ahead

The next chapter presents a synthesis of the literature on rent-seeking, market reforms, and the state which informs this study. The chapter also discusses telecommunications reform and how the changes in the industry provide a fertile source of empirical material for the theoretical issues that this book examines.

Chapter 3 presents background information on the type of state and business class that evolved in Malaysia and the Philippines. The discussion highlights the crucial events that gave birth to the type of state and business class in both countries. This refers particularly to the rise of a relatively strong state in Malaysia and an ethnically divided and defined business class, and that of a weak, penetrated state, and a dominant business class in the Philippines.

Chapter 4 discusses the condition of the telecommunications industry in both countries before the adoption of market reforms. The different situations of the sectors in both countries are discussed and contrasted, where one is a public monopoly while the other is a private monopoly. This discussion lays the stage for the adoption of similarly market-oriented reform policies.

The rest of the book analyses the telecommunications reform experience in Malaysia and the Philippines. Chapters 5–7 deal with the Malaysian case and Chapters 8–10 deal with the Philippines. The first chapters (5 and 8) of each case look into the dynamics of the policy reform process, focusing on the type of reform adopted, the role of various actors, and the power relations between them. The second chapters (6 and 9) provide a detailed account of the impact of liberalisation on the market structure, the economy, and society by identifying the beneficiaries of the market liberalisation. In addition, the types of rent outcomes and their impact on the industry are also assessed. The third chapters of each case (7 and 10) look into the state's regulatory responses after liberalisation.

Chapter 11 sums up the evidence and the similarities and differences in the experiences of both countries. The chapter considers what these cases say about the possibility and sustainability of reform in states with varying capacities, the role and type of coalition supporting reform, and the economic consequences of rent-seeking and rent-outcomes during market liberalisation. While it is the nature of case studies to be limited in the applicability of generalizations, this book argues that the conclusions are indicative of the dynamics of the political economy of both countries. We can thus draw insights from them, especially with regard to the way these states function, leading to a greater understanding of why one state has been more capable of pursuing developmental goals than the other.

NOTES

1. Paul Hutchcroft, *Booty Capitalism: The Politics of Banking in the Philippines* (Ithaca: Cornell University Press, 1998), p. 1. Hutchcroft borrowed the phrase from Ruth McVey. See also Ruth McVey, "The Materialization of Southeast Asian Entrepreneur", in *Southeast Asian Capitalists,* edited by Ruth McVey (Ithaca: Cornell University Press, 1992), p. 22.
2. *Second Malaysia Plan, 1971–1975* (Kuala Lumpur: Government Press, 1971), p. 1.
3. Mustaq Khan and K.S. Jomo edited a collection of essays that deal with this question, providing nuanced and empirically grounded arguments on the experience of Southeast Asian countries. See Mustaq Khan and K.S. Jomo, eds., *Rents, Rent-Seeking and Economic Development: Theory and Evidence in Asia* (Singapore: Cambridge University Press, 2000).
4. John Vickers and George Yarrow, "Economic Perspectives on Privatisation", *Journal of Economic Perspectives* 5, no. 2 (Spring 1991): 130.
5. Yet, market rents are not seen as wasteful but are regarded as productive because they are assumed to create more resources than the rent-seeking process uses up. See Ha-Joon Chang, *The Political Economy of Industrial Policy* (London: Macmillan Press, 1994), p. 44.
6. Hector Schamis, "Distributional Coalitions and the Politics of Economic Reform in Latin America", *World Politics* 51 (January 1999): 236–68.
7. Khan and Jomo, 2000, pp. 4–5.
8. Dani Rodrik, "Understanding Economic Policy Reform", *Journal of Economic Literature* 34, no. 1 (March 1996): 10.

9. See Schamis (1999); Kiren Aziz Chaundhry, "Economic Liberalisation and the Lineages of the Rentier State", *Comparative Politics* (October 1994): 1–25.

10. For literature on policy reform as crisis-led, see Rodrik (1996), pp. 9–41; Anne Krueger, *Political Economy of Policy Reform in Developing Countries* (Cambridge: MIT Press, 1993); John Williamson, ed. *The Political Economy of Policy Reform* (Washington: Institute for International Economics, 1994); and Gustav Ranis and Syed Akhtar Mahmood, *The Political Economy of Development Policy Change* (Cambridge: Basil Blackwell, 1992). For literature discussing the impact of technological developments on the telecommunications industry, see Roger Noll, "Telecommunications Reform in Developing Countries", *AEI-Brookings Joint Centre for Regulatory Studies Working Paper 9–10*, November 1999; Harvey Sapolsky, Rhoda J. Crane, W. Russell Neuman, and Eli Noam, *The Telecommunications Revolution: Past, Present and Future* (New York: Routledge, 1992); C.D. Foster, *Privatisation, Public Ownership and the Regulation of Natural Monopoly* (Oxford: Blackwell, 1992); Raymond Duch, *Privatizing the Economy: Telecommunications Policy in Comparative Perspective* (Ann Harbor: University of Michigan Press, 1991); Robert Britt Horwitz, *The Irony of Regulatory Reform: The Deregulation of American Telecommunications* (Oxford: Oxford University Press, 1989); Johannes M. Bauer. "Competitive Processes in Network Industries: Towards an Evolutionary Perspective". Prepared for presentation at the meeting of the Transportation and Public Utilities Group of the American Economic Association, January 1997 at http://www.bus.msu.edu/ipu/competit.htm; and Robin Mansell, *The New Telecommunications: A Political Economy of Network Evolution* (London: Sage Publications, 1993).

11. Stephan Haggard and Steven Webb, eds. *Voting for Reform: Democracy, Political Liberalization and Economic Adjustment* (Washington, DC and New York: World Bank and Oxford University Press, 1994), p. 3.

12. Executive Summary of the 2nd World Telecommunication Development Conference, 1998. http://www.itu.org (accessed 15 June 2000).

13. Ibid.

14. The telecommunications industry contribute about 2 to 3 per cent to the GNP of developing countries and constitute as much as 10 per cent of gross domestic investment. See Noll (1999).

15. See Noll (1999), p. 14. For additional background information see Joan Nelson, et al., *Fragile Coalitions: The Politics of Economic Adjustment* (Washington, DC: Overseas Development Council, 1989); Joan Nelson, ed., *Economic Crisis and Policy Choice: The Politics of Adjustment in the Third World* (Princeton, New Jersey: Princeton University Press, 1990).

16. Scott Wallsten, "Telecommunications Privatization in Developing Countries: The Real Effects of Exclusivity Periods", draft paper (12 May 2000), p. 3.

17. Wayne Sandholtz, "Institutions and Collective Action: The New Telecommunications in Western Europe", *World Politics* 45, no. 2 (January 1993): 246.

18. Noll (1999); Sandholtz (1993); Foster (1992); Duch (1991); Horwitz (1989); Bauer (1997); Mansell (1993).

19. Pekka Tarjanne, "Preparing for the Next Revolution in Telecommunications: Implementing the WTO Agreement". *Telecommunications Policy* 23 (1999): 52.

20. Mansell (1993).

21. Noll (1999).

22. Scott Wallsten, "Does Sequencing Matter? Regulation and Privatization in Telecommunications Reforms". *World Bank Policy Research Working Paper* (February 2002).

23. Laurel Kennedy, "Privatisation and Its Policy Antecedents in Malaysian Telecommunications. Ph.D. dissertation, Ohio University (November 1990), pp. 136–37.

24. K.S. Jomo, *Privatising Malaysia: Rent, Rhetoric, Realities* (Colorado: Westview Press, 1995), pp. 48, 81.

25. For a comprehensive account on the 1986 EDSA Revolution, see P.A. Daroy, Aurora Javate-De Dios, and Lorna Kalaw-Tirol, eds., *Dictatorship and Revolution: Roots of People's Power* (Metro Manila: Conspectus, 1988).

26. Fidel V. Ramos, "To Win the Future' Inaugural address as President of the Republic of the Philippines, 30 June 1992", in *Developing as a Democracy: Reform and Recovery in the Philippines, 1992–1998* (Hong Kong: Macmillan Publishers, 1998), p. 4.

27. *Philippine Daily Inquirer*, 20 June 2000.

2

REVIEWING THE LITERATURE: THEORIES AND PUZZLES

This chapter discusses relevant aspects of the literature on rent-seeking, market reforms, and the state which inform this study. After expounding on the theoretical background, the fourth section deals with telecommunications reform issues in order to provide an introduction to developments in the sector and demonstrate why the case is a rich source of data for the puzzles dealt with in this book. The final section links the puzzles and arguments with the pertinent theoretical issues.

Rents and Rent-Seeking

James Buchanan, one of the founders of public choice theory, defines rent as that part of payment to an owner of resources that is over and above what those resources could command in any alternative use.[1] Put another way, rents are incomes that are more than the income that an individual or a firm would have received in a competitive market.[2] Some scholars categorize rents according to their origin — whether they are economic (market rents) or political rents. For them, rent-seeking is usually reserved for actions that lead to the capture of political rents. Thus, rent-seeking is sometimes defined as resource allocation by the state or through political means, as opposed to resource creation in the market. The underlying assumption is that political rents are inefficient, but market rents are not. This classification, however, misses out on hybrid forms of rents: for instance, those that originate in market performance but are sustained by the government to encourage learning, innovation, or development, and rents that are politically created but are reaped in a competitive market.[3]

Khan enumerates no less than six types of rents: monopoly, natural resource, political transfer, Schumpeterian or innovation, learning, and monitoring.[4] Monopoly rents emerge as a result of entry barriers, allowing a firm in protected markets to charge higher prices for their products. Entry barriers can be natural (when the technology of production involves economies of scale) or state-produced (via the creation of exclusive production rights). Natural resource rents accrue to the owners of scarce natural resources. Political transfer rents not only help redistribute incomes, but also to create new property rights, and often entirely new economic classes in developing countries. Schumpeterian rents reward innovation. Rents for learning are artificially created by the state to encourage and accelerate learning in infant industries. Monitoring rents reward good management.

Khan argues that the discussion of rents in developing countries often assumes that all rents are monopoly rents and that the effects can be adequately analysed with a static neoclassical model. In reality, many monopolies are natural monopolies, which mean that they are created not by artificial entry barriers but by economies of scale in production that result in one producer dominating the market through lower costs. In cases of monopoly rents that are created by government protectionism to favour cronies, however, their dynamic effects are not always clear-cut. Khan argues that the task of analysis should be to look at precisely what these effects are.

While some rents signal innovation, others point to inefficiencies. Some types of rents are legal, while others are illegal. In whatever way they arise or are acquired, rents always generate incentives to create and maintain them because they represent higher than normal incomes. Activities that range from lobbying and advertising to bribing and coercion are usually called rent-seeking. Using this definition, Khan points out that almost all institutional change involves creating or destroying rents in so far as they result in change to the property rights regime. Thus, almost all distributive conflicts can be described as conflicts where both sides are seeking rents.[5]

Economists developed rent-seeking theory to show that state intervention in the economy leads to wasteful costs. Their aim was to strengthen the theoretical argument for freer markets.[6] The early literature on rent-seeking identifies political activities that aim to protect, maintain, or change systems of rights as "unproductive actions". Anne Krueger states that government restrictions give rise to a variety of forms of rent, which people often compete for. The general impact of these activities is welfare loss to society.[7] Jagdish Bhagwati describes actions by the private sector to bend public policy towards its favour as "'directly unproductive profit-seeking activities' which yield pecuniary return but do not produce or make available goods and services in the economy".[8] Clearly, the traditional view of rents is negative, with rent-seeking seen as welfare diminishing and thus, socially undesirable.

Politically created rents such as political transfers, entry restrictions, tariffs, quotas, and artificial creation of a monopoly position are the rents that neoliberals consider as costly and inefficient. These are the types of rents that neoliberals wish to eliminate when they talk about the removal of institutions and rights that protect rents, and not market-created ones like natural resource, Schumpeterian, learning, and monitoring rents. The removal of politically created rents is seen as the desirable path towards greater efficiency and economic performance.[9] Neoliberal market-oriented reforms are thus modelled to eliminate rents in order to promote economic growth.

The evidence from Southeast Asian countries, however, shows that the story is more complex. The dismissal of all politically created rents as undesirable and welfare-reducing ignores the fact that on closer examination, rents and rent-seeking activities have different growth and efficiency implications. Following Khan's lead, this book suggests that it is more interesting to ask why rents become socially desirable or undesirable, and to ascertain the conditions that determine the type of rents created. It is important to not merely look at rent-seeking costs but also rent outcomes. To do this, the theory of rent-seeking should be reinforced by insights from political science, political economy, and new institutional economics. In particular, it is important analyse the power relations between the state and social actors embodied in patron-client relationships.[10] In doing this, one must look at five interrelated questions: What types of rents are created? Are they competitively pursued or purposively allocated by the state? How are they captured? Who captures them? What are their outcomes? These questions are raised with regard to telecommunications reforms in Malaysia and the Philippines.

Market Reforms

When one speaks of economic reform, there is a general agreement today that this refers to the adoption of market-oriented policies and the withdrawal of state involvement in economic activities.[11] This persuasion is embodied in the so-called "Washington Consensus", which outlines the central role of the market and the supportive but minimal role of the state in attaining economic development.[12] As John Williamson, argues, "in most cases, market competition provides a better way than regulation of determining how much of a good will be produced and at what price ... Therefore the government should concentrate its policy attention on ensuring that the market is truly competitive, with regulations limited to those general rules that define the institutional infrastructure of a market economy ...".[13]

The neoliberal reform prescription includes liberalisation (the opening of the market to competition), privatisation (the transfer of the ownership of enterprises from state to private ownership), and deregulation (the removal of state regulations or barriers to market competition).[14] Collectively, these changes are hailed as the solution to the problems of inefficient state enterprises and poor economic performance. The dominant view today is that "governments perform less well than the private sector in a host of activities".[15] This claim disputes the assertions of development economists that state intervention is a necessary corrective to market failure.[16] Those who adhere to the neoliberal approach argue that correcting market failure through state intervention has led to the bigger and more costly problem of government failure. Apparently, politicians, bureaucrats, and state officials are more prone to pursuing policies in line with their personal or organizational interest than the national good. Meanwhile, societal groups engage in rent-seeking to influence public policy and sway it to their favour. Finally, direct state intervention via the establishment of state-owned corporations has proven ineffective because such corporations have acted as monopolies, were not faced with competitive pressures, and often pursued non-economic objectives that conflict with the goal of economic efficiency. Thus, the goal of economic reform is the reduction of government intervention in the economy, because "rent-seeking activity is directly related to the scope and range of governmental activity in the economy, to the relative size of the public sector".[17]

One of the bases of the neoliberal prescription of state withdrawal is traceable to the assumptions of collective action theory. Mancur Olson points out that groups of individuals with common interests usually attempt to further those interests when organizing costs are lower than the potential benefit. Due to free-riding problems, rational self-interested persons will not join a group that furthers their interest if they can enjoy the same benefits by not becoming a member.[18] However, when significant benefits accrue, individuals form coalitions that lobby the state to provide policies that are favourable to their interests. In contrast, politicians exchange policies for political support, thus providing rents to their constituencies, and prefer to allocate resources through political bargaining rather than the market. Hence, the more the state is involved in the economy, the more it has power to distribute economic benefits, the more that social groups will attempt to capture those benefits. Market-oriented reform, by reducing state involvement in the economy, is expected to lessen areas of state intervention, and consequently the government's capacity to favour family members, supporters, or friends. The introduction of market competition is anticipated to lead to improved efficiency for the liberalised industry in particular, and the economy in general. Neoclassical political economy literature expects that "the problem of distributional coalition is invariably resolved by liberalization",[19] where avenues for rent-seeking are removed as the state retreats from interfering in economic activities.

After almost a decade of neoliberal reform, even the World Bank has recognized that many of these policies have not yielded the expected results. In a 1995 study, the World Bank argues that reform can only be successful when it meets three necessary conditions: desirability, feasibility, and credibility. First, reform must be politically desirable to the leadership and its constituencies, whereby the political benefit outweighs political cost. This usually takes place during a regime change, a coalition shift in which those that oppose reform lose power, or an economic crisis. Second, leaders must have the means to enact reform and overcome opposition to it, either by compensating losers or compelling them to comply despite their losses. Third, investors must be convinced of the credibility of the government's commitment that the policy will not be reversed.[20] From this perspective, the emphasis is on the state — its capacity to enact, implement, and consistently demonstrate the credibility of its commitment to policy change. The state must take the initiative to carry out these changes. From the World Bank's point of view, influential social actors are expected to oppose change so that the emphasis should be on the state's capacity to quash such opposition when it arises. Social actors are perceived as supportive of the status quo and will not willingly let go of the benefits that they presently enjoy.

Yet, some scholars critical of the sweeping claims of neoliberal market reform point out that market failure is as prevalent as government failure.[21] A major critique raised is the absence of equivocal evidence supporting the superiority of the market over the state.[22] In contrast, the evidence of unprecedented high growth in East Asia provides proof on the positive role of government intervention.[23] In reality, state involvement in terms of building institutions — the legal system, property rights, capital markets, and regulatory bodies — that underpin a market economy is central to the market's effective operation. Thus, the leadership role of the state should not be ignored in the face of exuberant neoliberal advocacy for the market because a strong state is indispensable in setting up a working and properly functioning market. The hotly contested point is how big that role should be and in which areas should the state get involved.[24]

An area of concurrence between neoliberals and those who advocate state intervention is the centrality of the state in putting to place "a credible and effective regulatory framework that has the ability to encourage private investment and support efficient production and use of services".[25] Regulatory reform is found to be necessary after the introduction of market reforms because a competitive market does not naturally evolve on its own. Regulatory reforms involve setting up competition and pricing rules, and the establishment of an independent regulatory body that can ensure fair competition. The introduction of economic liberalization has not automatically led to competition in the market.[26] There is growing recognition that the extent of competition is largely influenced by government policy. Competition and the market itself have to be created. As Karl Polanyi puts it, "free markets could never have come into being merely by allowing things to take their course … *Laissez faire* itself was enforced by the state … The introduction of free markets, far from doing away with the need for control, regulation, and intervention, enormously increased their range … *Laissez faire* economy was the product of deliberate state action".[27]

Yet, most of the literature on the adoption of market reform focuses on how and why countries choose to implement these policies. Few scholars have studied the post-reform situation, leaving a gap in the literature, and not explaining "the reconstruction of institutions for market governance after the implementation of neoliberal reforms".[28] Richard Snyder, in studying the impact of reform on the Mexican coffee industry, found that re-regulation was needed soon after deregulation. Taking a step further, Stephen Vogel asserts that a strong state and a strong market are not mutually exclusive. Stronger markets actually require more state rules, and re-regulation is often desirable.[29] Thus, it is important to look at not merely the introduction of market reforms but also the essential regulatory reforms that follow them.

In considering the central role of the state in carrying out reforms, such role should not be overemphasized to the extent of ignoring other actors. The prevailing view in the literature stresses that governments adopt market reform policies as a response to economic crises or because of external pressure from multilateral institutions. Economists emphasize the importance of consistency and credibility of reform to be successful, while political scientists normally highlight the need for the political will and policy-insulation of bureaucratic elites to overcome opposition by those who benefit from the status quo. The implicit assumption of these frameworks is that reforms can only come from above, disregarding the possibility that social actors can organize in support of changes to the existing property rights regime.

Hector Schamis points out the central role of interest groups in organizing to pressure the state for benefits that could be enjoyed from policy change. This role of the private sector in the reform process has largely been neglected due to the overemphasis on the need for "relative autonomy of the state" and the importance of credible and consistent reform.[30] Schamis argues that some groups may organise to induce the government's withdrawal from the economy, in anticipation of market opportunities made available by liberalisation or privatisation. He provides empirical evidence to support this claim from the experiences of policy makers in Chile, Argentina, and Mexico where reformist governments succeeded in introducing liberalisation reform by forming alliances with prospective beneficiaries of the policy change.

The Latin American experiences demonstrate that the coalitions that organized in support of liberalisation are most appropriately described as distributional. They are rent-seekers who lobbied for state withdrawal in order to reap the benefits of reform. Policy makers built ties with these groups to ensure a constituency for the policies to be adopted. Schamis argues that while rent-seeking is expected of interventionist-regulatory regimes, the experience of market liberalisation reform in the three countries indicates that liberalisation can also generate rent-seeking activities, as potential beneficiaries of reform collude with policy makers to translate their preferences into policy.[31]

Emmanuel De Dios raises similar arguments for the Philippines. He observes that market liberalisation is now supported by some of the economic elite because they have experienced the benefits of a more open market, such as greater access to capital and technology, allowing them to diversify into other industries. "There is now a domestic constituency for the continuance of economic reform ... as elite interests [saw] some real opportunities of accommodation in a liberalizing environment. It is true, of course, that the liberalization process itself would remove the means for rent-seeking by narrowing the scope for government intervention. Nonetheless, some processes in a liberalizing agenda, especially deregulation and privatization, yield substantial rents as well, and no great technological shift is required on the part of dominant groups to use customary rent-seeking methods to win such battles."[32] De Dios, however, is talking about economic elites who now support reform, changing their position from opposition to it. This constituency is not necessarily the same as that signalled by Schamis in Latin America. Thus, two interesting questions to ask are, who constitutes the group supportive of reform, and to what extent does this constituency support the deepening of market reforms?

This study on telecommunications liberalisation looks into the role of state actors, and the composition of and the role played by distributional coalitions or rent-seekers in the adoption of market-oriented and regulatory reforms in Malaysia and the Philippines.

The State and Societal Actors

Scholars from varied theoretical persuasions have defined the state in different ways. While some portray the state as the instrument of the ruling class, others view it as embodying class relations and continually being shaped by class struggles. Still, others emphasise its role as an objective guarantor of the relations of production and economic allocation. In these definitions, the possibility of the state being autonomous is ruled out.

Several scholars, however, argue for the possibility of an autonomous state. Following Max Weber, Theda Skocpol defines states as compulsory associations claiming control over territories and the people within them. The state is composed of its administrative, legal, extractive, and coercive organizations that "attempt not only to structure relationships between civil society and public authority in a polity but also to structure many crucial relationships within civil society as well".[33] For Weber and neo-Weberians, the state can be potentially autonomous from dominant classes, and being the controller of the legitimate means of coercion and administration, may pursue goals that are at odds with the interest of dominant classes or any other social groups.[34]

Weber posits the ideal type of a centralized and fully rationalized bureaucracy as the form of a state that is most efficient and able to determine its will and implement it even in the face of social opposition. He argues, "the state's ability to support markets and capitalist accumulation depended on the bureaucracy being a corporately coherent entity in which individuals see furtherance of corporate goals as the best means of maximizing their individual self-interest. Corporate coherence requires that individual incumbents be to some degree insulated from the demands of the surrounding society".[35]

Revived academic interest in the state as an institution and social actor was signalled by the publication of *Bringing the State Back In* in 1985. Building on Weberian ideas, contributors to the volume analysed historically rooted examples of varying, uneven, and contradictory state capacities and autonomy. They propose that analyses of the state's role in economic development must look into concepts of capacity and insulation or autonomy. On the one hand, the state's capacities depend on the existence of a specific organizational structure that enables state authorities to undertake specific tasks. State autonomy, on the other hand, increases its ability to realize internally generated goals, while reducing the power of societal groups outside the state. However, it has to be pointed out that state strength, whether conceptualised in terms of the state's autonomy from social groups or in terms of its capacities to intervene on its own or other's behalf, can only be "fruitfully understood via dialectical analyses that allow for non-zero-sum processes and complex interactions between state and society".[36]

One type of a strong state that played a key role in economic development is called the "developmental state".[37] Chalmers Johnson proposes this concept in his analysis of the role of the Japanese state in its extraordinary post-war economic growth. He coined the term "developmental state" as a Weberian ideal-typical construct, based on his observation of the Ministry of International Trade and Industry (MITI), which was "plan-rational and capitalist, conjoining private ownership with state guidance".[38] Johnson argues that the credit for the Japanese economic miracle lies in the conscious and consistent government policies that were implemented from the 1920s onwards. He points out the central role of a powerful policy-making agency capable of developing a coherent economic policy and sufficiently insulated from societal demands to be able to pursue its development agenda.[39] Several other writers pursued the subject of the role of the state in economic development, focusing on the experiences of the NICs in Asia. These scholarly accounts provided further support for the case of a developmental state and the transformative role it can play in fostering economic development.[40]

Taking the theory of a developmental state a step further, Peter Evans introduced the concept of 'embedded autonomy' as an important characteristic determining the success of state intervention during late industrial transformation. Evans argues that a state's transformative capacity not only requires internal bureaucratic cohesion and sufficient autonomy to formulate developmental goals, but also connectedness to organized business and industrial networks that he terms "embedded autonomy". The idea of embeddedness is a relational concept that links an internally cohesive bureaucracy with the private sector. Embeddedness corrects the overemphasis placed on insulation,[41] as the state needs to rely on concrete networks of external ties to society to assess, monitor, and shape private responses to policy initiatives. Evans argues that embeddedness has value only in the context of autonomy. In the absence of a coherent, Weberian type of administrative structure, embeddedness will almost always lead to negative effects. He argues that it is the combination of embeddedness and autonomy that works, not either on its own.[42]

Yet, in most developing countries, a Weberian-type bureaucracy is rare, if not non-existent.[43] Scholars point to the absence of a "strong" or a developmentalist state in developing countries, which makes the project of economic development or instituting policy reforms problematic. The adoption of policy reform and the sustainability of the state's commitment in its implementation are perceived as lacking credibility in situations where a dominant social elite captures the state. Scholars argue for the necessity of the state developing relative autonomy, without which the feasibility of implementing credible and sustainable reform is doubtful. In cases where the state is permeated by rent-seekers, insulated policy-making technocrats can only draft and implement reforms during times of economic crisis.[44]

Another way of looking at the state is Joel Migdal's "state-in-society" approach. Migdal defines the state as "an organization composed of numerous agencies led and coordinated by the state's leadership that has the ability or authority to make and implement binding rules for all the people as well as the parameters of rule making for other social organizations in a given territory, using force if necessary to have its way".[45] This definition of the state encompasses policy making, administrative, policing, and military organizations under a central leadership. While "state" is a singular noun, the state is rarely a singular actor. The degree of coherence among organizations within the state varies over time, and from one state to another. Finally, while the state was created by society, it is different from society but is not separable from it. State and society engagements and disengagements mutually transform each other.[46]

Migdal views the state as intrinsically embedded in society and argues that it must be disaggregated for a better understanding of its power in developing countries. Far from being monolithic, internal tensions and competitions among state sectors take place. The state may help mould society, but is also moulded by the society within which it is embedded. The state and social forces may be mutually empowering, and not merely contending with each other in a zero-sum game. The state's capacities depend on the state's ties to other societal forces. Thus, one can fully elucidate what a strong or weak state really means by looking at the state as an organization within a societal arena where groups and individuals contest rules and norms.[47]

For Migdal, an important feature of state strength is social control measured in terms of its leaders' abilities to get people to do what state policies and laws require of them. High levels of social control enable a state to mobilize the population, collect taxes and other resources from society, have sufficient autonomy from other organizations to decide on and implement state programmes, monopolize coercion to prevent other groups in society from obstructing the state's authority, and stand up to external foes.[48]

Benedict Kerkvliet extends Migdal's argument by adding two other areas from which to assess state strength. Firstly, the state's strength can vary according to the policy area or the particular laws, programmes, or points of intervention. A state that is deemed an ineffective tax collector, for example, may be a strong implementer of policies against (or for) racial discrimination. Secondly, state strength should be analysed from a specific perspective — for instance, from the point of view of the favoured group or from those who oppose the policy agenda.[49] Adding these variables makes the analysis of a strong or weak state clearer and contextualised.

This study adopts Skocpol's and Migdal's definition of the state and disaggregates the state into its various — sometimes contending — parts, while contextualising it in the society in which it is embedded, and tracing its historical evolution and linkages to various societal actors. Kerkvliet's extension of Migdal's framework is also adopted in assessing the state's autonomy and capacity, by looking at specific policies and the perspective from which the state is seen as weak or strong. Adding these indicators to the analysis clarifies at a more definite level what is meant by a strong or weak state.

The level of state strength is tested in this study through an assessment of its interventions in the economy, and its capacity to formulate and implement a coherent policy, as the state is simultaneously connected to and affected by societal groups. In the two cases of states with varying capacities examined, the aim is to assess how policy reforms are adopted by identifying the major actors that pushed for reforms. That is, how are states faced with dominant vested interests and lacking a strong and insulated bureaucracy able to adopt and implement market-oriented reforms? Were the policy reforms adopted because a team of technocrats was able to insulate itself from societal pressures? Were these reforms undertaken during crisis periods? What are the outcomes of market reforms in situations where, as Evans has noted, embeddedness is not accompanied by state autonomy? Are the outcomes of the adoption of market-oriented reform in societies penetrated by patronage networks and rent-seeking necessarily or always deleterious? These issues are analysed in this book as it examines telecommunications reform in Malaysia and the Philippines.

Telecommunications Reforms

Among developing countries, the development of telecommunications services has gone through three phases. The first phase involved the establishment of a telephone company, which was owned, either by the colonial government or by foreign multinationals associated with the colonial regimes (such as AT&T, ITT, and Cable & Wireless). During this phase, telephone service was usually provided for the national capital and large cities, and was limited to government agencies and officials, large businesses, and wealthy individuals. The second phase occurred in the 1950s and 1960s, during the era of decolonisation, when these colonial or foreign-owned telephone companies were nationalised by domestic governments and run by a specific department or in rare cases, domestically owned private monopolies. Most post-colonial governments followed the policy model of their former colonizers: for instance, Malaysia followed the British model of state provision by a government department, while the Philippines followed the US model of a privately owned monopoly. The third and present phase began in the 1980s with the adoption of neoliberal reforms involving privatisation of nationalised corporations, and the introduction of competition to the industry.[50]

Countries that adopted neoliberal reform soon enough realised that privatisation and liberalisation were not enough to ensure fair competition in the sector. The need for regulatory reform consequently became evident. Reforming the telecommunications sector was thus a two-stage process that involved: first, the introduction of privatisation and liberalisation to reform the ownership and market structure of the sector; and second, the introduction of regulatory reforms. Although market opening greatly improved the availability of telephone services, regulations were needed to ensure the development of infrastructure in line with development needs, guarantee presence of competition, and the protection of the public interest. In fact, one of the areas of policy debate today is on the sequencing of reforms. In particular, are reform outcomes better if a regulatory framework is put in place before opening the market?[51]

Regulation takes places when the state acts to constrain private activity to promote the public interest. It is frequently associated with market failure, or the situation in which the market fails to produce goods and services at appropriate prices.[52] Regulation covers governmental activity that involves the formulation and promulgation of an authoritative set of rules, accompanied by a mechanism — usually a public agency — for monitoring and promoting compliance with these rules to mould private economic conduct. Rule making and monitoring or enforcement mechanisms are functions that may or may not be located in the same institution.[53]

Sam Breyer points out that the traditional and most persistent rationale for regulating a firm's prices and profits is the existence of a natural monopoly, as is often the case in telecommunications, where the industry cannot efficiently support more than one firm. An industry that is categorized as a natural monopoly is said to exhibit economies of scale, whereby the unit cost of a service would rise significantly if more than one firm supplied the service in a particular area. A good example is a telephone network system, in which it would be costly to have three companies rolling out separate cable systems in one area. It would be more efficient for the government to grant one firm monopoly status subject to the regulation of its prices and profits. The idea behind regulating a natural monopoly is the assumption that a monopolist, if left unregulated, would dictate prices. Higher prices mean less demand, but a monopolist is expected to willingly forgo sales to the extent that gaining more revenue through the increased price of the good that is sold will compensate for the loss in sales per unit. This results in social waste.[54]

Technological changes have challenged the assumption that telecommunications is a natural monopoly and has made competition in this sector possible. Nonetheless, the need for continued regulation became apparent as privatisation and liberalisation policies failed to create a competitive market and ensure the provision of efficient, affordable, and accessible communications facilities. Regulatory reforms were needed to promote the interest of consumers to contain market power, to foster competition, to create a favourable investment climate, and to narrow development gaps.[55] Former monopolists, even in countries like the United States, the United Kingdom, and Australia, which had introduced regulatory reform early on, continued to have dominant control over the market despite the regulator's interventionist role.[56] Indeed, the worldwide experience of telecommunications liberalisation called for a second stage of policy changes.

There are two critical issues in regulating the sector: the establishment of credible regulators and the creation of policies to resolve regulatory issues. The establishment of a regulatory agency with the autonomy, transparency, and resources to foster a competitive environment and protect consumer interests was of utmost importance. William Melody argues that the effectiveness of telecommunications markets depends on the creation of a solid legal foundation, and an independent, competent, and effective regulatory system. Regulation must be both independent of companies' and day-to-day government influence. The regulator's task is to implement government policy, ensure the performance accountability of telecommunications companies and other industry players in terms of economic and social policy objectives, resolve disputes between competitors and between consumers and operators, monitor changing industry conditions, and advise the government on developments that have bearing on policy.[57] Most importantly, Melody points out that competition among major industry players should be moved from "the arena of politics and bureaucracy to the marketplace" to achieve the industry performance objectives of government policy. Such a situation can only be achieved if regulatory decisions are made independently, on their own substantive merits, and not on the basis of political favouritism or the influence of the most powerful industry player.[58]

Brian Levy and Pablo Spiller outline what they call essential requirements in establishing a credible regulatory body: a strong administrative tradition, the credibility or the ability to undertake commitments that endure from one government to the next, immunity from government and political pressures, the presence of a substantial professional staff that is capable of handling complex regulatory concepts, and a judiciary that is impartial.[59] Levi and Spiller argue that a country's institutional endowments affect its regulatory capacity. Institutional endowments include the legislative, executive, and judicial institutions, customs and other informal but broadly accepted norms, the character of contending social interests within society, and the balance between them and the country's administrative capabilities. Finally, the structure and organization of a country's legislative and executive institutions also influence its regulatory options according to how well they restrain arbitrary government action.[60]

Aside from the establishment of a credible and independent regulator, five key regulatory issues need to be resolved. These are interconnection, pricing, universal access, convergence, and spectrum management.

The enforcement of interconnection is the single most important issue faced by regulators in developing a competitive marketplace for telecommunications services. According to Spiller and Carlo Cardilli, "interconnection is the right of a network operator to ensure that its users can make or receive calls to or from the users of another network. Without a right to interconnection ... the incumbent monopoly (is) likely to deny, delay, or overprice interconnection to attempt to preserve their dominant position, requiring entrants and small competitors to spend resources enforcing their interconnection rights."[61] Hence, it is crucial to decide which regulatory method to use to prevent the incumbent from creating a deadlock that has the effect of perpetuating its dominant position.

As will be discussed in Chapters Seven and Ten, new entrants were unable to connect to the national network or faced delays and difficulties, in both the Philippines and Malaysia. Indeed, experiences around the world shows that the incumbent does not freely make interconnection available to its competitors. Given the asymmetry in bargaining power and the urgency for the new entrant to interconnect, the incumbent usually slows down the process. In other cases, although the incumbent provides interconnection to a competitor, it does so at a high price. Thus, interconnection must be mandated and regulated.[62] Regulatory policies and decisions affect competitive outcomes in the market. Loosely defined interconnection rules can lead to delays that favour the incumbent.

Wholesale and retail pricing is the second regulatory issue that must be settled. Before the introduction of reform, rate of return regulation was the basis of telephone rates, wherein the regulatory body set a tariff ceiling for services based on a rate of return on capital needed by the company to run its operations, expand its network and services, and earn a moderate profit. Rate of return regulation was used to control the monopolist's profits. In a competitive environment, however, pricing is aimed at making sure the dominant carrier does not abuse its market position or adopt anti-competitive strategies.[63] Another pricing issue involves tariff re-balancing, whereby the cost of local calls is reassessed to reflect their true cost as subsidies from international or long-distance services fall.

The third regulatory issue is how to achieve "universal access" to telephones. In developed countries, the goal is universal service or the presence of a telephone in every home. In developing countries, however, household incomes are lower and the objective is to provide access to quality communications to all users at affordable prices[64] through the provision of village payphones, public calling offices, and urban and rural telecentres. In the Philippines, newly licensed telecommunications companies were required to install telephone lines in less economically viable areas. In Malaysia, universal service was a burden carried by Telekom Malaysia, but the establishment of a Universal Service Fund is currently being planned.

A fourth issue that regulators face is how to deal with the convergence of the information technology, broadcasting, and telecommunications industries. Most countries have laws that bar the cross-ownership of telecommunications, broadcasting, and print media, which is the situation in the Philippines. In 1998, Malaysia became the first country in the world to legislate a Communications and Multimedia Act, which recognized convergence and grouped together the three industries under one regulator. The fifth regulatory issue is spectrum management, or how to properly allocate scarce radio frequencies. Transparency and technology-neutral selection by the regulator serve to level the playing field for new entrants.[65]

As of 1 January 1998, 72 countries, including the Philippines and Malaysia, became signatories to the General Agreement on Trade in Services (GATS) and the World Trade Organization's Basic Telecommunications Agreement (BTA). These signatories made full or partial commitments to establish separate and independent regulatory bodies to act and decide in a transparent and non-discriminatory manner, establish competitive safeguards to prevent anti-competitive practices, provide for interconnection and dispute settlement mechanisms, apply universal service obligations in a neutral and transparent way, make licensing criteria publicly available, and fairly allocate scarce frequency resources. As it set down regulatory requirements that had to be complied with, the signing of the BTA was a major influence on the worldwide trend towards the standardization of regulatory agencies.[66]

This book examines not merely the adoption of market-oriented reforms, but also regulatory reforms in the telecommunications sector of Malaysia and the Philippines. In examining how the second stage of telecommunications reform took place, the role of state and business actors, and the interaction between them remain the focus of analysis. Regulatory reform is an area where international actors and consultants play an active role. Nonetheless, domestic political considerations determine the final outcome of what sort of regulatory regimes are adopted and established.

Puzzles and Implications

Building on the above literature, this book examines three interrelated issues on the possibility and sustainability of policy reform in states with varying capacities; the role and type of coalition supporting reform; and the consequences of rent-seeking and rent-outcomes during market liberalisation and their role in promoting or hindering economic development.

The adoption and outcomes of market-oriented reforms in the telecommunications industry of Malaysia and the Philippines are valuable cases to examine because they do not sit properly within existing conventional expectations of both countries.

The first puzzle pertains to the type of state and its capacity to adopt reform. Scholars assess the Philippine state as weak because it does not have the capacity to make consistent and coherent economic plans as it is overwhelmed by the particularistic demands of an influential elite. The literature on reform talks about the difficulty of introducing policy changes in weak states in the face of powerful vested interests. This body of literature predicts that reforms are usually adopted during times of crises or when international lending institutions pressure the government to adopt reform in exchange for aid or loans. Yet, despite the absence of these two "push" factors, liberalisation of telecommunications took place in the Philippines. The book argues that reform in the Philippines, despite expectations to the contrary, was made possible by the work of a coalition for reform, whose efforts were supported by then President Ramos. The occurrence of reform in the face of intense opposition from vested interest raises doubt as to the usual portrayal of the Philippine state as weak and incapable.

Meanwhile, scholars on the Malaysian state agree that it had the capacity to draw up and implement an ambitious economic development and restructuring programme. This capacity is oftentimes interpreted as proof that the Malaysian state is a strong, capable state. In this light, when market-oriented reform was adopted in the telecommunications industry, policy change could be viewed as another example of the decisiveness of a strong state responding to an economic crisis. Yet, it was plainly observable during privatisation and liberalisation of the telecommunications industry that the state under Mahathir purposively created rents and distributed them to a chosen few, who, in turn, actively sought the state-created market opportunities. In fact, while the state was considered strong and capable, and was becoming increasingly authoritarian under Mahathir, patronage and rent-seeking seemed to have intensified as a factor in the policy adoption and implementation process. This study argues that while there was a general policy framework for privatisation, the way in which the telecommunications industry was liberalised shows the active role of lobbying for market entry by rent-seekers who are politically well-connected. Thus, the role of rent-seeking and patronage in the reform process raises questions as to whether the Malaysian state is really as strong and insulated as it is usually portrayed.

An issue related to the first puzzle is the rationale behind the adoption of market-oriented reforms. In the Philippines, liberalisation was adopted because of the Ramos administration's goal of removing the monopoly control of an influential family over a specific industry. Such an objective is a straightforward expectation arising from the intended economic effects of market-oriented reforms. In Malaysia, however, while privatisation and liberalisation espoused the standard neoliberal rationale, the paramount objective that overshadowed the rest of the policy objectives was the creation of a bumiputera[67] business class. As the liberalisation of the telecommunications industry demonstrates, the Malaysian state deliberately adopted market-oriented reforms as a strategy to create rents that it could dispense to a group of favoured businessmen, with the ultimate goal of developing a bumiputra business class. Hence, while market reforms were adopted in the Philippines to break down monopoly rents, the Malaysian state used market reform to create rents, which were awarded to an influential few.

The second puzzle concerns the role and composition of the coalition for reform in both Malaysia and the Philippines. In Malaysia, those who lobbied for liberalisation of the telecommunications industry can be rightly termed as rent-seekers who advocated the reform because they were aware of the benefits they would gain from the change in rules governing the market. These rent-seekers nonetheless cannot be called "distributional coalitions" because they did not lobby as a group but rather sought market licenses in competition with each other. These licenses were secured via existing patronage networks based on racially organised political parties or through personal linkages to individual politicians. In contrast, the coalition for reform that supported the liberalisation of the telecommunications industry (and other key economic sectors) in the Philippines was composed of academics, professionals, and advisers to then President Ramos. The members of the coalition that moved, strategised, and pushed for liberalisation, in contrast to those who lobbied for it in Malaysia, were not the direct beneficiaries of market entry license to the industry.

The third puzzle pertains to the economic consequences of rent-seeking and rent outcomes. Often, the predominance of patronage and rent-seeking in the Philippines is considered as a major barrier to economic development. Yet, it seems that patronage and rent-seeking has not been a decisive hindrance to economic development in Malaysia. How and why this is so is an issue that this study explores. Following Mustaq Khan's lead, this study argues that the only way to understand the differences in the economic impacts of rents is to analyse the types of rents created, how they are captured, and identify their recipients. It will be argued that the types of rent outcomes are shaped by the characteristics and interactions of the state and business class in each country, and most importantly, the segment of society that supported the state's move towards market reform.

The Philippines and Malaysia had different industry status before reforms: the former was a private monopoly, while the latter was a state monopoly. Thus, pressures for initial reform and the specific policies adopted were different in each case, but were both market-oriented.

Table 2-1: Major Actors

Actors	Malaysia	Philippines
State actors	The Prime Minister, EPU, the METP/MECM, JTM, UMNO, and BN member parties	The President, presidential advisers, DOTC, NTC, Congress, and the Judiciary
Business class	Bumiputera and non-bumiputera businessmen	Economic elites, Cojuangcos and the new entrants to the industry
Civil society actors	Labour union, opposition politician, consumer organisations, academics	MORE Phones, Peoples' 2000
Foreign actors	Technical consultants, foreign investors	Donor influences (IMF, WB, ADB), technical consultants, competitive liberalisation pressures, and foreign investors

This study focuses on the roles played by four types of state and societal actors in the reform process. First, state executives (especially Mahathir and Ramos) played crucial roles in the policy-making process through their personal commitment to policy change that increased the feasibility and credibility of reform. Second, the support for liberalisation extended by pro-reform groups, which can be further categorized into business and civil society actors, was crucial. Identifying the exact composition of these pro-reform coalitions, the types of policies for which they lobbied, whether they included civil society or non-business organizations, and how different or similar they were from the earlier economic elites is an important focus of this study.

In the Philippines, support for liberalisation of the telecommunications industry came from a coalition of reformers who were not involved in the industry. The monopolist strongly opposed the reform, but support for liberalisation gained strength as the interests of established elites shifted in favour of reform. In Malaysia, while state support for privatisation came from the Prime Minister and his economic advisers, liberalisation of particular industries like telecommunications was supported largely by rent-seekers who were close to the Prime Minister and top politicians. Opposition arose from the corporatised but still government-owned company, the employees' union, the political opposition, and some academics. In both cases, reforms did not merely come from insulated policy-making teams of technocrats who designed and implemented policy autonomously from society. Rather, reform was implemented as a result of the link-up between state and societal actors. Finally, the role of international actors, for instance, through the diffusion of development prescriptions as part of loan or aid requirements from the International Monetary Fund, World Bank, and other international financial institutions was a factor in the Philippines. Such pressure, however, was largely absent in Malaysia.

Neoliberal economists see market-oriented reform as a way of removing politics from the market and reducing venues for corruption and rent-seeking. Yet, liberalisation and privatisation in the short term have not stopped patronage networks (by "cronies" or party-linked companies) from accessing privatisation contracts or licenses to enter the newly opened market. This situation has led some observers to dismiss the reform process as one that has been hijacked by dominant interests in society, and to claim that politics has merely adjusted to market reform.[68] However, one cannot simply ignore the positive economic impacts of reform, such as a decrease in the price of telecommunications services, increased productivity and investment, and improved teledensity, quality of service, and greater choice. In short, competition among the "politically well-connected" in both countries has proven to be better for consumers and the economy in general than the former monopoly situation.

At another level, however, this type of competition has not weeded out those who obtained market entry purely through state intervention. In Malaysia, reform seems to have been a domestic political strategy to further redistributive goals, as the state has continued to wield substantial power and prerogative. Indeed, even after market reforms, the Philippine and Malaysian states remain as involved as ever, although the level and quality of their intervention differ.

Finally, the persistence of economic and political structures where benefits are highly concentrated, despite reforms to the contrary, is an enigmatic consequence of liberalisation reforms. It is apparent that market liberalisation does not automatically solve the problem of rent-seeking. Rather, liberalisation creates new avenues for it. The type of rents created, who captures them, and their economic consequences are understandable only when one looks at the power relations between state and social actors. It remains an open question whether or not neoliberal reform can transform and ultimately eliminate structures of patronage and rent-seeking in the long term.

With the above theoretical backdrop, the next chapter considers the type of state, business class, and their interaction in Malaysia and the Philippines.

NOTES

1. James Buchanan, Robert Tollison and Gordon Tullock, eds., *Toward a Theory of the Rent-seeking Society* (College Station: Texas A&M University Press, 1980), p. 3.

2. Khan and Jomo (2000), p. 5.

3. Huizhong Zhou, "Rent-seeking and market competition", *Public Choice* 82, nos. 3–4 (1995): 225–41.

4. Mustaq Khan, "Rents, Efficiency and Growth", in Khan and Jomo (2000), pp. 26–33.

5. Khan and Jomo (2000), p. 6.

6. Gordon Tullock, "The Welfare Cost of Tariffs, Monopolies and Theft", *Western Economic Journal* 5 (1967): 224–32; Anne Krueger, "The Political Economy of Rent-Seeking Society", *American Economic Review* 64, no. 3 (June 1974): 291–303; Richard Posner, "The Social Cost of Monopoly and Regulation", *Journal of Political Economy* 83 (1975): 807–27; Buchanan, Tollison and Tullock (1980).

7. Krueger (1974), p. 291.

8. Jagdish Bhagwati, "Directly Unproductive, Profit-seeking (DUP) Activities", *Journal of Political Economy* 90, no. 5 (October 1982): 989–90.

9. Khan, "Rents, Efficiency and Growth", in Khan and Jomo (2000), p. 21.

10. Mustaq Khan, "A Typology of Corrupt Transactions in Developing Countries", *IDS Bulletin* 27, no. 2 (1996): 18.

11. Joseph Stiglitz, "Reflections on the Theory and Practice of Reform", in *Economic Policy Reform: The Second Stage,* edited by Anne Krueger (Chicago: University of Chicago Press, 2000), pp. 551–84; Michael Lipton, "Market, Redistributive and Proto-Reform: Can Liberalization Help the Poor?" *Asian Development Review* 13, no. 1 (1995): 2; Rodrik (1996).

12. John Williamson coined the term "Washington Consensus" in 1990 to describe what he calls "the lowest common denominator of policy advice" to Latin American countries by the Washington-based institutions such as the World Bank and the International Monetary Fund. The term, however, has since evolved to mean a set of neoliberal policy prescriptions based on free market principles or "market fundamentalism", See John Williamson, "What Should the World Bank Think About the Washington Consensus?" *World Bank Observer* 15, no. 2 (August 2000): 251–64.

13. John Williamson, "On Markets and Regulations" paper delivered at the University of California at Sta. Cruz Conference on 20–21 November 1998. For an important critique of the Washington Consensus and its limitations, see Joseph Stiglitz, "More Instruments and Broader Goals: Moving toward the

Post-Washington Consensus", The WIDER Annual Lecture, Helsinki, January 1998 in *Joseph Stiglitz and the World Bank: The Rebel Within* (London: Anthem Press, 2001).

14. The original ten policy prescriptions for Latin America contained in the "Washington Consensus" were fiscal discipline, redirection of public expenditure towards primary health care, education, and infrastructure, tax reform, interest rate liberalisation, competitive exchange rate, trade liberalisation, liberalisation of inflows of foreign direct investment, privatisation, deregulation, and secure property rights. See Williamson (2000), pp. 252–53.

15. Mary Shirley, et al., *Bureaucrats in Business: The Economics and Politics of Government Ownership* (Washington: World Bank, 1995), p. 3.

16. Market failure is said to occur when privately optimal decisions of utility maximizing rational individuals do not generate socially optimal outcomes due to the presence of the following conditions: (1) monopoly, (2) externalities, (3) public goods, (4) imperfect information, and (5) transaction costs. Furthermore, market systems rarely lead to equitable distribution of incomes. See Iyanatul Islam, "Political Economy and East Asian Economic Development", *Asia-Pacific Economic Literature* 6, no. 2 (November 1992): 69.

17. James Buchanan, "Rent-Seeking and Profit Seeking", in *Toward a Theory of the Rent-Seeking Society*, edited by James Buchanan, Robert Tolisson, and Gordon Tullock (College Station: Texas A&M University Press, 1980), p. 9.

18. Mancur Olson, *The Logic of Collective Action* (Cambridge: Cambridge University Press, 1965).

19. Schamis (1999), p. 244.

20. Shirley, et al. (1995).

21. John Ravenhill, "State and Market: Review Article", *Asian Studies Review* 16, no. 3 (April 1993): 113.

22. Ravi Ramamurti, "Why Haven't Developing Countries Privatised Deeper and Faster?" *World Development* 27, no. 1 (January 1999): 137–55.

23. The state in East Asia got involved through what economist call "price-distorting interventions" such as trade policy, industrial licensing and local content requirement, allocation of credit, establishment of state-owned corporations in capital-intensive industrial enterprises, and the introduction of discretionary investment incentives and labour policies. See Joseph Stiglitz, "The State and Development: Some New Thinking", International Roundtable on the

Capable State, Berlin, Germany, 8 October 1997, http://www.worldbank.org/html/extdr/extme/jssp100897.htm; Robert Wade, *Governing the Market: Economic Theory and the Role of Government in East Asian Industrialization* (Princeton: Princeton University Press, 1990); and Alice Amsden, *Asia's Next Giant: South Korea and Late Industrialization* (New York: Oxford University Press, 1989).

24. Douglass North, *Institutions, Institutional Change and Economic Performance* (Cambridge: Cambridge University Press, 1990). See also Joseph Stiglitz, "Redefining the Role of the State: What Should It Do? How Should It Do It? And How Should These Decisions Be Made?" Speech presented at the 10th Anniversary of MITI Research Institute, Tokyo, Japan, 17 March 1998.

25. Brian Levy and Pablo T. Spiller, "A Framework for Resolving the Regulatory Problem", in *Regulations, Institutions and Commitment: Comparative Studies of Telecommunications*, edited by Brian Levy and Pablo Spiller (Cambridge: Cambridge University Press, 1996), p. 7.

26. Paul Cook, "Competition and Its Regulation: Key Issues", *Centre on Regulation and Competition Working Paper Series Paper Number 2* (University of Manchester), October 2001, pp. 3–4. See also Scott Wallsten, "Does Sequencing Matter? Regulation and Privatization in Telecommunications Reforms", *World Bank Policy Research Working Paper*, February 2002.

27. Karl Polanyi, *The Great Transformation: The Political and Economic Origins of Our Time* (Boston: Beacon Press, 1957), pp. 139–41.

28. Richard Snyder, "After Neoliberalism: The Politics of Reregulation in Mexico", *World Politics* 51 (January 1999): 173.

29. Steven Vogel, *Freer Markets, More Rules: Regulatory Reform in Advanced Industrial Countries* (Ithaca: Cornell University Press, 1996).

30. Schamis (1999), pp. 236–68.

31. Ibid, p. 241.

32. Emmanuel De Dios, "Philippine Economic Growth: Can It Last?" http://www.asiasociety.org/publications/philippines/economic.html, pp. 18, 20 (accessed 10 October 2000).

33. Theda Skocpol, "Bringing the State Back In: Strategies of Analysis in Current Research", in *Bringing the State Back In*, edited by Peter Evans, Dietrich Rueschemeyer, and Theda Skocpol (Cambridge: Cambridge University Press, 1985), p. 7.

34. Peter Evans, Dietrich Rueschemeyer, and Theda Skocpol, "On the Road toward a More Adequate Understanding of the State", in *Bringing the State Back In*, edited by Peter Evans, Dietrich Rueschemeyer, and Theda Skocpol (Cambridge: Cambridge University Press, 1985), p. 350.

35. Peter Evans, "The State as Problem and Solution", in *The Politics of Economic Adjustment*, edited by Stephan Haggard and Robert Kaufman (Princeton, New Jersey: Princeton University Press, 1992), p. 146.

36. Evans, Rueschemeyer, and Skocpol (1985), pp. 353, 355.

37. Chalmers Johnson, *MITI and the Japanese Miracle: The Growth of Industrial Policy 1925–1975* (Stanford: Stanford University Press, 1982).

38. Meredith Woo-Cumings, "Introduction: Chalmers Johnson and the Politics of Nationalism and Development", in *The Developmental State*, edited by Meredith Woo-Cumings (Ithaca: Cornell University Press 1999), pp. 1–2.

39. Johnson (1982), pp. 314–20.

40. For instance, see Amsden (1989); Wade, (1990); Jung-en Woo, *Race to the Swift: State and Finance in Korean Industrialization* (New York: Columbia University Press, 1991); and Chang (1994).

41. Evans (1992), pp. 178, 179.

42. Ibid., pp. 179, 181. See also Peter Evans, *Embedded Autonomy: States and Industrial Transformations* (Princeton, New Jersey: Princeton University Press, 1995).

43. Ben Ross Schneider and Sylvia Maxfield, "Business, the State and Economic Performance in Developing Countries", in *Business and the State in Developing Countries*, edited by Sylvia Maxfield and Ben Ross Schneider (Ithaca: Cornell University Press), p. 5.

44. Anne Krueger, *Political Economy of Policy Reform in Developing Countries* (Cambridge: MIT Press, 1993), p. 109.

45. Joel Migdal, *Strong Societies and Weak States: State-Society Relations and State Capabilities in the Third World* (Princeton, New Jersey: Princeton University Press, 1988), p. 19.

46. Joel Migdal, "The State in Society: An Approach to Struggles for Domination", in *State Power and Social Forces: Domination and Transformation in the Third World*, edited by Joel Migdal, Atul Kohli, and Vivienne Shue (Cambridge: Cambridge University Press, 1994), p. 23. See also Benedict Kerkvliet, "Land

Regimes and State Strengths and Weaknesses in the Philippines and Vietnam", in *Weak and Strong States in Asia-Pacific*, edited by Peter Dauvergne (St Leonards, NSW: Allen and Unwin, 1998), p. 160.

47. Migdal (1994), pp. 7–37. It is worthwhile to note that Migdal's view of "state in society" is close to Evan's concept of "embedded autonomy", although they approach the issue in different ways.

48. Kerkvliet (1998), p. 160.

49. Ibid., pp. 161–62.

50. Noll (1999).

51. Wallsten (2002).

52. John Francis, *The Politics of Regulation: A Comparative Perspective* (Oxford: Blackwell Press, 1993), pp. 2, 5.

53. Robert Baldwin, Colin Scott, and Christopher Hood, eds., "Introduction", in *A Reader on Regulation* (New York: Oxford University Press, 1998), pp. 3–4.

54. Sam Breyer, "Typical Justifications for Regulation", *A Reader on Regulation*, edited by Robert Baldwin, Colin Scoot, and Christopher Hood (New York: Oxford University Press, 1998), pp. 61–79.

55. Peter Smith and Bjorn Wellenius, "Strategies for Successful Telecommunications Reform in Weak Governance Environments", draft paper (31 March 1999), pp. 2–3, http://www1.worldbank.org/wbiep/trade/papers_2000/BP2telcm.pdf (accessed 19 October 2000).

56. International Telecommunications Union (hereinafter ITU), *General Trends in Telecommunications Reform 1998 — World, Volume 1* (Geneva: International Telecommunications Union, 1998), pp. 3, 16.

57. William Melody, "Telecom Reform: Progress and Prospect", *Telecommunications Policy* 23 (1999): 12–13.

58. Ibid., p. 17.

59. Levy and Spiller (1996), pp. 4–5.

60. Ibid.

61. Pablo Spiller and Carlo Cardilli, "The Frontier of Telecommunications Deregulation: Small Countries Leading the Pack", *Journal of Economic Perspectives* 11, no. 4 (Autumn 1997): 128–29.

62. Spiller and Cardili (1997, p. 131) provide a survey of the varying experiences of Chile, New Zealand, Australia, and Guatemala in resolving interconnection and other regulatory disputes.

63. Two new methods of pricing computation are used today: the fully allocated cost (FAC) method, and the long-run incremental cost (LRIC) method. See ITU (1998), p. 14.

64. Helmuth Cremer, Farid Gasmi, Andrei Grimaud, and Jean-Jacques Laffont, *The Economics of Universal Service: Theory* (Washington: Economic Development Institute of the World Bank, 1998), p. 2.

65. ITU (1998), p. 14.

66. Ibid., pp. 5, 9.

67. Bumiputera literally means "son of the soil" or an indigenous person. It primarily identifies Malays, but also refers to non-Malay indigenous people, especially those in peninsular and East Malaysia. The Malays compose an overwhelming majority of the bumiputeras such that the two words are sometimes used interchangeably.

68. Shiela Coronel, "Monopoly", in *Pork and Other Perks: Corruption and Governance in the Philippines*, edited by Shiela Coronel (Manila: Philippine Centre for Investigative Journalism, 1998), pp. 148–49.

3

HISTORICAL OVERVIEW OF THE STATE AND BUSINESS IN MALAYSIA AND THE PHILIPPINES

This chapter deals with the development of the state and business in Malaysia and the Philippines, locates the reasons behind their characteristics, and outlines their interactions and power relations. The discussion emphasises crucial events that gave birth to the types of state and business class in both countries. In particular, the purpose of this chapter is to elucidate the rise of a relatively strong state in Malaysia, where a politico-administrative elite existed as early as the advent of independence, and the roots of the weak, penetrated state in the Philippines, with a poorly institutionalised and emasculated bureaucracy. While there are substantial differences between the two countries, patronage and rent-seeking are similarities that stand out but are not normally highlighted.

Malaysia is often depicted in scholarly accounts as a plural society where the concept of race is key to understanding its political, economic, and social life. Scholars often portray post-independence Malaysian politics as a form of consociationalism, whereby bargaining and consensus building among leaders of ethnic communities are the basis of government rule. However, the May 1969 racial riots marked the turning point towards a more pronounced Malay control of state power and the rise of a strong state that resolutely pursues its developmental goals. The launch of the New Economic Policy (NEP) signalled a strategy of active state participation in the economy to create a bumiputera business and entrepreneurial class through trusteeship guided by state institutions. This strategy of institutional trusteeship was replaced with individual trusteeship and ownership once the state believed that there were enough bumiputeras capable of building and running businesses of their own. Aided by state patronage and nurtured by the state under Prime Minister Mahathir Mohamad, a class of well-connected, mostly Malay, businessmen rose to prominence. This class of businessmen became the main beneficiaries of the state's privatisation and liberalisation policies.

By contrast, most scholars portray the Philippine state as weak because an oligarchic economic elite has continually raided the state and moulded state policy in accordance with its interests. The root of this economic elite is traceable to the landowning class that emerged during Spanish colonisation. This class was then able to capture the state through electoral victories during the American colonial period. In this view, the underdevelopment of the Philippines is traceable to the strength of this oligopolistic economic elite and the weakness of the state. Thus, one of the major tasks of post-Marcos administrations has been to lessen the monopolistic or oligopolistic control of this elite over the economy. Such goal was to be achieved through privatisation and liberalisation, with the introduction of competition and the lessening of the state's direct involvement in the economy.

The colonial experience greatly influenced both Malaysia and the Philippines, specifically in terms of the types of state that they developed, the types of business classes that emerged, and the bureaucratic traditions and institutions that linked the two. Thus, the impact of colonial rule in both countries is a logical starting point. Malaysia will be discussed in the first part of this chapter, followed by the Philippines in the second. The third part of the chapter will summarise the similarities and differences of the countries in terms of the state-business characteristics that are pertinent to this study.

HISTORICAL BACKGROUND OF MALAYSIA

The Legacy of British Colonisation

Most scholars view Malaysia's situation and experiences through the lenses of race or ethnicity. Many would agree with R.S. Milne's assessment that Malaysia's racial composition is an important key to understanding its economy, Constitution, democratic process, and party system.[1] The political, social, and economic life of Malaysia is dominated by racial discourse.[2]

Malaysia emerged as a multi-ethnic society from its colonial experience.[3] British colonial rule was formally established in Malaya in 1896 for the Federated Malay States, and effectively in 1919 for the entire Malay Peninsula.[4] The British worked with the indigenous aristocratic class whose members were given English education and recruited into the colonial bureaucracy. Ordinary Malays, who were traditionally peasants and fishermen, were not attracted to working in tin mines or rubber estates.[5] Thus, the Chinese merchants of the Straits Settlements of Singapore and Penang dominated these two industries as early as the mid-18th century until the British took over control in the 19th century. Between 1870 and 1940s, the British imported Indian and Chinese labour to work in tin mines and rubber plantations, regarding them as guest workers. At the same time, they enacted laws that ensured that rural Malays were not displaced. Thus, Malaya developed a dual economy where Europeans and non-Malays dominated the mining, plantation, and urban sectors, while the Malays were in the rural agricultural economy.[6] This policy of importing labour resulted in a multi-ethnic society in which almost half of the population was composed of immigrants.[7]

With no attempts at integration, the three communities developed separate forms of political awareness. From the late 1940s, the Indians looked to an India free of British rule for inspiration, the Chinese looked to developments in China, and the Malays experienced a nascent nationalism based on the fear of being overtaken by the immigrant races. Pre-war Malay nationalism was, in part, a reaction to the growing demands of non-Malay communities for citizenship and political rights. Thus, the dominant trend in Malay nationalism was the increasing awareness brought about by comparisons between the Malay community and the immigrant groups. Anti-colonial sentiment emerged but it was weak, when present. The first major show of force by Malay nationalism came in 1946 when the British proposed a Malayan Union that would greatly reduce the powers of the sultans and give citizenship to all inhabitants. Under intense pressure from the United Malays Nationalist Organisation (UMNO), the British withdrew the proposal. In contrast, anti-colonial sentiment in the Chinese community expressed itself in part through a communist rebellion that continued into the post-independence period.

Aside from the creation of a multi-racial society and a relatively well-functioning, merit-based bureaucracy, one of the lasting legacies of British colonial rule was the drafting of the country's Constitution in 1957, which provided the institutional basis of post-colonial government. The Constitution provided that Malaysia would adopt the British model of parliamentary government, where a prime minister and a cabinet are chosen from the party or coalition of parties that wins the majority of seats in parliament after a general election. The ministers are responsible to the elected lower house of parliament. An appointed upper house, the senate, with limited powers and functions similar to the British House of Lords, was also established. The Constitution also provided for an independent judiciary with the power of review over laws passed by parliament. Finally, an equivalent to the British monarch — who could serve as head of state and function to resolve constitutional conflicts — was established by setting up the office of the king. The hereditary sultans of the nine Malay states would choose a king from among themselves. The kingship would then rotate among the state rulers, each serving a five-year term. The existence of the state rulers prior to British colonisation led to the adoption of a federal instead of unitary government, with each state retaining the power to decide upon issues such as land and Malay customs and religion.[8]

Post-Colonial Malaysia

When Malaya gained independence in 1957, the society was composed of almost 50 per cent Malays and aborigines, 37 per cent Chinese, and 12 per cent Indians. The Malays were mostly located in the rural economy and largely involved in fishing and farming. Some, however, were employed in government service, especially in the higher levels of the bureaucracy. Although many non-Malays were also involved in agriculture, most Chinese were engaged in mining, commerce, and industry, and the Indians were largely employed in plantations or government service. The Malays were dominant in small-scale agriculture and the bureaucracy, while the non-Malays were located in commercial, professional, or working-class occupations.[9]

After independence, a government was formed by a coalition of ethnically based political parties. Known as the Alliance, this coalition drew together UMNO, the Malayan Chinese Association (MCA), and the Malayan Indian Congress (MIC). The Alliance had its genesis in a short-term electoral agreement between UMNO and the MCA in the 1952 Kuala Lumpur municipal elections. The addition of the MIC proved valuable for post-independence governance, as the British colonial government made it clear that independence would only be granted to a multi-ethnic leadership.[10] It soon became apparent that the appeal of a multi-racial coalition composed of distinct ethnic-based parties was greater than that of a multiracial party, with the Alliance winning all of the national elections that it contested from 1955 to 1969. Hence, early post-colonial Malaysia is usually described as a plural society in which consociationalism — the arrangement wherein the leaders of the main ethnic communities were represented in government and where decisions were reached by consensus — was practised.[11] The leaders of UMNO, the MCA, and the MIC met to reach compromises on sensitive ethnic issues (such as national language, citizenship, Malay "special rights", and education policy) instead of conducting public campaigns and open discussions that may provoke inter-ethnic tensions. The general idea guiding the government, called "the bargain", was that Malays would retain their political pre-eminence and "special position" as "indigenous people", while non-Malays would be allowed to maintain their economic position and practise their culture and traditions.[12] Right from the start, however, UMNO dominated the Alliance.[13]

The Economy and Business Class Development

Colonial Malaya was the most lucrative of Britain's colonies with its export of rubber and tin.[14] The British controlled international trade, while the Chinese were dominant in domestic retail trade and acted as intermediaries for British traders. The British, and to a lesser extent the Chinese, were the main beneficiaries of the export economy, while the Malays played a marginal role.

When Malaya gained independence, most writers characterise the Alliance's economic policy from 1957 to 1969 as laissez-faire and a continuation of those aspects of British colonial economic policy that coincided with the interests of its leaders.[15] The First and Second Malaya Plans (1955–65) and the First Malaysia Plan (1965–70) reflected the urban bias of economic policies. For instance, almost half of the budgetary allocation was for infrastructure development (power, transportation, and communications) in the urban areas, and only 7 per cent of the budget was allotted to rural areas.[16] However, a few measures were introduced to facilitate the entry of the Malays into the modern economy, with the establishment of the Rural and Industrial Development Authority (RIDA) and the Federal Land Development Authority (FELDA). RIDA was established in 1950 to provide Malays access to credit and business training. FELDA, established in 1956, was a land development scheme aimed at distributing land to the peasantry for cultivation of cash crops such as rubber and oil palm.[17]

The immediate post-independence economy was open and export-oriented. The Alliance government, however, soon adopted import substitution industrialisation (ISI) policies that offered infrastructure and credit facilities, as well as tariff protection to attract foreign investors. The ISI policies resulted in foreign companies establishing subsidiaries for the assembly, finishing, and packaging of goods produced with imported materials for sale to the protected domestic market. Foreign businessmen, mostly British, used the opportunity to consolidate their economic bases in the post-colonial regime. While Chinese entrepreneurs obtained some benefits, ISI policies did not encourage the development of a Malay business class.

The late 1960s was marked by a shift to export-oriented industrialisation (EOI). The government, nevertheless, established free trade areas for foreign investors and continued providing incentives such as tax holidays.[18] A key aspect of the Alliance government's economic policy was its reliance on foreign capital for the development of manufacturing and industrialisation while small and medium scale industries remained largely in Chinese hands. Therefore, foreign capital continued to play an important role, and reduced the possibility that economic growth would lead to the Chinese domination of the modern economy. Chinese businessmen were unable to make inroads into the large-scale manufacturing sector, not only because of government policy, but also due to the advantages of technology and material resources that foreign capital possessed. Thus, Chinese businessmen concentrated in trading, retail, and property.[19]

At the time of independence, Malay businesses were mostly small and located in traditional cottage industries. One study during this period found that only one Malay in every 623 was involved in business, compared with one in every 40 Chinese.[20] According to Harold Crouch, "Malays expected UMNO to use its political power not just to defend but also to advance Malay economic interests."[21] In the 1960s, a new small class of Malay businessmen emerged as a result of state support. These businessmen mostly benefited from close relations with political leaders and high-level bureaucrats.[22]

This new business class began pressuring the state for more pro-Malay economic programmes. It could influence officials because many of its members were past or present UMNO members or senior civil servants. In response, the government sponsored two Bumiputera Economic Congresses in 1965 and 1968, which recommended policies for increasing Malay economic participation. Following the first congress, Bank Bumiputera was established in 1965 to help Malay individuals and companies obtain credit. In 1966,

Table 3-1: Peninsular Malaysia: Ownership of Share Capital of Limited Companies by Ethnicity and Industry, 1970 (Percentages)

Sector	Malay	Chinese	Indian	Foreign
Agriculture, forestry, and fisheries	0.9	22.4	0.1	75.3
Mining and quarrying	0.7	16.8	0.4	72.4
Manufacturing	2.5	22.0	0.7	59.6
Construction	2.2	52.8	0.8	24.1
Transport and communications	13.3	43.4	2.3	12.0
Commerce	0.8	30.4	0.7	63.5
Banking and insurance	3.3	24.3	0.6	52.2
Others	2.3	37.8	2.3	31.4
Total	1.9	22.5	1.0	60.7

Table 3-2: Peninsular Malaysia: Ethnic Composition of Employment by Industry, 1970 (Percentages)

Sector	Malay	Chinese	Indian
Agriculture, forestry, and fisheries	68	21	10
Mining	25	66	8
Manufacturing	29	65	5
Construction	22	72	6
Utilities	48	18	33
Transport and communications	43	40	17
Commerce	24	65	11
Services	49	36	14

Note: Tables 3-1 and 3-2 are adopted from Edmund Terence Gomez and K.S. Jomo, *Malaysia's Political Economy: Politics, Patronage and Profits*, 2nd edition (Cambridge: Cambridge University Press, 1999), p. 20.

RIDA was expanded into the Majlis Amanah Rakyat (MARA, Council of Trust for Indigenous People), and received additional funding.[23] The two Congresses also helped Malay businessmen access businesses where the state controlled licensing such as in timber, mining, transportation, and contracting.[24] Malays welcomed these measures even though most outside the corporate sectors remained poor and marginalised. Many Chinese and Indian communities were also poor, but they did not expect the government to assist them.[25]

By the late 1960s, economic development was still not felt in the rural areas. The Malays remained poor and were concentrated in low-productivity peasant agriculture. In 1970, foreigners owned 61 per cent of all corporate capital, the Chinese owned 23 per cent, the Indians owned 1.0 per cent. The Malays, who constituted more than 50 per cent of the population, owned only 2.0 per cent.[26]

The 1969 Racial Riots

The "bargain" in its original form became untenable with the outbreak of racial riots after the May 1969 election, in which the Alliance won less than half of the votes.[27] Hostilities erupted in Kampung Baru in Kuala Lumpur where Malays armed with parangs and knives looted and burned Chinese shops and houses. The Chinese resisted and some launched counter-attacks. Rioting spread in Kuala Lumpur and in several states in peninsular Malaysia.[28]

Upon the request of government, the King immediately proclaimed an emergency and suspended the parliament and the Constitution. Executive power was placed in the hands of the National Operations Council (NOC), headed by then Deputy Prime Minister Tun Abdul Razak, and composed of seven Malays, one Chinese, and one Indian. Younger Malay nationalists criticised then Prime Minister Tunku Abdul Rahman's moderate policies as being pro-Chinese. These policies, they argued, were the main cause of the riots. Within UMNO, up and coming members called "Young Turks", like Mahathir Mohamad, publicly called for the Tunku's resignation. The riots exposed the vulnerability of the supposed multiracial unity.

Tun Abdul Razak replaced the Tunku as party president and government leader. This led to substantive changes in the government's form and style.

By 1971, Tun Razak had built a team that was willing to turn UMNO's pro-Malay preferences into government policy. Through an amendment of Constitution, the government made seditious the questioning of the Malay's special position. Also, Tun Razak expanded the Alliance membership to include 11 parties, including the opposition parties Parti Islam SeMalaysia (PAS), the Gerakan Rakyat Malaysia (Gerakan), and the People's Progressive Party (PPP). The Alliance was renamed Barisan Nasional (National Front). Yet, despite the expansion of membership, UMNO remained dominant.[29]

The New Economic Policy

The post-1969 Malaysian state can be characterised as a strong state that had the capacity to adopt a new set of economic policies that emphasised the bringing of Malays into the mainstream of the modern economy.[30]

The New Economic Policy (NEP), unveiled in 1971, became the centrepiece of post-1969 Malaysia. Its main goals were "reducing and eventually eradicating poverty by raising the income levels and increasing employment opportunities for all Malaysians, irrespective of race" and "accelerating the process of restructuring Malaysian society to correct economic imbalance, so as to reduce and eventually eliminate the identification of race with economic function" within 20 years. Poverty was to be reduced from 50 per cent in 1970 to less than 20 per cent by 1990. Malays and other indigenous people who owned less than 2 per cent of the nation's corporate sector in 1970 would own at least 30 per cent by 1990.[31]

The Department of National Unity prepared the Second Malaysia Plan under Tun Razak's leadership. These new interventionist economic policies were similar to the policies called for by two Bumiputera Economic Congresses.[32] Non-Malay participation in the formulation of the NEP came from the National Consultative Council, but only after the plan had already been drafted.

The NEP represented a shift in the government's role towards direct participation in commercial and industrial undertakings.[33] Also, its launch brought about a marked change in decision-making. Previously, the cabinet approved all major decisions involving economic policy. However, because the NEP was passed by parliament, it was not subject to debate and was implemented by a small group within the government. The leaders of non-Malay parties found themselves unable to participate in its implementation. Nonetheless, some of the leaders of the MCA and MIC were able to sway particular decisions in their favour by negotiating which company obtained potential business rewards.[34]

Tun Razak appointed Malays who were close to him to the Economic Planning Unit (EPU), and the Ministries of Finance, Trade and Industry, and Public Enterprises to guide the implementation of the NEP. From then on, this group of "economic bureaucrats determined the national economic policy," reflecting the priorities of the dominant Malay group.[35] In this situation, certain interest groups (Malay and occasionally, non-Malay) who had privileged access to particular political personalities benefited much from their close association. The Razak government's policy towards outsiders — businessmen, groups associated with the opposition, and worker and consumer groups — was to ignore them, and if they became influential, to restrict their activities. State insulation from one group was made possible because of support from another whose interest it actually served. The support from the latter group was substantial enough for the state's legitimacy not to be questioned.

James Jesudason documents how only those who concurred with the viewpoint of the new UMNO leadership were given authoritative positions in the bureaucracy's planning and decision-making processes.[36] In the 1960s, the officials of the Treasury and the Bank Negara (Central Bank), who emphasised balanced budgets and keeping a tight lid on the national and fiscal balance, were ascendant in economic policy making. However, by the 1970s, the development-minded EPU became the custodian of the NEP, overseeing the state's newly found urge to spend more. Increasingly, political considerations rather than fiscal balance became more important. From 1971, the power to decide on developmental projects rested with the EPU, despite the Treasury and Bank Negara's reservations.

Although the NEP had two goals, in reality, wealth restructuring received the most attention, especially the goal of creating a bumiputera business community. The often-repeated goal was the attainment of 30 per cent bumiputera ownership of the total commercial and industrial activities by 1990. Foreign- and Chinese-owned businesses were pressured to sell their shares or reorganise "voluntarily" to avoid state takeover. Two bureaucratic bodies played significant roles in this strategy. In 1974, the Foreign Investment Committee (FIC) drew together high-ranking officials from the Ministries of Finance, Trade and Industry, the Registrar of Companies, and the EPU. The FIC monitored foreign acquisitions of Malaysian companies, making sure that both large public and non-public corporations restructured their equity according to NEP requirements. Secondly, the Capital Issues Committee (CIC), originally established in 1968 to monitor the capital market, worked with the FIC to set the prices of shares to be issued to Malays (government enterprises or individuals), usually at below market price. These two bodies were important in providing the state with the power to influence and shape the economy according to the NEP goals.[37]

The state also created bodies such as Amanah Saham and Permodalan Nasional Berhad to hold the acquired assets of existing foreign or Chinese-owned businesses "in trust" for bumiputras until they could be turned over to capable bumiputra individuals.[38] Other state corporations, such as Perbadanan Nasional (Pernas), the Urban Development Authority (UDA), and the Heavy Industries Corporation of Malaysia (HICOM), were established to enter into construction, engineering, properties, securities, trading, and heavy industries. State Economic and Development Corporations (SEDC) were also formed at the local level. While there were only 54 government-owned enterprises in 1965, the figure reached 656 in 1980, and 1,010 in 1985.[39] Thus, for an essentially open market economy, Malaysia's SOE sector was among the largest in the world.[40]

Chinese Business Responses to the NEP

The NEP brought about drastic changes that constrained opportunities for expansion of Chinese capitalists.[41] In line with the NEP, the Industrial Coordination Act (ICA) of 1975 required manufacturing companies with above RM$100,000 capitalisation and more than 25 employees to have at least 30 per cent Malay equity. Chinese businesses initially protested against the NEP and the ICA and sought the help of the MCA and the Associated Chinese Chambers of Commerce in Malaysia (ACCIM). The government, however, ignored their protests. Depending on their size, Chinese businesses employed various strategies to stay afloat. The most common strategy for small businesses was the so-called Ali-Baba practice, wherein a Malay acted as a nominal partner for a Chinese business to facilitate licensing and paperwork approval. Along these lines, some sought linkages with Malay politicians. Secondly, some manufacturers restricted the size of their businesses or split them into two or more units to remain under the ICA threshold. Thirdly, big businesses incorporated their companies with Malay partners or sought links with senior members of the UMNO elite. Others responded by moving their businesses outside of Malaysia.

Jesudason found that that Chinese big business shifted its focus into short-term economic activities such as property development, banking, and finance rather than manufacturing, where controls were more stringent. His study also found that there was an increased reliance on mergers and takeovers rather than the establishment of new companies. A more assertive, controversial and, in the end, disastrous response to the NEP was the MCA's strategy of pooling Chinese resources to compete with large Malay corporations such as the case of the Multi-Purpose Holdings, which ended in deep financial troubles.

Chinese businessmen who established Ali-Baba relationships, especially those with UMNO politicians, became resilient and successful despite the NEP. Peter Searle categorises them into old and new groups, both needing close ties to the Malay elite to be successful. The old group was characterised by business success before the NEP, and developed special relations with UMNO politicians to keep their businesses afloat. In contrast, the new group rose in the corporate world via close ties with the UMNO elite during the NEP period, in much the same way as the Malay businessmen.[42]

Clearly, there were noticeable linkages between businessmen and politicians. The business community was racially divided, with the state playing a key role in shaping business development though its policies. First, it shaped the expansion of Chinese business, as it introduced equity limits and encouraged foreign participation in the economy. Second, the state served as a midwife to the development of a Malay business class through its interventionist policies.

The Rise of Dr Mahathir

Dr Mahathir Mohamed's ascendancy to the Prime Minister's office marks a turning point in Malaysian political economy. Under Mahathir, the strong, Malay-dominated state that was committed to achieving the NEP goals was increasingly characterised by authoritarianism, centralisation of power in the Prime Minister's office, and control over economic policy-making agencies. These changes that his administration brought about had important implications for how economic policies, such as privatisation and liberalisation, were drawn-up and implemented. Increasingly, executive patronage became important to accessing economic opportunities. Mahathir took a more active role in promoting development and compelled the bureaucracy to perform. Like Tun Razak, Mahathir intervened in bureaucratic decision-making and staffed the EPU and the planning agencies with individuals who shared his visions. He directed government resources to meet his strategies of development, and promoted politicians who were developmentalist and technocratically minded. However, in contrast to Tun Razak, who greatly relied on senior bureaucrats, Mahathir kept a close circle of confidants and associates who were not members of the bureaucracy, the UMNO, or the BN.[43] They acted as his personal advisers on important policy issues concerning development. In addition, the government established think tanks to provide policy analyses and recommendations. A network composed of special advisers, confidants, and technocratic politicians in the central planning agencies and economic bureaucracy surrounded the Prime Minister. Mahathir began to rely less on the bureaucracy and more on his personal advisers and outside consultants for development ideas, which would then be passed on to the bureaucracy to execute. Some civil servants complained about the level of political interference as Mahathir actively interfered with the workings of the bureaucracy, especially as regards the disbursement of government funds. Increasingly, government development funds and projects were used to reward individuals. Furthermore, politically appointed ministers and deputy ministers began to intervene more in the affairs of their ministries.[44]

Recent scholarship points to the development of authoritarian laws and institutions during the 1948 Communist Emergency, and the strengthening of those laws after the 1969 riots.[45] Thus, the Malaysian state had authoritarian instruments and qualities well in place before the 1980s. However, these authoritarian features and the centralisation of power in the hands of the Prime Minister became more pronounced during the Mahathir regime. First, the powers of the nine constitutional monarchs were diminished via amendments to the Constitution in 1984. Second, the sacking of the Lord President and three Supreme Court justices in 1988 undermined the judiciary's independence. In addition, a constitutional amendment removed the power of judicial review. Third, authoritarian laws such as the Internal Security Act were used against the opposition and people who were suspected of being threats to national security. The government rationalized these actions as necessary for the achievement of political stability, ethnic harmony, and economic development.[46] Constitutional and other legislative constraints on the powers of the government were weak because it has always commanded a two-thirds majority in parliament, thus enabling it to change the Constitution or pass laws at will.

The net result of these developments was the creation of a more centralised decision-making system focused on the powers and prerogatives of the office of the Prime Minister. Leong Choon Heng argues that Mahathir's increasing dominance in decision-making was a case of developing political insulation to enable the country to formulate and implement the policy instruments that were necessary for late industrialisation. While the Prime Minister indeed had the political will and determination to see the adoption and implementation of his policies, he was not insulated from the patronage demands of Malays and the UMNO. As will be argued in this study, the reform of the telecommunications sector was largely influenced by the active lobbying of politically well-connected businessmen who had close ties with top UMNO politicians or those considered to be personal confidants of the Prime Minister. The idea of insulation is relative, and in this particular case, although the executive was insulated from some interest groups, the patronage network within UMNO and of top politicians influenced it.

Other scholars argue that the so-called "multilateral elite consultation" in policy matters was abandoned and replaced by a system that depended primarily on the "good judgment and sense of equity and balance being exercised by one individual"[47] — Prime Minister Mahathir. This centralisation of power was a product of, and reinforced by, the use of authoritarian measures vested in the state. The exercise of power by the executive not only helped to channel state-created economic opportunities to well-connected businessmen, but also required them to deploy some of those rents in ways that supported the system, in particular by electoral funding. Thus, executive dominance had significant consequences in the implementation of key economic policies, particularly the NEP and later on, the privatisation policy, which was supposed to roll back state involvement in the economy and trim the overblown and inefficient public sector. Despite the seemingly contradictory means and objectives of these two policies, both involved political patronage in determining access to, and the allocation of, rents.[48]

Some observers contend that although real GDP growth was impressive and the standard of living of the Malays improved under the NEP, its implementation actually hindered economic growth.[49] They point to the capture of the state by patronage networks, in particular that of UMNO, which benefited from rentier economic opportunities. These opportunities were rationalised under the guise of wealth restructuring to attain greater inter-ethnic parity and to develop a bumiputera business class. The preferential treatment of political favourites was not a new phenomenon in Malaysian politics, but under Mahathir it reached rising proportions. Ordinary Malays and non-Malay entrepreneurs needed links to the executive to gain access to franchises, contracts, and projects.[50]

Thus, while the Malaysian state had been characterised by bargaining, the government became undeniably Malay-dominated and actively involved in the economy after 1969, as signalled by the introduction of the NEP. This Malay dominance of the state in turn was used to nurture the development of a Malay business class through preferential policies and state patronage. With Mahathir's ascendancy, observers of Malaysian politics agree that the state became increasingly authoritarian.[51] Diane Mauzy contends that Malaysian politics should now be characterised as "coercive consociationalism" where political bargaining is restricted. Power sharing and accommodation are gradually superseded. In reality, UMNO is hegemonic in the ruling coalition, and in UMNO Mahathir dominated.[52] No less significantly, institutions such as the judiciary, the bureaucracy, the legislature, the sultans, and the Constitution have been weakened by the centralisation of power in the hands of the executive.

To summarise, a strong, Malay-dominated state emerged after 1969, as shown by the dominance of UMNO in the expanded Barisan Nasional, and the enhanced power of the office of the Prime Minister and key economic policy-making agencies located under his department. These developments further reinforced the power of the Prime Minister's office under Mahathir. Thus, knowing where power in the state lies is crucial to understanding the type of economic policies that have been adopted and the ways in which they have been implemented. The privatisation and liberalisation of the telecommunications industry, as this study will demonstrate, is an illustrative example of these developments.

HISTORICAL BACKGROUND OF THE PHILIPPINES

In contrast to how scholars have portrayed the emergence of a strong and increasingly authoritarian state in post-colonial Malaysia, most scholars studying the Philippines assert that the state is weak in the face of a strong oligarchic class.[53] The state's weakness is seen in its inability to launch or implement a consistent economic development programme, as it is continually overrun by the particularistic demands of a powerful oligarchy. Through the capture of state institutions, this oligarchy has preserved and expanded its wealth by diversifying its economic base. The emergence of this elite is traceable to the indigenous landowning class that surfaced during the country's colonial experience under Spain. This class gained control of political power under the American colonial regime, as the Americans sought to co-opt the elites to accept its colonial rule. Paul Hutchcroft points out that the legacy of American colonialism is more of "oligarchy-building rather than state-building",[54] as elected officials in the Philippines learned to use public office for personal or familial wealth accumulation rather than the pursuit of the nation's interest.

Colonial Legacies

The Philippines was subjected to direct colonial rule for over three hundred years before Malaysia, which resulted in a different trajectory of state formation and business class development.[55] Spain's main commercial interest in the Philippines was its proximity to China, making it a useful trading post for Chinese silk and porcelain in the Galleon trade to Mexico.[56]

In 1834, with the collapse of the Galleon trade, Manila was opened to international trade. The Philippines was gradually integrated into the international economy due to the increasing demand in Europe for crops such as sugar, abaca, tobacco, and rice. Hacienda-based commercial agriculture was introduced in the country, from which emerged a landowning class who eventually formed the economic and political elites of the 20th century.

The Philippine revolution against Spain, the establishment of a Filipino government led by the Katipunan,[57] and its declaration of independence on 12 June 1898 were hijacked by the intervention of the United States of America.[58] Most of the landed elite and Manila-based, educated Spanish-Filipinos chose to collaborate with the Americans and campaigned for the acceptance of the new colonial rule. Several of them became members of the Philippine National Assembly, elected in 1907 by only 1.4 per cent of the country's 7.6 million population who were then qualified to vote.[59]

The three major legacies of American rule are the legal institutions patterned after the United States, the economic elite's control of the state, and the weak state. The American institutional legacies include a presidential system, a bicameral legislature, a multi-tiered hierarchy of elected local executives, and a bureaucracy that has little autonomy from elected officials.[60] The evolution of state institutions, which have been described as weak, captured, incapable of making consistent and coherent economic policies, and the strength of the economic elite outside of the state are traceable to how "democracy" was introduced to the country under colonization. Elections, another legacy of colonial rule, became an avenue for local and provincial elites to control local and national public offices.[61] This early entrenchment into public positions of local elites led to the fusing of economic power with political power even before democratic institutions took root.[62]

Secondly, by introducing elections and representative institutions from the bottom up, American colonial elections enabled local landlords to consolidate their hold on the national state, leading to the establishment of a "solid, visible, national oligarchy".[63] Meanwhile, American colonial rule created a minimal civil service. Many civil servants owed their employment to political patrons rather than their own merit. Hence, the primary loyalty of government employees was to the political patron who obtained them the job rather than to the agency head. Thus, the formal line of power in a government agency was often undermined by a more powerful tie of loyalty between a political patron and their client in the bureaucracy. Such situations led to a bureaucracy that was highly divided and weak. Thus, American colonial rule strengthened the oligarchy rather than built state institutions. The landed elites became a national force as they secured control of the central government in Manila. Having done this, they monopolised the new opportunities for wealth creation that the state provided.[64]

Post-Colonial Philippines, 1946–69

In contrast to Malaysia's communally-based political parties, the party system in post-colonial Philippines was dominated by two political parties that had identical policies and sources of support. The weak party system, in which intra-party solidarity was minimal and inter-party switching was prevalent,[65] and the subsequent weak presidency exacerbated the oligarchy's rent-seeking.

Since no common programme united party members, the President of the Republic, although head of his party, was actually weaker than the Constitution intended. Moreover, the presidential system of government meant that a president could not necessarily count on legislative support for his programmes.

Also different from Malaysia was the underlying dynamic in the political economy. Ethnic relations were scarcely a factor, and the landed elite of colonial era continued to control the economy using state institutions to protect their properties and help them to diversify into manufacturing. As Emmanuel De Dios has put it, access to the state machinery became the principal means of direct or indirect preservation and accumulation of wealth.[66] A key mechanism through which these dynamics were manifest were import substitution industrialisation (ISI) policies.

Under ISI, the Philippines enjoyed one of its highest post-war growth rates for the manufacturing sector and the economy in general.[67] In fact, during the 1950s, the Philippines had one of the highest per capita incomes in East Asia, higher than South Korea, Taiwan, Thailand, Indonesia, and China and was only below Japan, Malaya, Hong Kong, and Singapore.[68] Yet five decades later, the picture was starkly different, as its Asian neighbours left the Philippines behind.[69]

A partial explanation for this situation can be found in the state and the business class that emerged as an outcome of ISI policies of the 1950s and 1960s. The ISI class was composed mostly of landed elites who diversified from agriculture into the manufacturing without relinquishing their agricultural interests. Temario Rivera argues that because a significant section of the ISI capitalists were land-owning, they were faced with a "contradictory set of interests" that undermined their potential as an agency for industrial transformation. That is, since the landlords were also the major capitalists, the very class that could have benefited from the expansion of the internal market to absorb manufactured goods often blocked land redistribution, which in turn could have created a bigger domestic market. In reality, these landlord-capitalists were not interested in the development of one economic sector. Their loyalty lay in the profitability and expansion of their family conglomerates. Thus, the ISI capitalists who could have been important constituencies for industrialisation and for a coherent and consistent economic policy, as in other countries, did not play a similar role in the Philippines. Rivera contends that without a strong state to resolve the tensions between the traditional agrarian elite, local manufacturers, transnational capital, and global financial institutions, the transition to a higher stage of import substitution could not be pursued successfully.[70] Consequently, this situation reflected a structural constraint that made sustaining industrial growth in the country problematic, making it difficult for the state to formulate and implement policies independent of the powerful vested interests in society.

Martial Law

In 1969, Marcos became the first president to be re-elected, breaking the cycle of orderly exchange of power between the two dominant political parties. However, his second term in office came at a very high price, precipitating a balance of payments crisis by 1970 due to heavy government spending in the 1969 elections.[71] In 1971, Marcos convened a Constitutional Convention to debate the issue of changing the form of government from a presidential to a parliamentary system. Yet, before the Convention could finish its work, martial law was declared on 22 September 1972, ostensibly to protect the country from the growing threat posed by the Communist Party of the Philippines and the "exploitative traditional oligarchy".[72]

Under martial law, both houses of Congress were closed down. Civil rights were curtailed with the suspension of the writ of habeas corpus, and the arrest of real and potential opponents of the government such as opposition politicians, media commentators, and labour and student leaders. All media outlets were closed or taken over by the military. Marcos claimed that he declared martial law to "smash the self-serving power of the traditional elites or oligarchs", "reform the "US-style democratic processes that were unresponsive to the needs of Philippine society", and "to build a New Society".[73] He swiftly took over the companies of the so-called oligarchs, especially targeting the economic bases of his political rivals such as the Lopezes, Osmeñas, Jacintos, Elizaldes, and Aquinos-Cojuangcos. He jailed his political opponents on trumped up charges and those who escaped went into exile.

To weaken the "traditional feudal elite who rule the country", Marcos issued Presidential Decree 27 declaring land reform for the entire country, aiming to break up rice and corn haciendas and make tenants owners of the land they till. The government also took over the procurement and marketing of sugar and copra, which were the basis of the traditional elite's economic power, and issued decrees forcing large family-owned corporations to go public. Finally, Marcos centralised decision-making power in the presidency, increased the role of the military, and suspended elections. The former leverage of local elites vis-à-vis national politicians of delivering votes was removed, effectively centralising power in the hands of the executive.

While expropriating the wealth of his political opponents, Marcos also attempted to co-opt a part of the traditional economic elite that was not politically active.[74] Marcos proceeded to replace the dominant economic clans with a new set of business leaders who were loyal to him, more popularly known as the "crony capitalists". Most of the cronies were new rich, although some of them came from old landed families. They were conferred privileges such as the control of the assets taken over from the oligarchs (in broadcasting, newspapers, public utilities, agricultural land, and financial institutions), assigned trading monopolies, and received an array of other exclusive privileges, loans, and guarantees.

During the early years of martial law, the displacement of traditional clans with the new cronies seemed to mark a break from the old development path dominated by landed elite towards the rise of "new men" in Philippine society. Yet, as De Dios and Hutchcroft point out, the crony phenomenon was not a deviation but more of a logical extension and culmination of the pre-martial law process of using access to the political machinery to accumulate and preserve wealth. The main difference was that under martial law, one ruler had monopoly control of the state, which he used to further his personal ends.[75]

Marcos' consolidation of power during the 1970s was helped by a buoyant economy, brought about by high commodity prices and the availability of cheap credit from foreign private creditors. The availability of huge amounts of funds at low interest rates in private capital markets abroad meant that the Philippines' chronic current account deficits could be cheaply financed without adjusting economic policies. Thus, the inflow of funds allowed the country's gross national product (GNP) to grow at an average of 6.4 per cent from 1970 to 1983. Yet, the onset of the debt crisis in 1982 led to the collapse of this debt-driven growth strategy. Between 1980 and 1983, GNP growth dropped to 2.2 per cent annually. To keep the economy afloat, the Philippines switched to taking out short-term loans. The deepening of a worldwide recession, and its domestic impact in the Philippines was exacerbated in January 1981 with the flight of a well-known Marcos crony, Dewey Dee, who left behind huge debts that public financial institutions shouldered. In August 1983, the assassination of Senator Benigno Aquino brought the crisis of confidence in the Marcos regime to its highest levels. Sources of short-term credit dried up, and some lenders tried to call in their loans. This led to another balance of payment crisis that forced Marcos to declare a debt moratorium.[76] In a climate of a credit crunch and the increasing consolidation of opposition to the Marcos government, the country's GDP contracted by as much as 19 per cent between 1982 and 1986.[77]

Marcos rationalised his declaration of martial law as a way of dismantling the control of the traditional oligarchy in the county. Yet, he selectively prosecuted "the oligarchy", focusing only on his political opponents and created a new set of businessmen who temporarily displaced the traditional economic elite. Notwithstanding this, Marcos is credited for establishing an integrated economic policy-making agency, the National Economic and Development Authority (NEDA), which was staffed with technocrats, academics, and professionals.[78] Ironically, the rise of crony capitalism during the 1970s was covered up by the presence of credible technocrats in top government offices who served as link to international financial institutions. Their presence in the Marcos government provided the country access to international loans that financed the perennial balance of payments deficits, and enabled the government to launch debt-driven growth. Finally, the rule of Ferdinand Marcos, while in some ways was an aberration, can also be seen as consistent with the pattern of pre–martial law politics, especially with how he used the state for personal, familial, and crony enrichment.

The Aquino Administration

In February 1986, a "people power revolution" toppled the 20-year Marcos regime. The government of Corazon Aquino was supported by a broad coalition of people from varying ideological perspectives. In fact, a fraction of this coalition came from large business and landowning interests whose basic demand was the return of the pre-martial law property rights order.[79] Despite the presence of progressives, conservatives who represented big business and landowning interests dominated the Aquino government. Progressives lobbied Aquino to use the revolutionary power of her government to introduce redistributive policies such as land reform, as well as adopt a policy of selective debt repudiation of the massive foreign debt left by Marcos. Instead, Aquino restored the pre–martial law elite democracy with the 1988 elections that re-established Congress and the ratification of the 1987 Constitution.

The Aquino government trod carefully between the interests of the masses and those of the elite. Although it attempted to undo the Marcos legacy, especially in terms of economic plunder, little of Marcos and his cronies' ill-gotten wealth returned to the state. Moreover, the administration did not thoroughly investigate cases of crony-controlled companies that involved Aquino's relatives.

In the end, the Aquino government became more conservative and restorationist. The convening of Congress in 1987 opened the way for the return to power of traditional politicians and members of the elite, including former Marcos loyalists. Roughly 65 per cent of the restored Congress was composed of traditional politicians and about 35 per cent were from political families who were associated with the Marcos regime.[80] With the return to Congress of traditional landed elites, the prospect of redistributive measures such as agrarian reform dimmed. At the end of Aquino's term, the pre–martial law elites were back in their previous favoured economic and political positions.

The Ramos Administration: The Problem of a Weak State and Strong Oligarchy

Fidel Ramos ascended to the presidency in 1992 and declared the necessity of market reform to create a level playing field. Ramos' stated aim was to end the oligarchic control of a few who profit without being productive and reward economic productivity and efficiency. His rhetoric is best understood in the context of the weak Philippine state that has been raided by a powerful economic class. This class has been successful so far in using the state to preserve and continue accumulating wealth for its own benefit.

Scholars describe the weakness of the state and its capture by a strong oligarchic class as a developmental bog that the Philippines seems incapable of escaping. Hutchcroft argues that the word "oligarchy" is the appropriate term to use to describe the economic elite that dominates the Philippines because there is really no fixed aristocracy in the country. Though most of the oligarchs come from landowning backgrounds, there is a continuous creation of new rich that makes it big in business after gaining favourable access to the state apparatus.[81] Its type of capitalism has been described as booty capitalism wherein "business is born and flourishes or fails, not so much in the marketplace as in the halls of the legislature or in the administrative offices of the government".[82] The Ramos administration believed that the solution to this problem lay in attacking the structures that kept the economy underdeveloped through the introduction of market reform. Market-oriented policies were seen as the way to improve the economy, by removing state intervention in areas where the market would be best left alone, or by introducing competition in areas that were monopolised by politically influential people. As the case of liberalisation of the telecommunications industry shows, there were ways for reform to be implemented despite the dire situation the Philippines found itself in.

A SUMMARY OF KEY SIMILARITIES AND DIFFERENCES

Colonial Bequests: Institutional, Political, and Economic Legacies

The Philippines and Malaysia were fundamentally shaped by their colonial experiences. Malaysia emerged as a multiracial society from colonisation. The domestic business class was composed mostly of the Chinese, while political-bureaucratic power was in the hands of Malays. Malaysia's federal, parliamentary form of government with a constitutional monarchy, its bureaucracy, and its legal system were patterned after the British model. British colonisation left the Malaysian state with authoritarian instruments that had been developed to fight the communists in 1948. None of these repressive laws have been amended or repealed so far. In fact, they have become useful instruments for suppressing political opposition.

The political parties that evolved were racially organised, with the winning coalition composed of parties representing the three major races. Thus, the form of government in post-colonial Malaysia has been described as one characterised by consociationalism. Yet, after the 1969 racial riots, the state became categorically Malay-dominated, as shown by the creation of the New Economic Policy (NEP), which marked the rise of a strong state actively involved in the economy. A new, expanded alliance of 14 political parties, which at one point even included opposition parties, called the Barisan Nasional (National Alliance) controlled the government. Yet it is no secret that the United Malays National Organisation (UMNO) dominates it, and thus controls state power.

When Mahathir became Prime Minister in 1981, the Malaysian state is said to have become more authoritarian and pronouncedly Malay. With the adoption of privatisation and liberalisation policies, the state has created market entry opportunities for a chosen few, in the guise of creating a bumiputera economic class.

In the Philippines, the unitary, presidential system with a bicameral legislature and a judiciary is a legacy of American colonisation. While the British co-opted the Malay aristocracy by appointing them to the Malayan Civil Service, the Americans co-opted the Filipino elite by the early introduction of elections and the Filipinisation of the colonial bureaucracy.

With the early introduction of elections in the Philippines, economic and political power became fused in the same hands. This can be contrasted with the Malaysian case where economic power was in the hands of the Chinese while political power was in the hands of Malays. Thus, in the Philippines the economic elite's capture of state offices enabled it to use this state access as the primary means of wealth creation, expansion, and preservation. In Malaysia, state power was also used to dispense patronage, though in a different way. Following independence, state patronage benefited businessmen close to the Alliance, including the broader Chinese business community. After 1969, state patronage was extended to the Malays through the implementation of the NEP. Under Mahathir, state support was directed to Malay individuals, although some non-Malay businessmen also benefited.

In Malaysia, political power is firmly controlled by UMNO and the Barisan. In the Philippines competing factions of an economic elite organised in two nominally different parties up until 1970s controlled a weak state. The two parties were virtually undifferentiated in terms of ideology and policies. Party switching, which was rare in the case of Malaysia, was common in the Philippines. These two parties coexisted and alternated in power during the post-war period, until the pattern was broken in 1969 with the re-election of Ferdinand Marcos as president and his declaration of martial law in 1972. This two-party system changed into a multiparty system after 1986. However, political parties continue to be nominally differentiated and personality-centred.

The Post-Colonial State and the Economy[83]

The year 1969 proved to be a landmark year for both Malaysia and the Philippines. Like the Malaysian state, the Philippine state became highly involved in the economy. Marcos shrouded his actions with a discourse of breaking down the oligarchic control of the economy and creating a New Society. However, it became clear that Marcos only succeeded in replacing the old economic elite with a new set of favoured businessmen. In a way, Marcos' provision of privileges to his cronies is comparable to the special treatment that the privileged bumiputera businessmen enjoyed under Mahathir. Yet, while Marcos used the state to enrich himself and his cronies, Mahathir seemed to have believed that he was using the state to develop a group of businessmen for the nation's, or at least the Malay community's ends and not for his personal benefit.

The post-colonial economy of the Philippines was based on the export of natural resources and agricultural products. Similar to Malaysia, where foreigners controlled about 60 per cent of corporate wealth until 1970, Americans held a high stake in the Philippine economy until the early 1970s.

Malaysia was endowed with lucrative natural resource rents from oil and gas production that generated foreign currency revenues for the country and provided a soft budget constraint on the government.[84] This major difference provided the Malaysian state a leeway in making costly developmental decisions. The Philippines in its part had to resort to foreign borrowing.

Both states used their power to restructure the ownership and control of the economy. This was done decisively by the Malaysian state in its goal of creating a bumiputera business class, with the state intensifying its involvement in the economy. In the Philippines, however, the landowning elite captured state-created economic opportunities and diversified out of agriculture. Under Marcos these opportunities were appropriated for himself and his cronies.

In the 1970s, both economies grew rapidly, although debt borrowing propelled growth in the Philippines. Both countries were badly affected by the 1980s worldwide economic recession. However, Malaysia was not as badly affected, due to large foreign currency reserves coming from oil revenues and the entry of Japanese, Korean, and Taiwanese investments into the country. The Philippines meanwhile was severely affected as its foreign debt soared, coupled with a political crisis and decline in the Marcos regime's legitimacy. Malaysia recovered from the economic slowdown of the 1980s with the economy growing at an average of 6–7 per cent during the decade. In the Philippines, however, the 1980s are called "the lost decade", with the economy contracting as much as 19 per cent between 1982 and 1986, and the average growth rate for the entire decade was only 1 per cent.

Table 3-3: Average Growth of GDP in Southeast Asia, 1950–2000 (% p.a.)

Country	1950–60	1960–70	1970–80	1980–90	1990–2000
Malaysia	3.6	6.5	7.8	5.3	7.0
Philippines	6.5	5.1	6.3	1.0	3.2
Indonesia	4.0	3.9	7.6	6.1	4.2
Singapore	n.a	8.8	8.5	6.6	7.8
Thailand	5.7	8.4	7.2	7.6	4.2

Source: Arsenio Balisacan and Hal Hill, "An Introduction to the Key Issues", in *The Philippine Economy: Development, Policies and Challenges*, edited by Arsenio Balisacan and Hal Hill (New York: Oxford University Press, 2003), p. 7.

Patronage and Rent-Seeking Types

In Malaysia, state patronage was dispensed through racially organised political parties, in particular the members of Barisan Nasional, and most importantly UMNO.[85] Jomo and Gomez categorize state-created rents into two broad types. The first can be called developmental rents, which were created to diversify the economy, and promote industrialisation and technological development. The second type, called redistributive rents, is more political in nature and directly linked to the adoption of the NEP and its goal of the ethnic redistribution of wealth.[86] In the Philippines, family-based conglomerates and businesses secured state patronage. State-created rents were usually monopoly rents wherein entry barriers were established, thus generating protected markets for the rent-recipient whether it was in the extraction of natural resources, trade, or manufacturing. In contrast to Malaysia, family and factions, not political parties, were the main vehicles for wealth creation and accumulation in the Philippines.

Jomo and Gomez argue that the capture of state-created rents in Malaysia may not be competitive and costly because ethnicity limits the number of rent-seekers. Rent seeking by nature may not be very competitive because of the clandestine, exclusive, or protected nature of the rent-capture process, thus keeping the rent-seeking cost low. After 1969, most rents were redistributed along ethnic lines, and were rationalised as fundamental to maintaining political stability. Under Mahathir, access to executive patronage became increasingly essential to gain new market opportunities. While the outstanding reasons were still economic restructuring and redistribution, the select group of recipients of state created rents included non-Malays. Mustaq Khan describes this situation as "centralized clientelism" where the rent-seeking costs are kept low by the centralisation of power in the office of the Prime Minister.[87] Jomo and Gomez nonetheless make the argument that an analytical distinction should be made between "competitive rent-seeking" and "purposive rent allocation." The former refers to a situation wherein persons or groups actively seek out rents, while the latter refers to a situation where the state deliberately allocates rents to persons or groups who may exert little effort in capturing them. The rent-seeking cost is low when the state purposively allocates rents and when a small number of elites compete for them because of the a priori elimination of other potential rent-seekers due to the state's redistributive agenda. Both situations seem to coexist in the case of Malaysia.

In contrast to the predominance of purposive rent allocation in Malaysia, the Philippines has been characterised by competitive rent-seeking among members of an influential oligarchy who lobbied for privileged status from an incoherent state. The presence of two competing factions in the political elite that alternate in power ensured the continuity of this process, making the rent-seeking process highly variable, less calculable, and thus, more costly. Compared with the centralisation of power in the executive in Malaysia, the lines of formal authority in the Philippines were weaker, with a disjuncture between power and authority, as patrons were often outside of the formal structures of the state. In addition, government administrations had a short time frame that makes the political environment unpredictable and insecure, thus leaving little incentive for rent-recipients to adopt a long-run strategy of capital investment or use of the rents they receive. Compared with Malaysia, the Philippines lacks a state that deliberately created and allocated developmental or redistributive rents. Instead, the state was raided by the dominant economic elites who competed with each other to acquire a monopolistic or oligopolistic position in protected markets. Costly, unproductive monopoly rents were thus created by the state, which were secured by a strong oligarchy. Paul Hutchcroft describes this situation as a "politics of privilege" that was generally obstructive to growth and economic development.[88]

This chapter has traced the different trajectories and types of the state and business class, and the varying levels of economic development in Malaysia and in the Philippines. The types of patronage and rent-seeking networks have also been examined and contrasted.

Although the Malaysian state is depicted as strong because it is capable of drawing up a developmental plan and implementing it, this study argues that the Malaysian state is nonetheless highly penetrated by dominant Malay interests, which coincided with the developmentalist ideology of former Prime Minister Mahathir Mohamed. In Philippines, a consensus among observers is that the state is weak and penetrated by an influential oligarchy. Yet, this weak, penetrated state was able to introduce reforms that went against the established interests of the dominant economic class. These puzzles will be fully explained in the remaining chapters of this book.

The following chapter zeroes in on the case studies, focusing on the conditions of the Malaysian and Philippine telecommunications sector before discussing the adoption of market-oriented policies.

NOTES

1. R.S. Milne, *Government and Politics in Malaysia* (Boston: Houghton Mifflin Company, 1967), p. 3. Only a few scholars have tried to study the country using class analysis, such as Kwame Sundaram Jomo, *A Question of Class: Capital, the State and Uneven Development in Malaya* (Singapore: Oxford University Press, 1986); Hua Wu Yin, *Class and Communalism in Malaysia* (London: Zed Books, 1983); and Anne Munro-Kua, *Authoritarian Populism in Malaysia* (London: Macmillan Press, 1996).
2. K.J. Ratnam, *Communalism and the Political Process in Malaya* (Kuala Lumpur: University of Malaya Press, 1965), p. 2.
3. Gordon Means, *Malaysian Politics* (London: University of London Press, 1970).
4. Barbara Watson Andaya and Leonard Andaya, *A History of Malaysia* (Hong Kong: Macmillan Press, 1982), pp. 205–64.
5. Harold Crouch, *Government and Society in Malaysia* (St Leonards, NSW.: Allen and Unwin, 1996), pp. 17–18.
6. Diane Mauzy, "Malay Political Hegemony and 'Coercive Consociationalism'", in *The Politics of Ethnic Conflict Regulation*, edited by John McGarry and Brendan O'Leary (London: Routledge, 1993), p. 108.
7. Andaya and Andaya (1982), pp. 136–38, 175–81.

8. R.S. Milne and Diane Mauzy, *Malaysian Politics Under Mahathir* (London: Routledge, 1999), pp. 14–15.

9. Crouch (1996), pp. 14–15.

10. Means (1970), pp. 133, 153, 161–67.

11. Crouch (1996), pp. 152–53.

12. Ibid., p. 20.

13. Donald Horowitz *Ethnic Groups in Conflict* (Berkeley: University of California Press, 1985), pp. 418–19. For an opposing view, see John Funston, *Malay Politics in Malaysia: A Study of UMNO and PAS* (Kuala Lumpur: Heinemann Educational Books, 1980).

14. Edmund Terence Gomez and K.S. Jomo, *Malaysia's Political Economy: Politics, Patronage and Profits* (Cambridge: Cambridge University Press, 1999), p. 10.

15. Ibid., p. 14.

16. Munro-Kua (1996), p. 16; Gomez and Jomo (1999), p. 16.

17. Gomez and Jomo (1999), p. 15.

18. See the 1968 Investment Incentives Act and the 1971 Free Trade Zone Act.

19. Gomez and Jomo (1999), pp. 17–18; Munro-Kua (1996), p. 29.

20. James Jesudason, *Ethnicity and the Economy: The State, Chinese Business and Multinationals in Malaysia* (Singapore: Oxford University Press, 1989), p. 64.

21. Crouch (1996), p. 22.

22. Munro-Kua (1996), p. 30; Jesudason (1989), pp. 64–65.

23. Jesudason (1989), pp. 65–66.

24. Munro-Kua (1996), p. 30.

25. Crouch (1996), p. 22.

26. These statistics do not include government corporate ownership.

27. Crouch (1996), p. 74.

28. Gordon Means, *Malaysian Politics, the Second Generation* (Singapore: Oxford University Press, 1991), p. 7.

29. Means (1991), pp. 27–32; Andaya and Andaya (1982), pp. 280–82.

30. Means (1991), pp. 8–10; Crouch (1996), pp. 24–27.

31. *Second Malaysia Plan, 1971–1975* (Kuala Lumpur: Government Press, 1971), p. 1.

32. For a good background on the drafting of the NEP, see Mavis Puthucheary, "The Shaping of Economic Policy in a Multi-Ethnic Environment: The Malaysian Experience", in *Economic Policy-making in the Asia-Pacific Region*, edited by John Langford and K. Lorne Brownsey (Canada: Institute for Research on Public Policy, 1990), pp. 273–98.

33. *Second Malaysia Plan*, p. 7.

34. Puthucheary (1990), pp. 289–90.

35. Many of these officials were members of the two bumiputra congresses convened in 1965 and 1968 that lobbied for greater state intervention to aid Malays economically. See Puthucheary (1990), pp. 281–82.

36. Jesudason (1989), pp. 78–79.

37. Ibid.

38. Ozay Mehmet, *Development in Malaysia: Poverty, Wealth and Trusteeship* (Kent: Croom Helm, 1986); Mauzy (1993), p. 112.

39. Gomez and Jomo (1999), p. 31. Ironically, while privatization was formally introduced in 1983, the number of public enterprises continued to grow and by 1992 reached 1,149.

40. Christopher Adam and William Cavendish, "Background", in *Privatizing Malaysia: Rents, Rhetoric, Realities*, edited by K.S. Jomo (Boulder: Westview Press, 1995), p. 15.

41. This section draws from Jesudason (1989) and Crouch (1996).

42. Peter Searle, *The Riddle of Malaysian Capitalism: Rent-Seekers or Real Capitalists?* (NSW: Allen and Unwin, 1999), pp. 246–48. See also Edmund Terence Gomez, *Chinese Business in Malaysia: Accumulation, Accommodation and Ascendance* (Surrey: Curzon Press, 1999).

43. Leong Choon Heng, "Late Industrialization Along with Democratic Politics in Malaysia", Ph.D. dissertation, Harvard University, 1991, p. 276.

44. Leong (1991), pp. 270–71.

45. Tim Harper, *The End of Empire and the Making of Malaya* (Cambridge: Cambridge University Press, 1999).

46. Gomez and Jomo (1999), p. 3; Crouch (1996), pp. 94–95.

47. Means (1991), p. 112.
48. Gomez and Jomo (1999), p. 4.
49. Adam and Cavendish (1995), p. 15.
50. Andaya and Andaya (1982), p. 319.
51. Various scholars have described the increasing authoritarianism of the Malaysian state in various ways such as "quasi-democratic", "semi-authoritarian", "semi-democratic", or "neither authoritarian nor democratic". See Zakaria Haji Ahmad, "Malaysia: Quasi-Democracy in a Divided Society," in *Democracy in Developing Countries*, vol. 3, edited by Larry Diamond, Juan Linz, and Seymour Lipset. (Boulder: Lynne Rienner, 1989); Harold Crouch, "Authoritarian Trends, the UMNO Split and the Limits to State Power", in *Fragmented Vision: Culture and Politics in Contemporary Malaysia*, edited by Joel Kahn and Francis Loh Kok Wah (Sydney: Allen and Unwin, 1992); William Case, "Semi-Democracy in Malaysia: Withstanding the Pressures for Regime Change", *Pacific Affairs* 66, no. 2 (Summer 1993): 183–205; Harold Crouch, "Malaysia: Neither Authoritarian nor Democratic", in *Southeast Asia in the 1990s*, edited by Kevin Hewison, Richard Robison, and Garry Rodan (St Leonards, NSW: Allen and Unwin, 1993).
52. Mauzy's arguments on accommodation, however, should be reassessed in the light of the BN's increasing dependence on Chinese votes to win elections, as observed in 1999. Furthermore, it will be interesting to observe how power relations change within UMNO and Barisan, and how state power will be wielded following Mahathir's resignation as Prime Minister in October 2003.
53. Observers of Philippine government and politics have described its political system as cacique democracy, elite democracy, a patrimonial oligarchic state, an oligarchic democracy, boss politics, and an anarchy of families — to name just a few. See Benedict Anderson, "Cacique Democracy in the Philippines", in *The Spectre of Comparisons: Nationalism, Southeast Asia and the World* (New York and London: Verso, 2000), pp. 192–226; Francisco Nemenzo, "From Autocracy to Elite Democracy", in *Dictatorship and Revolution: Roots of People Power*, edited by Aurora Javate-De Dios, Petronilo Bn. Daroy, and Lorna Kalaw-Tirol (Manila: Conspectus Foundation, 1988), pp. 221–68; Paul Hutchcroft, *Booty Capitalism: The Politics of Banking in the Philippines* (Ithaca: Cornell University Press, 1998); John Sidel, *Capital, Coercion and Crime: Bossism in the Philippines* (California: Stanford University Press, 1999); Alfred McCoy, ed., *An Anarchy of Families: State and Family in the Philippines* (Wisconsin: Center for Southeast Asian Studies, University of Wisconsin, 1993).

54. Paul Hutchcroft, "Booty Capitalism: Business-Government Relations in the Philippines", in *Business and Government in Industrializing Asia*, edited by Andrew MacIntyre (Ithaca: Cornell University Press, 1994), p. 225.

55. Harold Crouch, *Economic Change, Social Structure and the Political System in Southeast Asia: Philippine Development Compared with the Other ASEAN Countries* (Singapore: Institute of Southeast Asian Studies, 1985), p. 7.

56. Anderson (2000), p. 194.

57. Reynaldo C. Ileto, *Pasyon and Revolution: Popular Movements in the Philippines, 1840–1910* (Quezon City: Ateneo de Manila University Press, 1979), pp. 75–103; Teodoro Agoncillo, *Revolt of the Masses: The Story of Bonifacio and the Katipunan* (Quezon City: University of the Philippines, 1956); Teodoro Agoncillo and Milagros Guerrero, *History of the Filipino People*, 7th edition (Quezon City: R.P. Garcia Publishing Company, 1987), pp. 149–65; O.D. Corpus, *The Roots of the Filipino Nation*, vol. II (Quezon City: Aklahi Foundation, 1989), pp. 206–69.

58. Agoncillo and Guerrero (1987), pp. 186–87, 208–9; Samuel K. Tan, *A History of the Philippines* (Quezon City: Department of History, University of the Philippines, 1987), p. 44.

59. Ruby Paredes, "The Origins of National Politics: Taft and the Partido Federal", in *Philippine Colonial Democracy*, edited by Ruby Paredes (New Haven: Yale University Southeast Asia Studies, 1988), p. 44.

60. Sidel (1999), p. 153.

61. Michael Cullinane, "Playing the Game: The Rise of Sergio Osmeña, 1898–1907", in *Philippine Colonial Democracy*, edited by Ruby Paredes (New Haven: Yale University Southeast Asia Studies, 1988), pp. 69–70.

62. Paredes (1988), pp. 7–8.

63. Anderson (2000), p. 203.

64. Hutchcroft (1994), pp. 224–25, 230–31.

65. Karl Lande, *Leaders, Factions and Parties: The Structure of Philippine Politics*, Monograph Series no. 6 (Michigan: Southeast Asia Studies Yale University, 1965), p. 1.

66. Emmanuel De Dios, "A Political Economy of Philippine Policy-making", in *Economic Policy-making in the Asia-Pacific Region*, edited by John Langford and K. Lorne Brownsey (Canada: Institute for Research on Public Policy, 1990), p. 111.

67. Temario Rivera, *Landlords and Capitalists: Class, Family and State in Philippine Manufacturing* (Quezon City: University of the Philippines Center for Integrative Studies and the University of the Philippines Press, 1994), p. 114.

68. Arsenio Balisacan and Hal Hill, "An Introduction to the Key Issues", *The Philippine Economy: Development, Policies and Challenges*, edited by Arsenio Balisacan and Hal Hill (New York: Oxford University Press, 2003), p. 3.

69. This was clearly indicated by the fact that by 2000, the Philippines' per capita GDP was about the same as its level in 1980. It was overtaken by Korea and Taiwan in the late 1960s, by Thailand in the 1970s, by Indonesia in the 1980s, and by China in the 1990s. The Philippines missed out on the Asian boom from the late 1970s to the mid 1990s. Specifically, the Philippines was bypassed when labour-intensive industries migrated to lower-waged countries in the ASEAN region such as Malaysia, Thailand, and Indonesia, and then to China.

70. Rivera (1994), p. 125–29. John Caroll, *The Filipino Manufacturing Entrepreneur: Agent and Product of Change* (Ithaca: Cornell University Press, 1965).

71. Manuel Montes, "The Business Sector and Development Policy", in *National Development and the Business Sector in the Philippines*, edited by Aichiro Ishii, et al. (Tokyo: Institute of Developing Economies, 1988), p. 32.

72. David Timberman, *A Changeless Land: Continuity and Change in Philippine Politics* (Singapore: Institute of Southeast Asian Studies, 1991), pp. 75–123; David Wurfel, *Filipino Politics: Development and Decay* (Cornell: Cornell University Press, 1988), pp. 114–54.

73. Rigoberto Tiglao, "The Consolidation of the Dictatorship", in *Dictatorship and Revolution: Roots of People Power*, edited by Aurora Javate-De Dios, Petronilo Bn. Daroy, and Lorna Kalaw-Tirol (Manila: Conspectus Foundation, 1988), p. 27.

74. Timberman (1991), pp. 87–88.

75. Emmanuel De Dios and Paul Hutchroft, "Political Economy", in *The Philippine Economy: Development, Policies and Challenges*, edited by Arsenio Balisacan and Hal Hill (New York: Oxford University Press, 2003), p. 50.

76. Montes (1988), pp. 32–33; Kenji Koike, "Dismantling Crony Capitalism under the Aquino Government", in *National Development and the Business Sector in the Philippines*, edited by Aichiro Ishii, et al. (Tokyo: Institute of Developing Economies, 1988), p. 209.

77. De Dios and Hutchcroft (2003), p. 50.

78. See Florian Alburo, Cesarina Rejante, and Charito Arriola, "Development Planning in the Philippines", in *Development Planning in Asia*, edited by Somsak Tambunlertchai and S.P. Gupta (Malaysia: Asian and Pacific Development Centre, 1993), p. 194.

79. De Dios (1990), p. 121.

80. Eric Gutierrez, Ildefonso Torrente and Noli Narca, *All in the Family: A Study of Elites and Power Relations in the Philippines* (Quezon City: Institute for Popular Democracy, 1992); Eric Gutierrez, *The Ties That Bind: A Guide to Family, Business and Other Interests in the Ninth House of Representatives* (Pasig, Metro Manila: Philippine Center for Investigative Journalism, 1994).

81. Hutchcroft (1998), p. 222.

82. Thomas McHale, *An Econecological Approach to Economic Development*, Ph.D. dissertation, Harvard University (1959) quoted in Hutchcroft (1998), p. 13.

83. See Appendix 1 for Selected Economic Indicators for Malaysia and the Philippines.

84. K.S. Jomo and E.T. Gomez, "The Malaysian Development Dilemma", in *Rents, Rent-seeking and Economic Development: Theory and Evidence in Asia*, edited by Mustaq Khan and Jomo (Singapore: Cambridge University Press, 2000), p. 288.

85. See Terence Gomez's studies on the business involvements of UMNO, the MCA, and the MIC: *Politics in Business: UMNO's Corporate Investments* (Kuala Lumpur: Forum, 1990), and *Political Business: Corporate Involvement of Malaysian Political Parties* (Townsville: James Cook University, 1994).

86. Jomo and Gomez (2000), p. 276.

87. Mustaq Khan, "Rent-seeking as a Process", in *Rents, Rent-seeking and Economic Development: Theory and Evidence in Asia*, edited by Mustaq Khan and K.S. Jomo (Singapore: Cambridge University Press, 2000), pp. 99, 117.

88. Hutchcroft (2000), pp. 209, 217, 220.

4

THE TELECOMMUNICATIONS SECTOR IN MALAYSIA AND THE PHILIPPINES BEFORE REFORM

This chapter describes the condition of the telecommunications industry in Malaysia and the Philippines before market reform by examining the condition of the telecommunications sector and the roles played by the state and private actors. The patterns of political patronage, the types of rents, the way they were obtained, who acquired them, and how they were used will also be analysed. Finally, the question why the Malaysian telecommunications sector was more efficient than that of the Philippines will be addressed.

Immediately after independence, Malaysia was left with a fairly efficient communications infrastructure. Telecommunications were originally the responsibility of a state department that had monopoly control of service delivery. The private sector's role was limited to equipment supply. With the introduction of the New Economic Policy (NEP), the state's monopoly over the telecommunications sector was used to allocate patronage to Malays through the award of licenses for some services and equipment supply contracts. The NEP goal of rapid economic development led to massive state investment for the expansion of the communications sector in the 1980s. The modernisation of the communications infrastructure, however, was also used to create business opportunities for Malays. State patronage of these businessmen resulted in increased costs and inefficiencies in the infrastructure expansion programme. Yet, the economic impact was generally expansionary and growth-enhancing.

Under American occupation, foreign-owned companies under state regulation provided telecommunications services in the Philippines. In 1967, a group of Filipino businessmen close to then President Ferdinand Marcos took over ownership of the Philippine Long Distance Telephone Company (PLDT). PLDT had sole authority to operate a national communications network. During the 1970s, PLDT consolidated its monopoly status, and was under very little state pressure to expand its network and improve its services. A few other businessmen were granted exclusive privileges by the Marcos regime to operate various communications services, and small provincial telephone companies proliferated due to unmet demand. Yet, the state protected the monopoly profit of PLDT, whose services were inefficient. Monopoly rent was captured for private profit, and was growth-hindering.

The case of Malaysia is presented first in this chapter, followed by that of the Philippines. Both cases are discussed chronologically, starting with the establishment of the sector under the respective colonial regimes. A second section focuses on the domestic take-over of the sector and events after independence. The third and fourth sections discuss the expansion and developments from the late 1960s up until the 1980s, before the introduction of market reform. A final section recaps the similarities and differences in the pre-reform telecommunications industries. The main argument of the chapter is that telecommunications was used for political patronage in both countries, although for different purposes and with different outcomes. One led to the growth of infrastructure and the economy, while the other stagnated and became an obstruction to economic development. The background presented in this chapter lays the groundwork for the discussion of privatisation and liberalisation reform in the succeeding chapters.

THE PRE-REFORM TELECOMMUNICATIONS SECTOR IN MALAYSIA

Colonial Developments

The British motivation behind developing telecommunications facilities in Malaya was twofold: to meet the communication needs of colonial businesses, particularly in tin mining, rubber plantations, and logging, and to monitor political and social unrest. In 1870, postal, telegraph, and telephone systems were introduced in the Malayan Peninsula with the construction of a submarine cable linking the Straits Settlement to London.

The Telecommunications Department was established in 1902 with responsibilities for postal, telegraph, and telephone services of Malaya.[1] The Telecommunications Department was patterned after the British telecommunications system and staffed by British civil servants. Malayan employees were confined to low-ranking positions such as technicians, assistants, and clerks. The British philosophy was to provide service to business first, while rural services were considered non-economical. This urban bias of telephone service was ingrained in the Telecommunications Department until the 1960s.[2]

The Telecommunications Sector After Independence, 1957–70

When Malaya became independent in 1957, it had a sound infrastructure base. According to a 1954 World Bank mission, the country's telecommunications system was well developed.[3] From 1947 to 1957, telecommunications services grew 16 per cent per annum. This slowed to 8.5 per cent per annum between 1960 and 1970. Teledensity, or the number of telephones available per 100 people, was 1.0 in 1970.[4] In terms of subscriber growth, there were 61,000 subscribers in 1957 compared with 103,000 in 1970.[5]

The Telecommunications Act of 1950 governed the operation of the Telecommunications Department. In 1956, 1,800, or two-thirds of the 2,650 departmental staff were British civil servants. The British held nearly all of the 64 senior positions.[6] Thus, the first reform

Table 4-1: Malaysian and Philippine Basic Telecommunications Indicators 1965–90

	Malaysia			Philippines		
Year	Total Population (Millions)	GDP per Capita (US$)	Teledensity	Total Population (Millions)	GDP per Capita (US$)	Teledensity
1965	9.5	1,164	.83	31.6	801	.25
1970	10.9	1,370	.95	36.6	867	.47
1975	12.3	1,712	1.37	42	998	.66
1980	13.8	2,297	2.87	48	1,172	.86
1985	15.7	2,586	6.14	54.2	974	.93
1990	18.2	3,104	8.92	61	1,083	1.00

Note: GDP per capita (constant 1995 US$).
Source: World Development Indicators Online, http://www.worldbank.org/publications/wdi.

of the Department after independence, as was the case throughout the bureaucracy, was its Malayanisation.[9] With the formation of the Federation of Malaysia in 1963, telecommunications service provision and regulation were placed under Jabatan Telekom Malaysia (JTM), a department subsumed under the federal Ministry of Works, Posts, and Telecommunications.

In line with the Malayanisation of the bureaucracy, a number of Malaysians were sent overseas to study engineering. From 1958 to 1965, the top 50 high school graduates each year were sent to Australia or New Zealand to study engineering under the Colombo Plan. JTM hired these graduates and rapidly promoted them. In 1965, Chew Kam Pok was appointed as the first Malaysian Director-General of Telecommunications. A telecommunications training centre was established in Kuala Lumpur in 1966, with the support of the United Nations and the International Telecommunications Union (ITU). This slow and orderly process of Malayanisation resulted in the inculcation of British civil service values to the Malaysian personnel, the preference for British-made equipment, and the continued use of British technology even after independence. Also, loans granted to the newly independent Malaysian government were tied to the purchase of British-made telecommunications equipment. Thus, the post-independence Telecommunications Department resembled its British counterpart in terms of structure, technology, and service orientation.[8]

The Investment Incentives Act of 1968, however, brought some changes, lessening the reliance on British telecommunications imports. The Act was part of an overall shift in economic policy towards export-oriented industrialisation (EOI) and encouraged the relocation of assembly operations of electronics companies to Malaysia. Among the electronics companies that relocated, a few were involved in telecommunications equipment manufacture. In early 1969, the government reached an agreement with Ericsson for the establishment of a telephone assembly operation. Eventually, Ericsson won a government contract worth RM6 million (US$1.9 million) to supply JTM with cross-bar switching equipment to be manufactured by its local subsidiary. In 1971, Philips secured a contract to provide JTM with locally manufactured VHF/FM radio sets. Japanese companies such as Fujitsu and NEC also established factories in Malaysia, and were awarded contracts to supply JTM with multiplexing and transmission equipment.

The NEP and Its Impact on JTM's Structure and Operations

As discussed in the previous chapter, the 1969 race riots were a watershed in the country's history, which led to the formulation of the New Economic Policy (NEP) with its twin goals of eradicating poverty and the restructuring of society.

The impact of the NEP on JTM was twofold: an increase in the demand for the provision of communications infrastructure necessary for economic growth, and a demand for business opportunities and employment for bumiputeras. The following discussion will provide instances of how JTM's efforts to modernise its operations were hampered by the NEP's second goal. This conflict played out in the telecommunications sector, where the primary goal was to help create a bumiputera entrepreneurial class rather than to develop the telecommunications network in the most efficient and cost-effective way.

In 1971, JTM's management was reorganised with the early retirement of its first Malaysian Director-General, Chew Kam Pok. His deputy, Lee Chye Watt, was bypassed for promotion to the top post. Instead, a Malay, Buyong Abdullah, was appointed to the position. From that time, a Malay has always held the post of Director-General.[9]

In its attempt to expand the telecommunications network, JTM's main problem was the lack of qualified technical staff. In line with the NEP, the government promoted bumiputera employment in the civil service. This obliged JTM to recruit under-qualified, mostly Malay, personnel. The staff was given minimal training, and JTM became top- and bottom-heavy as management and unskilled positions swelled. The total number of employees rose from 7,000 in 1957 to 8,000 in 1969, and to 21,000 in 1979. Management and unskilled positions grew by 300 per cent in the 1970s, but technical positions only grew by 50 per cent. Consequently, JTM's inefficiency became a regular subject of critical letters to newspapers.[10]

JTM also faced budget difficulties. According to JTM officials, from 1966 to 1980 the department barely received half of its original budgetary requests. Furthermore, its budgetary allocation was closely linked to the country's five-year plans and their mid-term reviews, which were often revised. This constrained equipment purchases and prevented long-term planning.[11]

In line with a World Bank loan in 1971, the Telecommunications Act was amended to introduce a commercial system of accounting to the Telecommunications Department. JTM adopted a new system of accounting designed by officials of the Australian Post Office. Financial statements were to be submitted to the Auditor-General on the first of March each year. In his first report, the Auditor-General noted numerous discrepancies, and irregularities in JTM's accounts, which were submitted for auditing three years after they were due.[12] Furthermore, unrecovered telephone arrears, lack of financial control over the issuance, and collection of bills, improper documentation of equipment purchases, unvouched payments, thefts, frauds, and losses in departmental cash and stores were not taken seriously, even though these were tabled in parliament. These issues were still unresolved when JTM's assets and liabilities were transferred to the corporatised Syarikat Telekom Malaysia (STM) on 1 January 1987.[13]

In 1978, the government reorganised JTM and the Ministry of Works, Telecommunications, and Posts. The Ministry was transformed into the Ministry of Energy, Telecommunications, and Posts (MEPT), and a separate ministry was created for public works. Operating under the MEPT, JTM was revamped into seven regional offices. Each regional office was given planning, implementation, maintenance, and operational management authority. The reorganisation meant that more staff was needed to fill the newly created offices. By the end of 1981, there were 32,423 approved posts, but only 26,965 were filled. Most of the vacancies occurred in technical areas.

Aside from its effect on personnel, the NEP pressured JTM to help nurture the development of bumiputera business by awarding contracts for equipment and services to bumiputera companies and trust agencies. The awarding of these contracts was not based on competitive bidding, but on the ethnicity of the company's owner.[14] Thus, while subcontracting was supposed to enable JTM to meet the increasing demand for telecommunications, it was implemented in such a way that it provided business opportunities for bumiputras.

The first contracting out of telecommunications service was the franchise to install, maintain, and operate public payphones in urban areas awarded to Uniphone Sendirian Berhad in 1972. Uniphone was a subsidiary of United Motor Works (UMW), a local dealer of heavy automotive equipment owned by Eric Chia.[15] The idea to take over JTM's coin-operated payphone business came from Shamsuddin Kadir, one of Chia's Malay recruits to UMW. Shamsuddin was a British-educated telecommunications engineer who resigned from JTM in 1971. Having worked in JTM for 11 years, Shamsuddin gained insider's knowledge of how the department operated. The five-year contract for the payphone business involved a 50:50 division of revenues between Uniphone and JTM. In 1974, Shamsuddin purchased Uniphone from Chia, and in January 1975 formed his own holding company, Sapura Holdings, to serve as Uniphone's parent company.[16] Shamsuddin's swift success points to the state's provision to Malays of exclusive business opportunities in the telecommunications industry.

A second area for bumiputera business development emerged from the 1974 federal procurement policy. The policy required at least 20 per cent local content, with additional preference given to bumiputra-owned companies. Some bumiputeras who were former JTM employees established their own companies in the hope of winning JTM contracts. In 1973, Shamsuddin Kadir formed Teledata Sdn. Bhd. to manufacture mainframes in Malaysia. In 1975 he also established Premier Communications and Engineering Sdn. Bhd. to manufacture telephone sets, equipment, and payphones. In 1976, Uniphone became the local assembler and distributor of Siemens product and was awarded a new contract to supply telephone sets to JTM.

Like Shamsuddin, Karim Ikram also left JTM, and established Zal Enterprises and Binafon Sdn. Bhd. The former company went into the manufacture of cable casing used by both JTM and the National Electricity Board (NEB), while the latter installed telephone cables. Soon, Zal Enterprises won a contract to supply cable casing to JTM and the NEB. In 1975, both Uniphone and Binafon were awarded a RM3 million contract to install cables for JTM.

Another former JTM official, Zahari Hassan, formed Sri Communications Sdn. Bhd. In 1975, his company was awarded a contract to install switching equipment for telephone exchanges. Two other notable companies were Electronscon Kinki Communications, established in 1978 by Dato Haji Mohktar Mohiddin in partnership with Ericsson, and Syarikat Edani Snd. Bhd, of former JTM Director, Buyong Abdullah.[17] As will be discussed in detail in the next section, these companies owned by former JTM employees played key roles in the modernisation of the telecommunications infrastructure.

A third area of bumiputera business opportunities was from the Jadual Kadar Harga (JKH), the Schedule of Standards and Rates. This schedule established rates for public works projects and signalled the state's decision to subcontract all such segment of JTM operations. Contractors had to register with the Treasury, which then allocated them jobs based on a round-robin system. With the JKH, bumiputera companies owned by ex-JTM employees flourished as they cornered JTM contracting jobs. In 1979, the JKH system was modified to classify contractors according to technical capacity and ownership (that is, that they were certified bumiputera companies). Firms were classified into three types: Class A, those that had the capacity to undertake contracts valued over RM1 million; Class B, those capable of doing contracts worth up to RM1 million; and Class C, those that could do jobs valued at no more than RM500,000. Shamsuddin's Uniphone, Karim's Binafon, and Mokhtar's Electroscon were classified as Class A. They dominated the JKH system and secured the most lucrative contracts.[18]

The final area of business for bumiputeras was the production of telecommunications equipment. Due to their lack of technological capacity to manufacture electronics and cable equipment, bumiputera-owned companies formed joint ventures with foreign producers. JTM's Supplies and Stores Division cautioned that it would end up paying unreasonably high prices for substandard equipment. However, JTM pushed through with the policy of sourcing equipment from bumiputera firms. This development led to the growth of bumiputera-owned companies in joint ventures with foreign companies.[19]

In addition, two bumiputera trust agencies, the Lembaga Tabung Angakatan Tentera (LTAT, or the Armed Forces Provident Fund) and Perbadanan Nasional Bhd. (Pernas, National Corporation), entered into joint ventures to take advantage of the state-created opportunities. In 1974, LTAT and Ericsson formed Ericsson Telecommunications Sdn. Bhd. (later known as Perwira Ericsson), with 30 per cent equity ownership by LTAT. The company was awarded a ten-year telephone supply contract. Also, the government announced new import duties on telephones that Ericsson was producing locally. In the same manner, Pernas and NEC of Japan formed Pernas Engineering Sdn. Bhd., in which Pernas held 60 per cent of the equity. The company won a contract worth RM6 million to supply JTM with multiplex equipment, and another valued at RM7.6 million for satellite earth station equipment.

Learning from the experiences of Ericsson and NEC, foreign firms discovered that the best route to break into the Malaysian market was via partnership with a bumiputera company. Thus, Siemens formed a partnership with Shamsuddin's Sapura, NKF Kabel formed another with Binafon, and Alcatel formed a venture with the Fleet Group, UMNO's first holding company. Although these partnerships provided market access to foreign firms, they also allowed bumiputera companies to gain manufacturing licences, bank guarantees, and bigger contracts from JTM.[20]

In sum, during the first decade of NEP implementation, JTM was reorganised, placing Malay personnel in the top ranks and unskilled Malays in the lower levels. Secondly, JTM contracted out its public payphone operation in urban areas and its public works and cable-laying to Malay-owned companies. JTM also awarded equipment supply contracts to joint ventures between Malay-owned companies and foreign suppliers, which were required to build factories in Malaysia. Although the official rationale for the policy of contacting out was that JTM could not cope with the expanding demand for its services, the way that the contracts were awarded — based on the ethnicity of company owners and not on the basis of competitive bids — indicates the primacy of providing patronage to Malays rather than improving telecommunications services.

Expansion of Infrastructure via Turnkey Contracts

Given the changes in JTM's structure and operations during the 1970s, the Department was unable to service the demand for telephones during the 1980s. JTM was not able to expand its services in part because it only received roughly half of its requested budget, and in part because it had to hire unskilled workers.

Yet, Rita Hashim documents that JTM's financial status from 1976 to 1984 actually showed a positive revenue stream, growing at an average of 21 per cent during the period. These revenues were enough to cover operating expenditure, but not development expenditure, which grew at a faster rate. Given this qualification, it would be incorrect to claim that JTM was a loss-making enterprise before privatisation, as is normally the reason for privatisation.

In line with the state's shift towards expansionary spending to revive the economy, JTM was allocated a budget of RM1.4 billion or 3.56 per cent of the total development spending allocation under the

Table 4-2: Malaysian Telephone Demand, 1970–84 (Before Privatisation)

Year	No. of Current Subscribers	No. of Applicants on Waiting List	Total Demand[a]	Annual Change (%)	Unsatisfied Demand (%)[b]
1970	103,763	13,704	117,467	—	13.2
1971	110,019	11,924	121,943	4	10.8
1972	121,603	13,674	135,277	10	11.2
1973	136,494	20,591	157,085	16	15.1
1974	149,458	35,085	184,543	17	23.5
1975	169,538	48,306	217,844	18	28.5
1976	194,259	65,303	259,562	19	33.6
1977	227,564	76,438	304,002	17	33.6
1978	271,010	84,247	355,257	17	31.1
1979	325,154	105,699	430,853	21	32.5
1980	395,640	133,609	529,249	22	33.8
1981	488,675	149,945	638,620	21	30.7
1982	585,387	189,808	775,195	21	32.4
1983	700,097	199,831	899,928	16	28.5
1984	849,127	190,542	1,039,669	15	22.4

[a] Total Demand = No. of Current Suscribers + No. of Applicants on Waiting List.
[b] Unsatisfied Demand = No. of Applicants on Waiting List ÷ No. of current Subscribers.
Sources: Rita Hashim, "Privatisation of Telecommunications in Malaysia", MPA thesis, University of Malaya (1986), p. 52; Laurel Beatrice Kennedy, "Privatization and Its Policy Antecedents in Malaysian Telecommunications", Ph.D. dissertation, Ohio University (1990), p. 110. The statistics were taken from the Annual Reports of the Department of Telecommunications, 1970–84.

Table 4-3: Revenue and Financing of the Telecommunications Department, 1976–84 (RM Millions)

Year	Revenue	Operating Expenditure	Surplus	Development Expenditure	Deficit	Expenditure Funded by Revenue (%)
1976	282	145	137	210	–73	79.9
1977	301	189	112	288	–176	63.1
1978	358	205	153	359	–206	63.5
1979	433	240	193	381	–188	69.7
1980	524	299	225	638	–413	55.7
1981	637	363	274	697	–423	60.1
1982	772	427	345	749	–404	65.6
1983	931	655	316	1,037	–721	57.4
1984	1,231	723	508	1,603	–1,095	52.9

Source: Rita Hashim, "Privatisation of Telecommunications in Malaysia", MPA Thesis, University of Malaya (1986), p. 51. The statistics were taken from the Annual Reports of the Department of Telecommunications, 1977–84.

Fourth Malaysia Plan (4MP). The 1983 Mid-term Review of 4MP almost doubled this allocation to RM2.9 billion.[21] Ostensibly, the increased budgetary allocation was designed to meet the increasing demand for telephone services.[22]

Under the 4MP, the Implementation and Coordination Unit (ICU) of the Prime Minister's Department formulated a telecommunications modernisation and expansion program worth around RM5 billion.[23] It is important to note four things about this modernisation program. First, an office under the Prime Minister, and not JTM, devised the plan. Second, private contractors, not JTM, would undertake the project. The modernisation of the telecommunications infrastructure was an example of the Mahathir government's commitment to limiting the public sector's role through contracting and his strategy of nurturing the rise of bumiputera businesses by opening business opportunities for local firms. Third, none of the contracts involved competitive tenders. Instead, they were awarded to chosen bumiputera companies owned by well-connected Malays. Fourth, these companies were encouraged to find foreign joint-venture partners. The modernisation of the telecommunications infrastructure demonstrated the increasing politicisation of business enterprises development and the growing centralisation of decision-making power in the office of the Prime Minister.

The three-stage expansion program was undertaken through the awarding of contracts without a bidding process. The first stage involved the awarding of a ten-year contract to supply and install computerised digital switching equipment to Pernas-NEC and Perwira Ericsson. Pernas-NEC's contract was worth RM1 billion, while Perwira Ericsson's deal was valued at RM700 million. The second stage, which will be discussed in detail below, was the awarding of four turnkey contracts worth RM2.54 billion for the installation of 1.7 million telephone lines. The third stage was the upgrading and expansion of microwave relay switches and fibre optic cable, where two contracts valued at RM510 million were awarded to Standard Elektrik Lorenz, a West German subsidiary of ITT (which set up a local factory to supply the contract), and Sapura Holdings in a joint venture with Fujitsu.[24]

Of these, the biggest was the RM2.54 billion turnkey project. It was considered the largest build-and-transfer contract that the government had ever awarded to the private sector during this time. In 1983, four bumiputera-owned companies controlled by former senior JTM officials were awarded the project: Uniphone, Binafon, Electroscon, and Sri Communications. These four were reportedly chosen because of their proven track record and were the only companies that qualified under JKH's Class A category. As in the case of earlier contracts, there was no competitive bidding involved. Instead, the four firms were invited to submit proposals for the project, and the job was divided equally among them.

Each turnkey contract, valued at RM636 million, entailed the installation of 444,100 lines of cable pairs throughout the country from 1983 to 1988. The country was divided into four parts, with Binaphone covering Kelantan, Terengganu, and Negri Sembilan; Uniphone covering Selangor, Sarawak, and a major portion of Kuala Lumpur; Sri Communications covering Penang, Kedah, Perlis, and Sabah; and Electroscon covering Pahang, Melaka, Johor, and portions of Kuala Lumpur. There was no competition among the contractors in terms of price or area of service, but only in terms of quality and the time to finish the installations. The contracts included penalty fees if work was not completed by the end of 1988, with 68.8 per cent of the work to be completed by the end of 1985.[25] The task was so far the biggest that the four bumiputera companies had undertaken. Before this, they only received contracts worth around RM2 million. The contracts were said to contain provisions for the use of smaller subcontractors registered under the JKH system to make sure that Class B and C contractors would still receive their share of business.

Laurel Kennedy documented how the idea of the turnkey project, also known as build and transfer type of privatisation, originated from a similar project undertaken by the Saudi Arabian government in the 1970s to expand its telecommunications system. One of the contractors in the Saudi project was NKF Kabel, which was a venture partner of Binafon. NKF reportedly encouraged Karim Ikram, Binafon's owner, to propose the idea to JTM, but JTM was non-committal. Reportedly, Karim brought the idea to MEPT Minister Leo Moggie and Daim Zainuddin.[26] The latter was then recognised as Prime Minister Mahathir's closest and most influential adviser. JTM allegedly objected to the contracts because the prices were inflated, and claimed that it could do the work more cheaply. It also argued that the project would render its 28,000 employees redundant, further raising its cost. Finally, the contracts gave all authority for the planning and installation of the network to the contractors. Yet, no local company had any experience in planning because they had all relied on JTM in the past. Despite JTM's objections to, and misgivings about, the project, the contracts were signed. The capacity to decide about technical matters such as telecommunications modernisation did not rest with the department involved in its provision. Instead, the ministry encouraged the contractors to find a foreign partner that could provide the expertise that was needed. Karim Ikram's Binafone continued its joint venture with NKF Kabel. Shamsuddin's Uniphone formed a joint venture with Sumimoto Electric for the project. Mokhtar's Electroscon S.B. formed a joint venture with Ericsson, and Zahari's Sri Communications decided to employ foreign expertise in its design and planning.[27]

The project's development was monitored by an inter-ministerial Steering Committee headed by the Minister of EPT and composed of other senior staff members of the Ministry, the Ministry of Finance (MOF), the ICU, the Prime Minister's Department and JTM. The Committee's composition is further proof that JTM did not have much influence in the planning and decision to modernise the country's telecommunications infrastructure.

By the midpoint of the contract period, all of the contractors had performed miserably, having installed only an average of 29 per cent of their targets. In the 1985 JTM annual report, the contractors' achievements without specifying names were as follows: 23.8 per cent, 19.4 per cent, 29.1 per cent, and 43.5 per cent. JTM and the contractors blamed each other for the delays in the project. JTM criticised the contractors for the delays and their poor work quality. Meanwhile, smaller subcontractors complained about the monopoly of the four bumiputra companies, who instead of subcontracting to the smaller JKH registered contractors, set up their own subsidiaries and awarded parts of the work to these firms, thus cornering the multimillion contracts. The four contractors lashed out at JTM for causing delays in approving the plans at various stages. They alleged that JTM officials resented how their former colleagues had taken over the more "glamorous" aspect of their jobs.[28]

By 1991, three years after the supposed end of the contracts, 1.4 million new telephones lines had been installed, but about 18 per cent of the project had still not been completed. It is not known if the project was ever completed or whether penalties for delays were ever levied on these companies. Moreover, the total cost of the project was never revealed. The whole idea was quietly forgotten, as the corporatisation and privatisation of JTM became the major issue.

The massive telecommunications modernisation program embarked upon during Mahathir's first term showed an evolution in the NEP redistributional goals away from the use of bumiputera trustee institutions to nurturing individual bumiputeras. These businessmen were identified as being close to top-ranking UMNO politicians. The expansion programme had ambitious targets for telecommunications development but the awarding of contracts was non-competitive. In fact, the turnkey project itself seems to have been the result of the lobbying of those who were eventually awarded the contracts. The state required these companies to take in foreign equipment suppliers as partners, bringing with them the needed technology and experience. This suggests that the state also incorporated a performance criterion in its award of contracts. On the

Table 4-4: Turnkey Project Outcomes

Contractor	Target No. at End of 1985	No. of Installed Lines 1985	1986	1991
Binaphone	304,669	72,624	220,918	333,330
Electroscon	304,669	59,060	167,728	336,176
Sri Communications	304,669	88,768	213,475	292,063
Uniphone	304,669	132,610	255,277	485,277
Total	1,218,676	353,062	857,773	1,446,846

Sources: Vincent Lowe, "Malaysia and Indonesia: Telecommunications Restructuring", in *Telecommunications in the Pacific Basin: An Evolutionary Approach*, edited by Eli Noam, Seisuke Komatsuzaki, and Douglas Conn (New York: Oxford University Press, 1994), p. 124.

part of foreign equipment suppliers, a condition for their entry was that they took up local partners and committed to establishing local manufacturing outlets. Thus, pressure from foreign equipment suppliers who wanted a share in the expanding telecommunications market in Malaysia corresponded with the lobby of the four big bumiputera-owned companies and Mahathir's modernisation program.[29] More importantly however, the state was pursuing a dual goal: the expansion of telecommunications infrastructure and the creation of a class of bumiputera businessmen.

With these developments, JTM was sidelined. The Prime Minister's Department took over both the planning and the awarding of contracts. JTM's protests about the cost and inefficiency of the contracts were ignored. Under Mahathir, the authority even in the planning and awarding of contracts for the modernisation of telecommunications moved from JTM to the Prime Minister's Office. Four handpicked bumiputera companies were awarded the contracts, to the detriment of smaller bumiputera-owned companies. This indicates the increasingly political basis of contracts awarding and the centralisation of power in the hands of the Prime Minister. Under Mahathir, the nexus of politics and business, and the importance of executive patronage, became more pronounced and decisively moulded the country's development. This is even more evident in the introduction of privatisation and liberalisation reforms in the telecommunications industry, as will be later examined.

THE PRE-REFORM TELECOMMUNICATIONS SECTOR IN THE PHILIPPINES

This section examines the development of the telecommunications sector in the Philippines in four parts. The first discusses the establishment of the sector during colonial times. The second focuses on the domestic take-over of the biggest telecommunications company and discusses the other private players in the industry after independence. The third concentrates on developments during martial law, when state policy benefited a single company. The last examines events under the Aquino administration, and looks into why attempts to liberalise the industry failed.

Telecommunications in the Colonial Period

Compared to Malaya, private companies developed and offered telecommunications services in colonial Philippines. Two companies provided international and domestic communications. The first was Eastern Extension Australasia and China Telegraph Company Ltd. (Eastern Extension), a subsidiary of Britain's Cable and Wireless. The Spanish colonial government authorised Eastern Extension in 1880 to construct and operate the first submarine cable linking the Philippines and Hong Kong.[30] A second firm, the American-owned Philippine Islands Telephone and Telegraph Company, started its operations in Metro Manila in 1905. In 1928, it merged with the Cebu, Panay, and Negros Telephone and Telegraph Companies, to form the Philippine Long Distance Telephone Company (PLDT).[31] In 1932, the colonial Philippine Congress granted PLDT a 50-year franchise to operate a national telephone system.

To regulate the industry as well as other public utilities, the colonial government established the Public Services Commission. Also, a Bureau of Posts was created to operate telegraph services nationwide.[32]

The Telecommunications Sector After Independence, 1946–69

Takeover of PLDT by Filipinos[33]

The initial management control of PLDT lay in the hands of Americans, although the major stockholder was a Canadian company, British Columbia Telephone. In 1956, British Columbia Telephone sold its stake to an American company, General Telephone and Electronic Corporation (GTE).[34]

In March 1967, a few years before the expiration of the Laurel-Langley Agreement, which gave Americans economic rights in the Philippines, GTE decided to dispose of its 28 per cent controlling interest in PLDT. Several groups of Filipino businessmen were interested in buying GTE's stake in PLDT. Foremost was the Ayala Corporation, then the second largest shareholder of PLDT. However, it decided to sell its stake in the company when it found out that Marcos was interested in PLDT.[35]

Pedro Cojuangco and his brother-in-law, Benigno Aquino, then-Governor of Tarlac province, led a second interested party. They were reportedly close to finalising an agreement when President Marcos intervened and identified a more "suitable" group of businessmen to take over the company. Ramon Cojuangco, a first cousin of Pedro and of Aquino's wife, Corazon, led the group that obtained Marcos' blessing.[36]

On 7 November 1967, the Philippine Telecommunications Investment Corporation (PTIC) was registered with the Securities and Exchange Commission (SEC) to buy GTE's controlling interest in PLDT. Ramon Cojuangco, his wife Imelda, Alfonso Yuchengco, Leonides Virata, and Antonio Meer were the main incorporators. PTIC formally took control of PLDT on 1 January 1968. Ramon Cojuangco, his wife, and brother in law Luis Tirso Rivilla, held 57 per cent of PTIC. GTE held 22.5 per cent of the company as partial payment for its PLDT stock, Yuchengco owned 7 per cent and Leonides Virata, Gregorio Romulo, and Antonio Meer were each given 3 per cent stakes in PTIC for helping to organise the deal. Minor stockholders held the rest of the shares. The details of the ownership transfers to Filipino businessmen, which will be discussed in greater detail below, reveal not only Marcos' intervention in the sale,

Figure 4-1: PLDT Ownership Structure, 1967

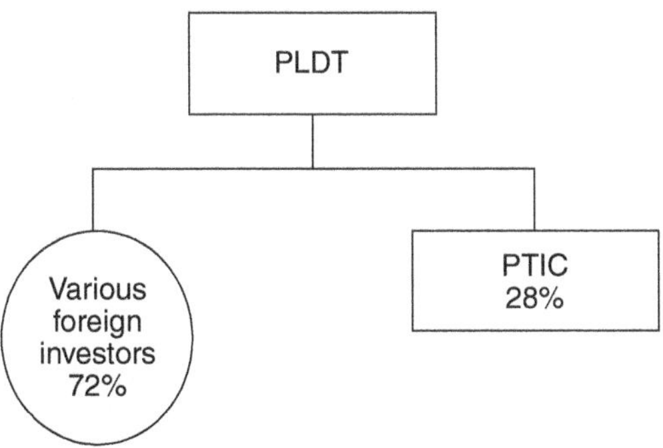

Table 4-5: PTIC Ownership, January 1968 (Percentages)

Stockholders	Ownership
Ramon Cojuangco, Imelda Cojuangco, and Luis Tirso Rivilla	57.0
GTE	22.5
Alfonso Yuchengco	7.0
Leonides Virata	3.0
Gregorio Romulo	3.0
Antonio Meer	3.0
Others	4.5
Total	100.0

but also his ultimate ownership of a substantial percentage of PLDT stock. This issue is still contested today. Understanding the question of ownership of PLDT sheds light on how government policy under martial law bolstered the company's monopoly status.

Other Players and Rural Telephone Companies

PLDT became the dominant player in telecommunications because it was the only company authorised to operate a national network. However, company officials dispute the monopoly criticism because of the existence of a government telephone system, about 60 provincial companies, four international, two domestic data carriers, and two satellite services companies. At first blush, PLDT's defence seems reasonable, but a deeper investigation proves otherwise.

In 1947, the government created the Bureau of Telecommunications (Butel) to operate a government telephone system and take over the telegraph system from the Bureau of Posts. Butel's function was expanded in 1972 to pioneer telephone and telegraph services in unserved areas.[37] By 1975, Butel had 34,643 operational telephone lines, which constituted 10.2 per cent of the total telephone capacity of the country, and was operating in nine of the eleven regions.

Four companies were licensed to offer international data communication. The oldest of these was Eastern Extension. By virtue of Marcos' Presidential Decree 484, its franchise was transferred in 1974 to Eastern Telecommunications Philippines Incorporated (ETPI).[38] ETPI restructured its ownership, selling 60 per cent of its stock to Filipino businessmen, while Cable and Wireless kept 40 per cent.[39] The Filipino shareholders were Roberto Benedicto, Jose Africa, and Manuel Nieto, each owning 20 per cent stakes. All three businessmen were well-known associates of President Marcos. According to a then-senior executive of Cable and Wireless, the choice of local partners was made purely out of political expediency. The decision certainly paid off, because ETPI gained special treatment from the Marcos government in the form of tax holidays, duty-free imports, and the exclusive right to the ASEAN communications market.[40]

The second international data carrier was Globe Mackay Cable and Radio Corporation (GMCR). The company, established in 1928, was owned by the Ayala family and the American company ITT Communications Services. The Ayala family was probably the oldest and most established of the country's economic elite, and owned the tract of land that is today the Philippines' central business district, Makati City.

The Santiago family owned the third international record carrier, Capitol Wireless Inc. (Capwire). Capwire was established in 1962 and was part of a group that included Retelco, then-second largest telephone company, and PT&T, a nationwide telegraph and telex company. The fourth international record carrier, Philippine Global Communications (Philcom), was established in 1977 with RCA Global Communications as partner. Philcom was a data carrier, but President Marcos gave it exclusive rights to handle calls to Japan, Australia, Korea, Guam, and Thailand. The company's principal stockholders were then Defence Minister Juan Ponce Enrile and Energy Minister Geronimo Velasco.

Finally, two companies were licensed to operate satellite services: the Philippine Communications Satellite Corporation (Philcomsat) and the Domestic Satellite Company (Domsat). Philcomsat was established in 1967 to operate voice and data telecommunications using the International Telecommunications Satellite Consortium's (Intelsat) services. Philcomsat's biggest client was actually the US

Table 4-6: Major Pre-Reform Telecommunications Firms in the Philippines

Companies	Franchise Date	Local Owner	Foreign Partner	Type of Service
Voice carriers				
PLDT (Philippine Long Distance Telephone Company)	1928	PTIC (Ramon Cojuangco), PHI	Institutional investors	Local, long distance, and international calls
Bureau of Telecommunications		Government department		Government telephone system, telegraph and telex; telephone in unserved areas
RCPI (Radio Communications Philippines Inc.)	1950	Filipino-owned	None	Telegraph and telex
PT&T (Philippine Telegraph and Telephone Company)	1962	Jose Santiago	None	Telegraph, domestic voice
Provincial companies (Paptelco)				Local voice service

International Record Carriers				
ETPI (Eastern Telecommunications	1967, successor of Eastern Extension, 1880	Roberto Benedicto Jose Africa Manel Nieto	Cable and Wireless, 40%	International data and common carrier
GMCR (Globe MacKay Cable and Radio Corp)	1928	Ayala family	ITT Communications Services	International data carrier
CAPWIRE (Capitol Wireless Inc.)	1962	Jose Santiago	Korea Telecoms	International data carrier
Philcom (Philippine Global Communications)	1967	Juan Ponce Enrile	RCA Global Communications	International data carrier
Satellite				
Philcomsat (Philippine Communications Satellite Corp)	1967	Benedicto, Enrile	GTE Plessey (UK)	Satellie service via Intelsat
Domsat (Domestic Satellite Corp)	1977	Benedicto, Cojuangco	Marubeni	Domestic relay of satellite broadcasting

Military, and a study argues that the company was originally established for the use of the US Military during the Vietnam War. Meanwhile, Domsat was established in 1979 for domestic television broadcasts using channels leased from Indonesia's Palapa Satellite.[41]

Aside from the above players, small provincial telephone companies that were owned by local businessmen proliferated in the 1960s due to PLDT's focus on Metro Manila and a few regional cities. By 1975, there were around 60 small telephone companies providing services in 101 of the 346 most populated cities and municipalities nationwide. These companies operated 11.7 per cent of the total telephone capacity at that time. Of the 101 towns and municipalities, only in 14 places was there duplication of the presence of Butel or PLDT.[42] This meant that these small private companies were established due to local needs.

In 1976, the Philippine Association of Private Telephone Companies (Paptelco) was organised to protect the interests of these

Table 4-7: Coverage of Local Telephone Operators by Region, 1973–74

	No. of Telephone Lines by Firms Operating in the Area				
Region	BUTEL (1974)	PLDT (1973)	Small Operators	Regional Total	As a % of National Total
I — Ilocos	1,200	3,100	6,774	11,074	3.26
II — Cagayan Valley	600	0	952	1,552	0.46
III — Central Luzon	1,700	3,100	7,822	12,622	3.72
IV — Southern Luzon	24,743	222,250	9,600	256,593	75.56
V — Bicol	1,000	0	3,070	4,070	1.20
VI — Western Visayas	1,200	14,800	1,065	17,065	5.03
VII — Central Visayas	2,500	16,800	1,100	20,400	6.0
VIII — Eastern Visayas	1,200	0	1,920	3,120	0.92
IX — Western Mindanao	0	0	1,300	1,300	0.38
X — Northern Mindanao	500	0	3,500	4,000	1.18
XI — Southern Mindanao	0	5,200	2,590	7,790	2.29
Total	34,643 10.2%	265,250 78.1%	39,693 11.7%	339,586	100%

Table adapted from Task Force on Human Settlements, *Infrastructure/Utilities Development in the Philippines Technical Report Part IV: Telecommunication* (Quezon City: Development Academy of the Philippines, March 1975), p. 25.

Table 4-8: Distribution of Local Telephone Operators by Region, 1973–74

	No. of Firms Operating in the Area			
Region	BUTEL (1974)	PLDT (1973)	Small Operators	Regional Total
I — Ilocos	1	1	9	11
II — Cagayan Valley	1	0	4	5
III — Central Luzon	1	1	16	18
IV — Southern Luzon	1	1	15	17
V — Bicol	1	0	8	9
VI — Western Visayas	1	1	6	8
VII — Central Visayas	1	1	3	5
VIII — Eastern Visayas	1	0	4	5
IX — Western Mindanao	0	1	5	6
X — Northern Mindanao	1	0	12	13
XI — Southern Mindanao	0	1	8	9
Total	9	9	90	106

Table adapted from Task Force on Human Settlements, *Infrastructure/Utilities Development in the Philippines Technical Report Part IV: Telecommunication* (Quezon City: Development Academy of the Philippines, March 1975), p. 24.

small telephone companies and negotiate collectively with PLDT. The two most contentious issues between Paptelco members and PLDT were interconnection and access charges. The small telephone companies were dependent on PLDT to place inter-provincial and overseas calls. PLDT used interconnection, the process by which subscribers of one company can call subscribers of another, as a tool to maintain its control over the national network in two ways.

First, it could allow, slow down, or deny interconnection at will, despite the regulatory body's order to allow access. Some of the companies that found it financially impossible to operate without interconnection decided to sell their companies to PLDT.[43] Second, PLDT dictated the interconnection access rates, which meant that it effectively cornered most telecommunications revenues.[44]

Because the Philippine telecommunications industry, in contrast to the state monopoly sector in Malaysia, involved numerous private players with distinct and captive market segments, PLDT consistently claimed that it was merely the largest of the private players. This, however, ignores the fact that it was the only company with a franchise to operate a national telephone network. In 1975, PLDT owned and controlled 78 per cent of all telephone lines in the country and about 94 per cent of the total lines by 1991. This dominant market position was the result of state policy that segmented market sectors based on type of network and technology. More importantly, PLDT used political influence to maintain this situation and ward off the entry of competitors.

Developments under Martial Law

Ferdinand Marcos declared martial law on 21 September 1972, purportedly to rid the Philippines of oligarchic control over its economy and politics and to establish a new society. Along with the centralisation of power in the Presidency, Marcos used state power to create a new set of favoured businessmen who enjoyed exclusive privileges. An illustrative example of a company that benefited from this set-up is PLDT. Through state nurturing, PLDT consolidated its dominant position in the telecommunications industry. The previous section argued that PLDT's dominance in the sector occurred despite the presence of various private players. This section analyses how PLDT consolidated its virtual monopoly status.

The Subscriber Investment Plan (SIP)

The first indicator of PLDT's privileged position was demonstrated by a presidential edict. Marcos announced Presidential Decree (PD) 217 in 1973 mandating all PLDT subscribers to invest in PLDT to raise equity and finance its expansion program.[45] This law, which is still in effect today, required all PLDT subscribers to buy preferred non-voting shares in the company as a prerequisite for a telephone connection. In theory, the Subscribers Investment Plan (SIP) would broaden public ownership of the company while raising the capital that was needed for network expansion. But because the shares carried no voting rights, mandatory investors held about 85 per cent of the total company equity shares but had no say in how the company was run. Section 5 of PD 217 allowed for the conversion of these preferred shares into common voting stock.

Yet, PLDT managed to avoid such conversion by liberally interpreting the time frame in which they were required to take place. Thus, PLDT theoretically had a wide public ownership, but only a small group of businessmen controlled the company.

The Role of the World Bank

Aside from the SIP, a source of funding for PLDT was its access to international loans approved by, or came from, the World Bank. In contrast to the minimal role that it played in Malaysia, the World Bank played a more direct role in shaping telecommunications development in the Philippines because of the country's reliance on Bank funding.

The Bank's lending for telecommunications in the Philippines began in 1969 when the International Financial Corporation (IFC) invested in an US$11 million bond issue to help PLDT finance an expansion program.[46] As it did in other developing countries, the Bank tied its loans to the maintenance of a local telephone monopoly and a centralised regulatory body. The prevailing idea during this time, which the Bank subscribed to, was that a monopoly would best facilitate the development of telecommunication infrastructure. Telecommunications was thought to be a natural monopoly. Yet, it is vital to note that World Bank funding for telecommunications expansion went to PLDT and not to Butel.

Until the early 1990s, the Philippines was the sixth largest recipient of World Bank loans. The telecommunications sector was one of the principal areas of Bank lending and was considered a high priority sector for public investment.[47] In 1973, a World Bank mission evaluated the state of the country's telecommunications infrastructure and proposed a plan for its long-term development. The mission called for the integration of all telecommunications services into one network and proposed that a monopoly franchise be issued.

PLDT became the main beneficiary of World Bank advice and lending. In 1977, the company initiated an US$870 million joint-venture project with Siemens. The project, also known as Expansion Phase 4, was to be financed by internal earnings and a 20-year unsecured loan of US$307 million from a consortium of foreign banks led by the European Asian Bank. This loan was the largest-ever private loan without government guarantees.[48] In 1985, the World Bank again extended a US$4 million loan to PLDT.[49] In 1986, the IFC lent PLDT US$30 million to finance its stake in the first trans-Pacific fibre optic cable. In 1988, the IFC again extended PLDT a US$74 million loan to help upgrade its long-distance infrastructure. During the same year, the Asian Development Bank (ADB), which was planning to locate its central office in Metro Manila, provided PLDT with a US$48 million loan to improve its facilities.[50]

One study documented that at the end of 1984, PLDT owed US$1.5 billion to foreign banks.[51] Evidently, World Bank lending bolstered PLDT's dominance, as it became the single largest private recipient of foreign loans to the Philippines. Thus, the agency known to be the principal advocate of liberalisation and market reform did not play that role in the telecommunications sector of the Philippines. Rather, by the 1990s local rumblings for competition overtook World Bank's policy advice of supporting a monopoly.

Telecommunications Sector Consolidation

Echoing the World Bank's counsel, a technical report on telecommunications by the Task Force on Human Settlements in 1975 called for the sector's integration. The report's most important recommendation was the development of a Telecommunications Masterplan "to avoid overlapping of activities and diseconomies of scale."[52] Although this was sensible advice, it ignored the fact that the proliferation of small provincial telephone companies resulted not only from the lack of a national policy guiding the industry's development, but more importantly from the failure of PLDT and Butel to provide telephone services in areas that needed them.

In 1978, the country's first 10-year telecommunications master plan was issued. It aimed to upgrade the telecommunications infrastructure by constructing an integrated communications network, with a projected budget of P4.6 billion (US$650 million). However, in 1979, the inflation rate reached 20 per cent and foreign currency reserves shrunk. This forced the Marcos government to eliminate public investment in the sector. The government, instead, encouraged private sector investment by providing them access to letters of credit. Butel, which was supposed to receive the bulk of the P4.6 billion budget for 1978–1980 was only allocated P87 million, with even less disbursed.[53] Thus, in contrast to the case of Malaysia, the development of telecommunications infrastructure was left in the hands of the private sector. This led to the concentration of expansion in major cities with high demand and assured profit, such as in Metro Manila and the emerging export-processing zones, to the detriment of rural areas.

During the later part of 1981, a National Telecommunications Development Plan was released. Arthur D. Little, an international consultancy company, drew up the plan and was funded by the World Bank. The 11-volume document proposed a 20-year plan to expand the country's telephone capacity to 3.56 million lines in 2002. A section of the plan recommended the integration of all private telephone companies under one monopoly, PLDT.[54] This consolidation took place with the state supporting PLDT's takeover of its smaller competitors, including the government telephone system. Marcos even issued a presidential directive to direct Republic Telephone Company (Retelco), PLDT's main competitor in Metro Manila, to merge with PLDT.[55] Despite the objection of the owners of Retelco, PLDT bought out its main competitor, after Marcos threatened to withdraw the companies' franchises. The merger was described as a shotgun wedding that led to a bigger and stronger PLDT.[56]

The above discussion demonstrates that PLDT consolidated its dominant position during martial law because of state policies that favoured the company and its privileged access to international funding. Yet despite the funding and loans, the telecommunications network in the Philippines remained underdeveloped. As can be seen in Table 1-2, teledensity remained at one telephone per 100 people for over 10 years. The next section explains why Marcos nurtured PLDT during martial law, and how the Aquino administration failed to change this situation.

Developments Under Aquino

The Question of PLDT Ownership and Failed Sequestration by the PCGG[57]

Until today, the actual ownership of PLDT is still contested. Four pieces of evidence are analysed in this section that strongly support the claim that former President Marcos was a majority owner of PLDT. That would explain why he nurtured the company during his administration. Marcos' actions point to how the state was used for personal profit. Yet, the failure of the Aquino administration to sequester the company reveals how the weak Philippine state continued to be captive to strong vested interests.

On 12 January 1977, the US Securities and Exchange Commission (SEC) filed charges against GTE, PTIC, and Stamford Trading Company (STC). The lawsuit charged Ramon Cojuangco, Alfonso Yuchengco, Luis Tirso Rivilla, and Antonio Meer for violating US federal securities law.[58] This lawsuit exposed two things: anomalies in the 1967 deal and Marcos' ownership of PLDT. It was revealed that PTIC did not have enough money to buy the GTE shares. However, in an incongruous turn of events, GTE helped PTIC buy its own shares in PLDT by agreeing to lend money to PTIC officials in exchange for an exclusive long-term contract to supply equipment to PLDT. The US SEC also found that GTE sold overpriced, sometimes second-hand equipment to PLDT. The price of equipment depended on the commission paid to PTIC officials. In effect, GTE agreed to sell its shares in PLDT to PTIC in exchange for a contract that secured its position as PDLT's equipment supplier. The Cojuangco group quietly agreed to an out-of-court settlement with the US authorities, and consented to court imposed penalties. Yet, no similar prosecution took place in the Philippines as PTIC continued to control PLDT.

A second revelation in the lawsuit pertains to the actual owners of PLDT. The lawsuit stated that GTE had already agreed to sell its stake in PLDT to a group of Filipino investors when "officials at the highest levels of the government of the Philippines urged GTE for political and purported security reasons not to sell its interests to the first group and told GTE to deal with another group, the PTIC group."[59] Theodore F. Brophy, then Executive Vice President for Finance and General Counsel of GTE, admitted that when he visited the Philippines he met with Mr. Cojuangco's principal, whom he identified as no other than President Marcos.[60]

While the US lawsuit was ongoing, GTE abandoned its 22.5 per cent stake in PTIC without any explanation. It was surprising that GTE would voluntarily dispose of its shares in a very profitable company. Equally surprising was how none of the PTIC stockholders were interested in the free shares, except Ramon Cojuangco, who ended up owning an additional 22.5 per cent stake in PTIC. This peculiar turn of events would become understandable in the light of who eventually gained control of the said shares. On 5 May 1978, Ramon Cojuangco and Luis Tirso Rivilla transferred 46 per cent of the PTIC shares to Prime Holdings Incorporated (PHI). Jose Campos and Rolando Gapud, well-known Marcos cronies, established PHI in 1977. When Marcos fled the country in 1986, Jose Campos executed an affidavit acknowledging that he and his financial adviser, Rolando Gapud, incorporated PHI on behalf of President Marcos. On 8 April 1986, Campos executed a second affidavit turning over to the government all shares and assets that he claimed were only under his custody for the former president. Campos' testimony is the second piece of evidence that revealed Marcos' hand in PLDT.

When Corazon Cojuangco Aquino became President in February 1986, her first Executive Order was to establish the Presidential Commission of Good Government (PCGG) to investigate and recover the Marcoses' ill-gotten wealth.[61] Aquino's second Executive Order was to "freeze all the assets and properties in the Philippines of the former president and his wife, their close relatives, subordinates, associates, dummies, agents or nominees."[62] PLDT became one of the first companies that PCGG took over on the basis of the Campos affidavit. A team of PCGG investigators led by lawyer Luis Sison was appointed to uncover the extent of Marcos' ownership of the company.

A third evidence that substantiated claims of Marcos' involvement in, and ownership of, PLDT emerged through the testimony of Alfonso Yuchengco. Yuchengco had been a minority investor in PLDT since the 1960s and was also interested in taking control of the company. In an affidavit executed to block the PCGG's sequestration of PLDT, Yuchengco claimed that not all of the PHI and Cojuangco shares were rightfully theirs. He revealed that Marcos coerced him to give up his shares in the company and instructed him to cooperate with Ramon Cojuangco.[63]

The PCGG investigating team presented the fourth piece of evidence. On 21 April 1986, Sison issued a preliminary report to the PCGG revealing that Prime Holdings Investment (PHI) was a Marcos dummy company. PHI owned a controlling share of 46 per cent of PTIC, which in turn held 24.95 per cent of the total PLDT voting shares at that time. In effect, Marcos was PLDT's single biggest stakeholder. (See Figure 4-2 and Tables 4-9). Based on Sison's initial report, PCGG Commissioner Mary Concepcion Bautista ordered the sequestration of the PHI shares in PLDT.[64]

Yet, on 2 May 1986, the PCGG unconditionally ended the sequestration of PLDT, except for the PHI shares. The unconditional lifting of sequestration orders was issued even before the Sison team concluded their investigation. Sison's final report on 27 May 1986 was very telling. His team discovered a list of anomalous and illegal practices within PLDT. The investigators discovered that PLDT accrued US$594.5 million worth of foreign loans from 1969 to 1982, but there was no evidence of network expansion that resulted from these loans. The report concluded by listing recommendations for reclaiming Marcos' stake in PLDT, freezing company assets, and initiating criminal prosecution against PLDT officials for fraud and the misuse of company funds. Sison also recommended the sequestration of the Cojuangco family's stake in PLDT because Ramon Cojuangco was clearly a business associate of Marcos, as defined by Aquino's EO 2.

None of these recommendations were heeded. Instead, Sison and his team were dismissed from their tasks and replaced by two new PCGG representatives, Mario Locsin and Benjamin Guingona.

Figure 4-2: Distribution of PLDT Equity Shares, 1987[66]

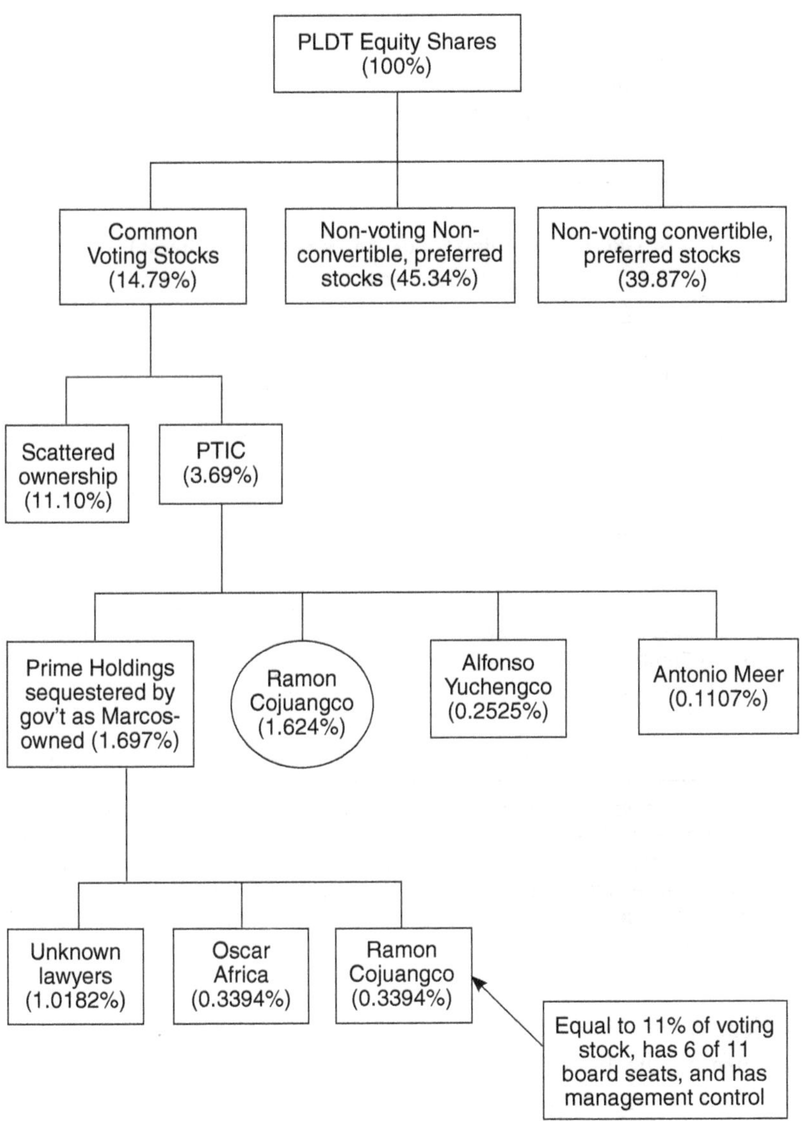

Table 4-9: PLDT Ownership Structure, 1987[67]

Type of Shares	Percentage
Voting Common	14.79
Non-voting, non-convertible, preferred	45.34
Non-voting, convertible, preferred	39.87
Total	100.00

Breakdown of Voting Stock (14.79 % of Total Shares)

Stockholders	Percentage
PTIC	3.69
Scattered Ownership (foreign)	11.10
Total	14.79

PTIC Share Ownership

Shareholders	% in PTIC	As a % of PLDT's Total Shareholding	% of Voting Power
Prime Holdings	46	1.69	11.43
Cojuangco Family	44	1.62	10.95
Alfonso Yuchengco	7	0.25	1.70
Antonio Meer	3	0.11	0.74
Total	100	3.69	24.95

Note: PTIC controls only 3.69 per cent of the total stock in PLDT. However, these shares represent 25 per cent of the total voting stocks in the company. Through these shares, the Cojuangco family, which controls PTIC, ended up controlling PLDT. Meanwhile, the subscribers who are required to invest in the company, controls 85 per cent of the total stocks in the company but these stocks are non-voting stocks.

In September 1986, Locsin and Guingona released a new report overturning the findings of Sison's team. The new report concluded that there was no proof that Ramon Cojuangco was a Marcos crony. It also denied the existence of evidence of management anomalies. Locsin and Guingona became PCGG representatives to the PHI, PTIC, and the PLDT board, but instead of protecting the public's interests they became close to Antonio (Tonyboy) Cojuangco, Ramon Cojuangco's son, who took over the management of PLDT when Ramon died in 1984.[65]

Based on the revelations of the 1977 US SEC lawsuit, the affidavits of Campos and Yuchengco, and the findings of the Sison team, it can be concluded that Ramon Conjuangco gained control of PLDT through Marcos' intervention and that Marcos himself held a substantial stake in PLDT. According to President Aquino's Executive Orders 1 and 2, the PHI and Ramon Cojuangco and family's shares in PLDT should have been sequestered. Yet, the Cojuangco shares in PLDT were not sequestered. Reportedly Antonio Cojuangco used his familial connections to stop the PCGG from taking over the company.[68] The Aquino government's double standards was made more evident by the unreserved sequestration of alleged cronies' shares in ETPI, Philcom, and Philcomsat, while the Cojuangco shares in PLDT were untouched. Former PCGG Chairman Jovito Salonga remarked in 1993 that President Aquino, a first cousin of Antonio's father Ramon, "treated the PLDT case with benign neglect, although she made a show of being neutral."[69]

The establishment of the PCGG indicated the Aquino government's sincerity in its initial efforts to rectify Marcos' misgovernance. Yet, this resolve slowly gave way to various competing demands from influential segments of society who had a large stake in maintaining the status quo. This situation made the prosecution and recovery of the Marcos wealth difficult, if not impossible. The failure to fully settle the question of PLDT's ownership and the government's lack of determination to resolve the matter was illustrative not merely of the failure of the PCGG, but also the failure of the Aquino government to lay to rest the injustices that had been committed under martial law. The PLDT case demonstrates the weakness of the state and its penetration by influential vested interests in society.

Failed Attempts to Liberalise

If "benign neglect" is how President Aquino treated the attempts to reclaim Marcos' ownership of PLDT, a more direct interventionist approach from her family members was observed when the Department of Telecommunications and Transportation (DOTC) and the industry regulator, the National Telecommunications Commission (NTC), attempted to liberalise the industry. The Aquino government's policy with respect to telecommunications was two-pronged: to increase public spending in underserved or unserved municipalities and to allow the entry of new players. The efforts to allow new players into the lucrative international gateway facility business and the cellular mobile telephony threatened PLDT's monopoly. This section discusses how PLDT blocked liberalisation in telecommunications by using its political and familial connections in the three branches of government. The success of PLDT's actions demonstrates the weakness of the state. Reforms to break down the monopoly control in the economy were selectively applied, indicating a lack of political will to introduce real changes to the economy and politics of the country.

The efforts to liberalise the telecommunications industry officially began in 1987 when the DOTC, in consultation with a Government-Industry Relations Committee, adopted a series of policies aimed at rationalizing the development of the industry.[70] These policy statements, contained in the DOTC's Department Circular (DC) 87-188, reversed the Marcos administration's push towards the integration of the telecommunications system under a monopoly. DC 87-188, issued on 22 May 1987, affirmed that a coherent development of the national telecommunications system could take place through the introduction of competition and regulated entry into the market. This policy statement was the first indicator that liberalisation and the introduction of competition would be pursued. Yet, the circular made no further mention of how the department would actually go about introducing competition.

In line with the government's objective of providing telephone service in rural areas, the DOTC launched the National Telecommunications Plan (NTP) in 1988. The Plan, worth US$389 million, consisted of three turnkey projects in Luzon, the Visayas, and Mindanao. The aim was to install 138,000 digital lines in 85 areas by 1992. After the construction of the network, the operation would then be privatised.[71] Japanese, French, and Italian firms would construct these networks, funded by bilateral concessional loans from the aforementioned countries. Additional funding for the project was expected to come from the World Bank.

When the Luzon segment covering 27 provinces was completed in 1990, the DOTC opened bidding for a 30-year lease to operate and manage the infrastructure. Digital Philippines Incorporated (Digitel), a subsidiary of ETPI, won the bidding with an offer of P40 billion. In contrast, PLDT's bid was only P7.46 billion. Digitel was to be awarded the contract in September 1991, when PLDT objected, arguing Digitel lacked the proper legislative franchise to provide telecommunications services. It also claimed that the high bid would ultimately be passed on to subscribers. The new DOTC Secretary, Nicomedes Prado, agreed with PLDT's position, and called for rebidding.[72] Digitel went to Congress to ask for a franchise, but legislators took time deliberating on its application. While Digitel's franchise application was stuck in Congress, the management of an already operational telephone network came to a standstill. In the meantime, PLDT moved quickly. In 1992 it secured a World Bank loan to install telephone lines in precisely the same areas where Digitel was supposed to operate.[73]

Aside from the policy of government involvement in constructing infrastructure in unserved areas and eventually privatising them, liberalisation also entailed the issuance of new licenses to operate international gateway facilities (IGF) and cellular mobile telephone systems (CMTS).

First, on 8 May 1988, the NTC authorised the Express Telecommunications Company (Extelcom) to operate a second cellular mobile telephone system (CMTS) and an alphanumeric paging system in Metro Manila and Southern Luzon. The NTC also mandated Extelcom's interconnection with PLDT. Extelcom's franchise to operate maritime radio services was issued to Felix Alberto in 1958. Control of the company eventually shifted to Ruby Tiong-Tan, in a joint venture with Motorola and Comvik of Sweden. Extelcom's services would compete with PLDT's subsidiary, Pilipino Telephone Company (Piltel).[74]

PLDT protested the issuance of Extelcom's license, arguing that the company did not have the proper franchise, that it was not the original franchise holder, and that the NTC had gravely abused its authority in mandating interconnection. In a close eight-seven vote, the Supreme Court affirmed the NTC's jurisdiction over all telecommunications entities, and its ability to issue provisional authority to operate services. The Court also ruled that the NTC had the power to mandate interconnection between authorised telecommunications companies. The media heralded this decision as one of the most important steps towards breaking up the PLDT monopoly.[75]

Despite the Supreme Court decision, PLDT deliberately postponed interconnection with Extelcom. In July 1992, Extelcom's major shareholder, Ruby Tiong-Tan, reportedly sold her shares in the company out of exasperation. Antonio Cojuangco was widely believed to be the ultimate owner of the buyer, Marifil Holdings Corporation. These suspicions were given more credence when almost immediately after Marifil Holdings bought into Extelcom, PLDT agreed to interconnect.[76]

Second, in October 1989 the NTC, headed by Commissioner Jose Luis Alcuaz, approved Philcom's application to operate a second international gateway facility (IGF). A month later, the NTC was set to issue a third IGF license to ETPI, when Alcuaz was fired. Alcuaz declared that he was removed from office because of interventions by relatives of President Aquino in government affairs.[77] Nonetheless, the newly appointed Acting Commissioner, Josefina Lichauco, affirmed Alcuaz's decision on the ETPI license. PLDT responded by lowering its international call rates and by suing the NTC and ETPI.

In the Supreme Court, PLDT called for the reversal of the NTC decision on ETPI.[78] PLDT lawyers argued that ETPI's franchise allowed it to transmit data messages but not voice messages.[79] Subsequent arguments focused on the entry of ETPI as a wasteful duplication of facilities. PLDT further challenged the NTC's authority to compel interconnection between PLDT and ETPI. PLDT claimed that interconnection could only be mandated between two telecommunications systems, but ETPI was not a telephone company. In a surprising turn of events that contradicted its earlier ruling on the Extelcom case, the Supreme Court voted ten-four in favour of PLDT. The decision, issued on 27 August 1992, accepted all of

PLDT's arguments. The Court ruled that ETPI had first to obtain the correct franchise from Congress, although it opined that a third international gateway was not really necessary.[80]

In a dissenting opinion, Justice Florentino Feliciano and three other Justices chided the decision for uncritically accepting PLDT's "eccentric interpretation" of the law, which had "absolutely no legal basis except PLDT's own bare assertions."[81] Justice Feliciano argued that PLDT's claim of a supposed distinction between voice and data messages had no basis. He pointed out that even PLDT's franchise provided no support for such highly selective definition. Moreover, the NTC had the authority to mandate the interconnection of public communication networks. He concluded that PLDT's petition was merely an attempt to protect its lucrative IGF monopoly and lamented that the Court's majority decision was an attack on the government's principle of free competition.

To summarise, PLDT used familial ties under the Aquino administration to evade government sequestration, block the entry of competition by filing legal challenges in the Supreme Court, and — with the help of its allies in the Congress — stall the awarding of franchises to its competitors. The NTC was an ineffective regulator due to its lack of independence from political pressures. The Supreme Court issued contradictory decisions whereby it allowed competition in the case of Extelcom but supported PLDT's monopoly in the case of ETPI. Finally, Congress obstructed reform by using its franchising power in favour of the monopolist and sitting on the franchise applications of potential PLDT competitors. Thus, by the end of Aquino's presidential term in 1992, the state of Philippine telecommunications was very poor. Despite the attempts of well-meaning executives in the DOTC and the NTC to break-up the PLDT monopoly, the company's political influence prevailed.

SUMMARY

Both the Malaysian and Philippine telecommunications sectors were monopolies before the introduction of reform, though one was a public monopoly in the British mould and the other a private monopoly in the US mould. In Malaysia, JTM, a government department, was the provider and regulator of all telecommunications services. In the Philippines, one company was a virtual monopoly although other private companies existed. Malaysia's state monopoly was relatively efficient and provided reasonable telecommunications service. The Philippine dominant player, meanwhile, was inefficient, and had poor service quality.

After independence, Malaysia was left with a fairly efficient communications infrastructure. With the adoption of the New Economic Policy (NEP), the state used the sector to allocate patronage to Malays in two ways. First, Malays were recruited to staff JTM at both the executive and lower levels. Second, some of JTM's services were licensed out and equipment supply contracts were awarded to bumiputera-owned businesses. These practices were rationalised on the basis of the NEP's redistributive goals. In the early 1980s, the state launched an ambitious plan to modernise the telecommunications infrastructure. Turnkey contracts were awarded to Malay businessmen who lobbied for the project's approval. This demonstrates the coincidence of the goals of the state with the rent-seeking actions of some bumiputera businessmen. That is, while the Malaysian state was determined to achieve both NEP goals — the creation of a bumiputera business class and the expansion of the economy through active state investment — there was support from individual businessmen for these goals to be attained in a particular manner. Under Mahathir, four handpicked bumiputera companies were awarded contracts despite the objections of smaller bumiputera companies. This indicates the shift in the strategy of state patronage from trustee institutions to patronage of individual bumiputras. However, these companies were required to have joint-venture partners who would bring in technology and expertise, suggesting that the distribution of patronage had a performance requirement. Thus, Mahathir's goal of rapid economic development, the creation of a Malay business class, and the goal of attracting foreign investments came together in the turnkey project. Under Mahathir, the nexus of politics and business, and the importance of executive patronage became more pronounced and decisively moulded the country's development. Such change will become more evident with the case of reforming the telecommunications sector that is presented in the next chapter.

In the Philippines, the state regulated foreign-owned private companies which provided telecommunications services. In 1967, a group of Filipino businessmen close to then President Marcos took over ownership of the private company with the sole authority to operate a national communications network. Under martial law, this company consolidated its monopoly status, and was under very little pressure to expand its network and improve its services. Marcos granted other businessmen exclusive privileges in the industry, while small provincial telephone companies proliferated due to unmet demand. The state under Marcos protected the monopoly profit of PLDT. This was allegedly because Marcos was himself a substantial stakeholder in PLDT. Under the Aquino administration, PLDT's owners used familial ties and its political influence in Congress and the Supreme Court to protect its position. Monopoly rent was captured for private benefit and was growth-hindering.

In both countries, the telecommunications sector was used for political patronage, but for different purposes and with different outcomes. The provision of state patronage to Malay businessmen was coupled with the requirement of having a foreign joint-venture partner. Having a foreign partner in a capital- and technology-intensive industry, while not an infallible mechanism in ensuring success, indicated the existence of a performance criterion. As the state invested massively in the telecommunications sector, the economic impact was generally expansionary and growth-enhancing. The coverage of Malaysia's telecommunications sector expanded considerably despite its use for political patronage purposes. This can be contrasted with the inefficient industry in the Philippines, where the dominant player was more interested in earning monopoly profits. PLDT was the recipient of many foreign loans for the expansion of its network but there were no indications of any service improvement. Loans dissipated in questionable deals that did not result in any infrastructure expansion. Despite its non-performance, the state favoured this company allegedly because Marcos himself owned shares in it. Under Aquino, the company owners used their familial links and their influence over other state institutions to block attempts to liberalise the industry.

In the Philippines, the World Bank was a major source of funding for communications infrastructure development, supporting the inefficient operation of PLDT. In contrast, Malaysia was not as reliant on external sources for funding. Thus, the impact of World Bank advice was not as influential in Malaysia as it was in the Philippines.

In Malaysia, infrastructure investment grew rapidly on the back of the NEP. The 1983 turnkey project, even though it was criticised for political favouritism, resulted in the installation of 1.4 million lines by 1991. This project considerably increased Malaysia's landline network. In contrast, telecommunications in the Philippines stagnated, with teledensity remaining at 1 telephone per 100 persons for over 10 years until 1993. The Philippine government left the modernisation of telecommunications infrastructure in private hands when its planned expansion program was overtaken by an economic crisis. Being a virtual monopoly, PLDT had neither the incentive nor pressure to expand and improve its services. Thus, while the state in Malaysia led investments for the expansion of its telecommunications infrastructure starting in the 1980s, telecommunications in the Philippines stagnated as the sector was used by Marcos to enrich himself and his cronies.

Some of the inefficiencies of Malaysia's telecommunications sector were brought about by the pre-eminence of attaining ethnic restructuring above economic growth goals. Yet, despite the use of sector for patronage purposes, the way it was done led to the growth in the telecommunications network and had an expansionary economic impact. In the Philippines, telecommunications was run not for the attainment of public goals but for the benefit of the President and his cronies, to the detriment of the country. In this way, one can conclude that rents were expansionary in the telecommunications sector of Malaysia before reform. Meanwhile, rents in Philippine telecommunications were monopolised and growth obstructing.

This chapter has discussed the state of telecommunications in Malaysia and the Philippines before market reform, and highlighted the roles of the state and private actors and the pattern of state patronage that emerged in the sector. The types of rents created, who reaped them, and their economic impacts have been identified. Building on this as background, the subsequent chapters will examine the experiences of Malaysia and the Philippines in reforming their telecommunications sectors.

NOTES

1. Fong Chan Onn, "Malaysia", in *Telecommunications in the Pacific Basin*, edited by Eli Noam, Seisuke Komatsuzaki, and Douglas Conn (Oxford: Oxford University Press, 1994), p. 134.
2. Laurel Beatrice Kennedy, "Privatization and Its Policy Antecedents in Malaysian Telecommunications", Ph.D. dissertation, Ohio University (1990), pp/ 57–59, 62.
3. G. Naidu and Cassey Lee, "The Transition to Privatization: Malaysia", in *Infrastructure Strategies in East Asia: The Untold Story*, edited by Ashoka Mody (Washington, DC: World Bank, 1997), p. 27.
4. Ben Petrazinni, *The Political Economy of Telecommunications Reform in Developing Countries: Privatization and Liberalisation in Comparative Perspective* (Westport, Connecticut: Praeger, 1995), p. 144.
5. Fong (1994), p. 139.
6. Ibid., p. 136.
7. The term "Malayanisation" was used until 1963, after which the term was changed to Malaysianisation
8. Kennedy (1990), p. 85.
9.. Ibid., 1990, pp. 109–10.
10. Fong Chan Onn, "The Malaysian Telecommunications Services Industry: Development Perspective and Prospects", *The Columbia Journal of World Business* (Spring 1989), p. 86.
11. Kennedy (1990), pp. 107–8.
12. *Report of the Auditor-General on the Account of the Telecommunications Department for the Year Ended 31st December 1971*, pp 1, 31–33, 47–48. Hereafter, it will be cited as *Report of the Auditor-General.*
13. See various *Reports of the Auditor-General,* 1971, 1972, 1974–81, 1985–86.

14. The use of ethnicity as a basis of contract award, however, does not preclude the possibility of competition among bumiputra-owned business if more than one applied for the contract. In reality, contracts were awarded to bumiputras without competitive bidding.

15. Eric Chia, a Straits-born Chinese or "Baba", was an example of Chinese businessmen who prospered under the NEP through their links to Malays. See Peter Searle, *The Riddle of Malaysian Capitalism: Rent-seekers or Real Capitalists?* (NSW: Allen and Unwin, 1999), pp. 161–64.

16. See Searle (1999), pp. 170–74. See also Sally Cheong's *Bumiputera Controlled Companies in the KLSE*, 2nd edition (Kuala Lumpur: Corporate Research Services, S.B., 1993), pp. 297–98; and *Bumiputera Entrepreneurs in the KLSE*, vol. 2 (Kuala Lumpur: Corporate Research Services, 1997), pp. 251–54.

17. Kennedy (1990), pp. 115–22.

18. Ibid., pp. 112, 122–26.

19. Ibid., p. 129.

20. Ibid., pp. 129–33.

21. Government of Malaysia, *Mid-term Review of the Fourth Malaysia Plan 1981–1985* (Kuala Lumpur: Government Printers, 1984), p. 327.

22. For a discussion on the 4th Malaysia Plan and the changes after its mid-term review under Dr Mahathir, see Lai Yew Wah and Tan Siew Ee, "Towards Effective Planning in Malaysia: Some Strategic Issues", in *Development Planning in Mixed Economies*, edited by Miguel Urrutia and Setsuko Yukawa (Tokyo: United Nation University, 1988), pp. 107–55.

23. Rita Hashim, *Privatization of Telecommunications in Malaysia*, MPA thesis, University of Malaya (1986), pp. 53–54.

24. Fong (1994), p. 142; Kennedy (1990), p. 152–53, 155.

25. Hashim (1986), pp. 60, 64

26. Kennedy (1990), p. 155–58.

27. Hashim (1986), p. 61.

28. Vincent Lowe, "Malaysia and Indonesia: Telecommunications Restructuring", in *Telecommunications in the Pacific Basin: An Evolutionary Approach*, edited by Eli Noam, Seisuke Komatsuzaki, and Douglas Conn (New York: Oxford University Press, 1994), p. 124. See also Hashim (1986), p. 69–73.

29. Malaysia's experience is similar to the opening of the US telecommunications market, where the lobbying of equipment suppliers eventually led to market opening. See William Melody, "Telecom Reform: Progress and Prospect", *Telecommunications Policy* 23 (1999): pp. 7–34.

30. Thomas Aquino, "Philippines", in *Telecommunications in the Pacific Basin,* edited by Eli Noam, Seisuke Komatsuzaki, and Douglas Conn (Oxford: Oxford University Press, 1994), p. 179.

31. *Philippine Long Distance Telephone Annual Report* (1997).

32. Hadi Salehi Esfahani, "The Political Economy of the Telecommunications Sector in the Philippines", in *Regulations, Institutions and Commitment,* edited by Brian Levy and Pablo Spiller (New York: Cambridge University Press, 1996), p. 163.

33. See Appendix 2 on the PLDT Ownership Issue Timeline, specifically points 1–4.

34. Aquino (1994), p. 180.

35. Stella Tirol-Cadiz, "Monopoly Unmasked: The Story of the PLDT", *Malaya,* 22–24 March 1993.

36. At that time Ramon Cojuangco was president of a family-owned bank, the Philippine Bank of Commerce, and was on good terms with Marcos. Ramon's wife, Imelda Ongsiako Cojuangco was part of Imelda Marcos' Blue Ladies Corps, who frequented Malacanang. See Shiela Coronel, "Monopoly", in *Pork and Other Perks: Corruption and Governance in the Philippines,* edited by Sheila Coronel (Manila: Philippine Center for Investigative Journalism, 1998), p. 120.

37. Task Force on Human Settlements, *Infrastructure/Utilities Development in the Philippines Technical Report Part IV: Telecommunication* (Quezon City: Development Academy of the Philippines, March 1975), p. 6. Hereafter cited as *Infrastructure Report.*

38. G.R. 94374, "Philippine Long Distance Telephone Company (PLDT) vs Eastern Telecommunications Philippines Incorporated (ETPI) and National telecommunications Commission (NTC)", 27 August 1992, p. 14. This document was found at the Philippine Center for Investigative Journalism Library in Quezon City.

39. http://www.etpi.com.ph/aboutus/index.asp (accessed 1 October 2002).

40. Gerald Sussman, "Telecommunications Transfers: Transnational Corporations, the Philippines and Structures of Domination", Third World Studies Center, University of the Philippines Dependency Series no. 35 (June 1981), p. 10.

41. Ibid., pp. 7, 12–13.

42. *Infrastructure Report*, pp. 91–103.

43. Examples are Zamboanga Telephone and Telegraph, Pineda Telephone System, Manang Telephone Company, Filipinas Telephone Company, Belle Telephone System, Republic Telephone Company, the Cotabato Telephone Company, and the Davao City Telephone System. See *Business World*, 12 October 1992.

44. For long-distance calls, the revenue sharing formula was 30:40:30, with the calling company receiving 30 per cent of the revenue, the national backbone owner receiving 40 per cent, and the company with which the call terminated receiving 30 per cent. This effectively meant that PLDT cornered 70 per cent of the revenue. For international calls, PLDT and the foreign telephone company received 50 per cent of the revenue. PLDT shared 12 per cent of the international call revenue to a provincial telephone company only if the international call originated locally. However, 82 per cent of international calls were incoming and only 17 per cent outgoing. Thus, PLDT effectively cornered the revenues of the lucrative international call market. See *Business World*, 12 October 1992.

45. *Presidential Decree 217*, "Establishing Basic Policies for the Telephone Industry, Amending for the Purpose the Pertinent Provisions of Commonwealth Act 146, as Amended, Otherwise Known as the Public Services Act No. 3436, as Amended, and All Inconsistent Legislations and Municipal Franchises Including Other Existing Laws", 16 June 1973.

46. *Far Eastern Economic Review*, 11 October 1969, p. 139; World Bank, *Philippines Private Sector Assessment*, vol. I (12 July 1994), p. 18.

47. Gerald Sussman, "Telecommunications for Transnational Integration: The World Bank in the Philippines", in *Transnational Communications: Wiring the Third World*, edited by Gerald Sussman and John Lent (California: Sage Publications, 1991) p. 42, 47.

48. *Far Eastern Economic Review*, 12 December 1980, p. 57.

49. Robert Saunders, Jeremy Warford, and Bjorn Wellenius, *Telecommunications and Economic Development* (Baltimore: JohnsHopkins University Press, 1994), p. 420.

50. *Far Eastern Economic Review*, 3 July 1988.

51. Sussman (1991), p. 140.

52. *Infrastructure Report*, p. 51.

53. Sussman (1981), p. 5.

54. Sussman (1991), pp. 60–61.

55. *PLDT Corporate Profile* (Research Department, Makati Stock Exchange, May 1991), p. 3.

56. *Far Eastern Economic Review*, 7 May 1981, p. 46.

57. See Appendix 2 for Timeline of PLDT Ownership Question, specifically points 1–12.

58. The US SEC had jurisdiction over PLDT because it was listed and actively traded on the New York Stock Exchange. This section draws information from Ricardo Manapat, *Wrong Number: The PLDT Telephone Monopoly* (no place and date of publication); and PCGG case no. 3, Alfonso Yuchengco versus Republic of the Philippines, Presidential Commission on Good Government, Ferdinand Marcos, Imelda Marcos, and Prime Holdings, Inc. 2 August 1988, p. 4. Herein cited as "Yuchengco versus PCGG et al".

59. Manapat, n.d., p. 2.

60. Internal Memorandum to President Joseph Estrada on "The Participation of Prime Holdings in the Philippine Long Distance Telephone Company (PLDT) and the Government's Right of Recovery", 26 October 1998. Hereafter cited as *Memo to Estrada*. The document was found at the Philippine Center for Investigative Journalism Library in Quezon City.

61. *Executive Order No. 1*, Creating the Presidential Commission on Good Government, 28 February, 1986.

62. *Executive Order No. 2*, Presidential Commission of Good Government Rules and Regulation, 12 March 1986.

63. Yuchengco versus PCGG et al., pp. 9–10.

64. *Memo to Estrada.*

65. Coronel (1998), p. 127.

66. Figure 4-2 was adapted from Hilarion Henares, "The Trouble with PLDT and Emperor Tony", *Philippine Daily Inquirer*, 27 March to 7 April 1989, reprinted in "The Case of Lawyers Against Monopolies of the Philippines (LAMP) versus Philippine Long Distance Company (PLDT) for Quo Warranto" filed at the office of the Solicitor General, 29 May 1989, p. 31.

67. These tables are constructed from information contained in the *Memo to PCGG Chairman Ramon Diaz from Victor S. Limlingan*, 23 March 1988. Some differences exist in the sums due to rounding off.

68. *Far Eastern Economic Review*, 6 October 1988, p. 69.

69. Coronel (1998), p. 126.

70. Department of Transportation and Communications, *National Telecommunications Development Plan 1991–2010: Executive Summary*, July 1993 Update, p. 1.

71. *Far Eastern Economic Review*, 6 October 1988, p. 72.

72. The DOTC under Aquino was headed by a succession of secretaries: Rainerio Reyes, Oscar Orbos, and Nicomedes Prado.

73. *Far Eastern Economic Review*, 4 June 1992, p. 42.

74. *Newsday*, 26 October 1990.

75. *Newsday*, 26–28 October 1990.

76. *Philippine Daily Inquirer*, 17 March 1991.

77. *Daily Globe*, 15 November 1989.

78. Curiously, although PLDT blocked ETPI's IGF application, it accepted without any legal challenge the entry into the IGF market of Philcom. This was perhaps because PLDT and Philcom had interlocking boards of directors.

79. ETPI's franchise states that it is authorized to "operate telecommunications by cable or other means known to science or which in the future may be developed for the transmission of messages between any points in the Philippines to points exterior thereto". See *Republic Act 5002*, 17 June 1967.

80. Justice Hugo Gutierrez, Supreme Court of the Philippines Decision on GR 94374, PLDT vs ETPI and NTC, 27 August 1992.

81. Justice Florentino Feliciano, Supreme Court of the Philippines Dissenting Opinion on GR No. 94374, PLDT vs NTC and ETPI, 27 August 1992.

REFORMING THE MALAYSIAN TELECOMMUNICATIONS SECTOR

This chapter examines the first phase of Malaysian telecommunications reform. Three main arguments are presented. First, Prime Minister Mahathir played a central role in the adoption of privatisation in Malaysia. Second, while the Malaysian state is strong and determinedly adopted privatisation as an economic policy, its implementation was greatly influenced by the active lobbying of Malay businessmen. This points to the fact that, a strong state that is able to formulate policies independently is not necessarily insulated from lobbying that shapes implementation. Third, the implementation of telecommunications privatisation mainly benefited politicians, their political allies, and those within the Barisan Nasional patronage network.

The discussion is divided into three parts. The first looks into how and why privatisation was adopted as a state policy. The second focuses on the privatisation of the telecommunications sector by analysing its stages, its implementation, and the roles of major actors. A final section synthesises the chapter's evidence and main arguments.

PRIVATISATION IN MALAYSIA

Malaysia was one of the first among developing countries to adopt privatisation as part of its economic reform program. In the early 1980s, the total or partial sale of state enterprises to the private sector was viewed as *the* solution to numerous problems confronting the state including inefficient state enterprises, high levels of public debt, and poor economic performance. In March 1983, Prime Minister Mahathir Mohamed spoke of privatisation as a cornerstone of his economic policy. He said:

> Privatisation means the opposite of nationalisation. The objectives of nationalisation is for government to take over the ownership of private enterprises while privatisation means the transfer of government services and enterprises to the private sector. Normally, the companies and services owned and managed by the government have been less successful or have run at a loss because the government's management methods differ greatly from those of the private sector. On the other hand, private businesses and enterprises are usually profitable ... In view of this possibility, there is a need to transfer several public services and government-owned businesses to the private sector.[1]

Along with announcing privatisation, Mahathir spoke of Malaysia Incorporated, a concept inspired by "Japan Inc." The phrase encapsulates cooperative practices between the government and business leading to economic growth. Mahathir explained, "the Malaysia Incorporated concept means that Malaysia should be viewed as a company where the government and the private sector are both owners and workers ... Only through the success of the company will the owners and workers' well being be safeguarded and improved ..."[2]

Initially, there was confusion about whether privatisation and Malaysia Inc. meant the same thing. In a speech to the Malaysian Economic Society in November 1983, Abdullah Badawi, then Minister in the Prime Minister's Department, maintained that privatisation and Malaysia Inc. were two different concepts. He defined privatisation as the transfer of selected government interests or investment to the private sector, and Malaysia Inc. as cooperation between government and business for the mutual benefit of both the public and private sectors, and ultimately the nation.[3] Thereafter, government-sponsored think tanks organised various conferences to explain the policy options and debate their merits.[4]

The Economic Planning Unit (EPU), the country's highest economic policymaking body, pointed out that the adoption of privatisation as a state policy was in line with the international ideological shift following the poor performance of many state-owned enterprises.[5] The Malaysian public sector had ballooned after the introduction of the NEP, and the dismal financial performance of public enterprises drained public funds.[6] Privatisation was seen as a solution to these problems. In 1985, the EPU issued its Guidelines on Privatisation, outlining the policy objectives, forms of privatisation, the identification and selection of entities to be privatised, privatisation's consistency with the NEP, and the organisational, legal, and valuation aspects of privatisation. The rationale for privatisation in Malaysia were to: relieve the financial burden of government; reduce the size and presence of the government in the economy; accelerate economic growth through private investment; to promote competition, raise efficiency and increase productivity; and achieve the distributional objectives of the NEP.[7] The government was very careful to point out that all privatised projects had to have at least 30 per cent Bumiputera participation, which rendered them tools for the continued restructuring of capital ownership in the economy.[8] The close connection between privatisation and affirmative action goals for the Malays made Malaysian privatisation unique. Ultimately, the goal of redistribution, in line with the NEP, dominated the other four objectives.

The government's concept of privatisation was much broader than the usual transfer of government-owned businesses and public services to the private sector to improve efficiency. Various forms of privatisation were adopted, including the complete, partial, or selective transfer of ownership to the private sector through the: sale of assets; commercialisation of public enterprise structures; management buy-outs, contracts or purchasing of management expertise; contracting out of existing services; leasing; and construction projects based on the build-operate-transfer (BOT), build-operate-own (BOO), build-operate (BO), and build-transfer (BT) models.[9] The licensing of private competition to enter an activity or industry that had been previously monopolised by the public sector, such as telecommunications, television broadcasting, or tertiary education, which is called liberalisation in other countries, was also considered to be a mode of privatisation.[10]

After 1983, the government privatised over 400 SOEs, ranging from ports to telecommunications, power utilities, highways, water provision, sewerage services, shipping, car distribution, petrol retailing, and television stations. The policy was extended beyond the federal government as state governments and departments were encouraged to prepare their own privatisation plans. Various state government-owned or -controlled entities that managed water, ports, dockyards, abattoirs, hotels, cement factories, parking and commercial vehicle inspection, city buses, and even rubbish collection were privatised.

A distinctive feature of privatisation in Malaysia was how the Guidelines invited the private sector to submit proposals, although the final decision to accept or reject a tender rested entirely with the government.[11] This provision basically enshrined in policy the "first come first served" principle wherein the first group to submit a proposal was awarded the privatisation contract, without the benefit of any public bidding. The EPU rationalised this practice as recognising private sector initiative and rewarding innovation, ingenuity, and entrepreneurial spirit.[12]

The government established an Inter-Departmental Committee on Privatisation chaired by the EPU Director-General to plan, monitor, and evaluate the progress of privatisation. The Committee was composed of the Directors-General of the Implementation Coordination Unit (ICU) and the Public Services Department, the Attorney General, and the Secretaries-General of the Ministries of Finance, Trade and Industry, Transport, Works, Energy, Telecommunications, and Post, and Home Affairs. Four technical sub-committees were established on manufacturing and commerce; transport; energy, telecommunications, post and information; and social and other services. The EPU served as the Chair and Secretariat of each sub-committee. Private sector proposals for privatisation were initially handled by the relevant technical committees, after which successful applications were sent to the Inter-Departmental Committee. The Committee then assessed private sector proposals before sending them to the

Cabinet for approval. While this was the ideal process, the practise however, was extremely ad hoc and lacked transparency. Moreover, almost all of the approved proposals came from well-connected and influential businessmen or the relatives of politicians. Even Bumiputera businessmen voiced their concern that only political favourites were benefiting from privatisation.[13]

A few months after the release of the Guidelines on Privatisation in 1985, the government commissioned J. Henry Schroder Wagg and Company of London and the Arab-Malaysian Merchant Bank to draw up a privatisation master plan to lend coherence to the process.[14] The consultants identified 424 enterprises with a market capitalisation of around RM15–20 billion as candidates for privatisation over five years. However, the result of the study, the Privatisation Masterplan (PMP), was only released six years later, in 1991. The PMP was announced, along with the Sixth Malaysia Plan, the Second OPP and the New Development Plan (NDP), making privatisation a central aspect of the post-NEP economic program. A Privatisation Action Plan (PAP) accompanied the PMP, identifying viable privatisation candidates and ranking them according to the ease of privatisation.

The PMP was a response to the criticism that privatisation decisions had been made in an ad-hoc, "first come first served" manner. It was designed to provide a clearer definition of policy objectives, guidelines, and methods, to develop an overall strategy to facilitate the privatisation of the remaining SOEs, and to place privatisation as a strategy within the broader context of economic liberalisation. The PMP also provided a review of privatisation's achievements in the previous years, and claimed success in terms of efficiency gains, providing growth stimulus, reducing governmental responsibilities, and economic redistribution. Despite the PMP, many of the earlier practices of opacity in the awarding of contracts were unchanged. The case of telecommunications liberalisation provides support to this claim.

Although both the Guidelines and the PMP cited the standard neoliberal reasons for market-oriented reform, the role of Mahathir's political will and commitment to the policy cannot be overemphasised. Amidst a severe recession during the mid-1980s and declining terms of trade[15] exacerbated by his own heavy industrialisation drive, Mahathir considered privatisation as a crucial means of sponsoring the emergence and consolidation of a Malay business class that could become internationally competitive.[16] Mahathir had consistently argued that private, not public, enterprises should be the main conduit for the creation of a Bumiputera capitalist class and for national economic development, as he articulated in his *The Malay Dilemma*.[17] He and his then-closest ally and Finance Minister, Daim Zainuddin, thought that there were already enough Malays capable of acquiring privatised entities and managing them profitably if given the chance.[18] Consequently, it is probable that even without the economic recession of the mid-1980s and international prompting, Mahathir would have actively pursued privatisation, if only as a central tool to promote Bumiputera capitalism. Nevertheless, the presence of these two factors certainly gave him an additional push. Mahathir's influence was so significant that when he underwent an operation for a heart ailment in 1989, privatisation proposals and meetings at the EPU came to a standstill. His recovery reinvigorated the process.[19]

In December 1986, when criticisms of the implementation of privatisation were increasing, the parliament amended the Official Secrets Act (OSA). While the amendment was not primarily a response to these criticisms, it extended the definition of official secrets to include, among other things, "government tender documents and any other documents or materials which ministers and public officials may arbitrarily deem secret or confidential." The decision to classify a document or material as an official secret could not be challenged in any court of law. Any public official found guilty of breaking the law would receive a mandatory one-year jail sentence.[20] Thus, although there were attempts to initiate public discussion on policy matters via conferences and seminars, the manner of privatising and the amendments to the OSA point to the lack of public accountability and transparency, and the obvious attempt to douse criticisms.

With power centralised in the hands of the Prime Minister, the implementation of policies such as privatisation increased opportunities for rent creation and appropriation and advanced private business and patronage interests. Private interests were nurtured, especially those of the politically well-connected. The privatisation and liberalisation of the telecommunications industry clearly demonstrates this development.

PRIVATISING TELECOMMUNICATIONS IN MALAYSIA

The reform of the Malaysian telecommunications sector took place in three stages: the corporatisation of the Telecommunications Department in 1987, its partial privatisation through public listing in 1990, and the liberalisation of the industry through the introduction of competition in 1993.

Moves to Privatise JTM and Early Liberalisation

JTM was the first government department identified for privatisation. This was ironic for two reasons. First, compared to the telecommunications systems of other developing countries, JTM had consistently earned profits since the 1960s, despite charging among the lowest calling rates in the world. Even the World Bank held up JTM during the 1960s as a model for other nations to emulate, and credited it with an "A" financial rating.[21] In addition, as discussed in the previous chapter, JTM consistently had a positive revenue base that grew by an average of 21 per cent per year until its corporatisation in 1984. It was not a loss-making enterprise, despite the inefficiencies that the implementation of the NEP imposed on its performance. Second, JTM had a reputation for adopting technological innovations and providing modern business communication services in the commercial centres of the country while maintaining a program of rural telephony.[22]

Nevertheless, consumer dissatisfaction with JTM's service grew during the 1970s, centering on contested billing and unmet demand. The unsatisfied demand for telephone lines rose from 13,704 in 1970 to 190,542 in 1984, which reflected a growing economy and an upwardly mobile population. Charges of corruption against those allocating new facilities also became an issue. An Ombudsman's office set up in the early 1980s mainly received complaints about JTM's service. The major daily newspapers also regularly printed letters from unhappy subscribers. Furthermore, the reports of the Auditor-General from 1975 to 1987 consistently criticised JTM for questionable practices, despite the commercialisation of its operations.[23]

Responding to the demand for better telecommunications services, the government increased spending in both the 4th Malaysia Plan and its Mid-Term Review particularly for turnkey and equipment contracts with Bumiputera companies. While this increased funding meant the expansion and modernisation of the infrastructure to help meet the increased demand for telecommunications services, the decision to use private contractors pointed to the successful lobbying of Malay businessmen, despite JTM's resistance on the basis of cost effectiveness. As was the case of the turnkey contract discussed in the previous chapter, competitive bidding was dispensed with as the Prime Minister handpicked the contractors for the RM2.54 billion project.[24]

On another front, the international trend towards privatisation inspired some people to think about following the precedent of British Telecom, which was privatised in 1984 via the public sale of its shares. While the EPU was in the process of crafting a general framework for privatisation, there was active business pressure to privatise JTM and liberalise certain market segments. Shamsuddin Kadir, the founder and owner of the Sapura Group of Companies and one of the turnkey contract awardees, reportedly commissioned Arthur D. Little, a US-based consultancy firm, to research on "The Advantages and Feasibility of Privatising Jabatan Telekom Malaysia." The study was completed in November 1983, and presented to the Prime Minister and the Cabinet. It argued that a privatised JTM would be more responsive to the nation's needs by eliminating the inability to hire qualified staff, to install sophisticated management systems, inappropriate financial regulation, and to control data processing and resources. Telecommunications services had to be separated from civil service administration to solve the problems of inflexibility, non-commercial orientation, and the incapacity to respond to the demands of high technology.[25] The study also called for the liberalisation of value-added network services (VANS) such as radio paging, while asserting that JTM's monopoly control of basic network services should be maintained due to natural monopoly considerations.

Within a year, the VANS segment was liberalised. Komtel, a company that was 60 per cent owned by Sapura Holdings, received the first radio-paging license to serve the Klang Valley. Three other licenses to operate in Klang were issued to Electroscom Bumi Engineering (one of the turnkey contractors), Kilatcom (a subsidiary of Karim Ikram's Binafon, also a turnkey contractor), and Telesistem (a company owned by Tan Sri Hassan Wahab, a former Director-General of JTM). By 1987, 32 licenses had been issued, and this number reached a height of 38 by 1992. Kennedy observes that the most lucrative, urban areas were granted to the most politically influential firms.[26] All licensees were Bumiputera firms, though the Ministry of Energy, Telecommunications, and Posts declared that the awarding of licenses and the criteria used for the evaluation of applications was not a matter of public record.

The ultimate measure of the success of Shamsuddin's commissioned study was how it helped to convince the government to privatise JTM. The Cabinet agreed to adopt the recommendations, but did not formally accept the study because it was privately commissioned. In August 1984, the government hired a consortium composed of the Arab-Malaysian Merchant Bank, Klienwort Benson, (the UK bank that helped in the privatisation of British Telecom) and local accountants Hanafiah Raslan Mohamad and Associates. They were expected to make recommendations on the feasibility and financial procedures of privatising JTM. This second study, which remained confidential, was submitted to government on 28 December 1984, and was said to have set forth procedures on how to go about privatising JTM.[27] It is important to note that the second study was completed in December 1984, but a new company to take over from JTM was already incorporated in October 1984. This indicates that the decision to privatise had already been taken, perhaps influenced by Shamsuddin's commissioned report.

On 12 October 1984, Syarikat Telekom Malaysia (STM) was incorporated under the Companies Act as a wholly-owned government company under the Ministry of Finance. Next, various legal constraints had to be resolved. Three laws and the Constitution were amended for this to take place. First, the Telecommunications Act of 1950, which provided that establishing, maintaining, and operating telecommunications services in the country was the exclusive responsibility of the government, was amended to allow the transfer of that function to a successor company. JTM retained regulatory authority over the successor company, and any future operators of telecommunications services had to be licensed by the Ministry. Second, the government enacted the Telecommunications (Successor Company) Act of 1985 to create a new operating company, STM, to assume the business from JTM. The Act was closely patterned after the British Telecom Act of 1984, whereby all assets, liabilities, rights, and privileges were transferred to the successor company. However, in comparison to British Telecom's immediate sale of shares, the Malaysian government was not categorical about when exactly it would privatise STM. Then Minister of Energy, Telecommunications, and Posts (MEPT), Leo Moggie, stated that the company would initially be fully owned by government via the Ministry of Finance. He also said that the government intended to allow the entry of private capital at some future date, selling a majority of shares to the private sector, with the government retaining an approximately 30 per cent stake.[28] Furthermore, a distinctive feature of the Act was its provision for the future of JTM employees. It provided that staff members who opted to join the new company would be offered terms and conditions of service "not less favourable" than those they enjoyed as government employees. Third, a special provision was incorporated into the Pension Act of 1980 (the Pension Act Amendment of 1985) to preserve the retirement benefits of staff members who chose to retire or join the new company. Finally, the Constitution had to be amended to allow the disposal of state land on which JTM assets and facilities were located.[29]

Opposition to Privatisation

While the government was obviously committed to privatising JTM and wanted to do so swiftly, opposition arose from various sectors, including from legislators, civil servants, employees, unions, academics, and consumer groups. The government, however, was not keen to listen to those whom it could afford to ignore.

On 23 July 1985, the government tabled in Parliament three bills pertaining to JTM's privatisation. The Members of Parliament were only given two days notice about the vote on the upcoming bills, with legislators obtaining draft copies of the bills a day before the vote. Despite the complexity of the bills and the far-reaching implications for the country, there was no debate on them. The Malaysian Technical Services Union (MTSU) and the National Union of Telecom Employees (NUTE) complained that they were not even consulted.[30] Lim Kit Siang, the head of the opposition Democratic Action Party (DAP), raised a motion to postpone consideration of the bills so that deliberation and analysis could be made. Lim criticised the lack of consultation and discussion on the matter and asked whether the people or the "UMNOputras" would be the ultimate beneficiaries of privatisation. The motion, however, was defeated and the bills were passed without much ado.[31] This was possible because the BN controlled more than two-thirds of the seats in parliament.

Consumer groups such as the two Penang-based associations, the Consumers Association of Penang and Aliran, launched public information campaigns on the issue through the publication of newsletters that aimed to make consumers aware of the costs and issues of privatisation.[32] Meanwhile, some academics who attended government organised conferences and symposia on privatisation also raised concerns, but their voices were barely heard.[33]

The most vocal and organised opposition to the privatisation of JTM came from the labour sector.[34] Of the 28,000 JTM employees, NUTE claimed 22,000 members, and 2,500 were members of the MTSU.[35] Since the announcement of privatisation in 1983, what was left of the weak union movement in Malaysia[36] consistently expressed opposition to the policy. Unions argued that privatisation would negatively affect the welfare of workers and the people, as it would lead to increases in the cost of services that were traditionally the responsibility of government.[37] In 1984, NUTE publicly rejected privatisation. It argued that JTM was profitable and that it could be made more so if the constraints set by the Treasury and the Public Services Department were removed. NUTE called for strikes and organised picketing and work stoppages to protest JTM's privatisation. In the same year, the MTSU also publicly stated its opposition to privatisation, pointing to the lack of transparency in the process.[38]

The government could not easily dismiss the 30,000 JTM employees, who threatened to launch a national boycott and work stoppages to paralyse JTM's operations. To appease them, the Telecommunications Successor Act contained a provision that JTM staff members who chose to join STM would be offered terms "not less favourable" than the conditions they had previously enjoyed as public servants.[39] On 18 May 1986, Minister Leo Moggie announced that letters would be sent to all employees detailing the available options. Employees were given 60 days to decide whether to join STM or stay with government. The government threatened that those who did not respond by the set date would be considered to have retired from service.[40] NUTE, however, insisted that outstanding issues of claims and problems of pay and pensions should be settled first. Hence, the plan to turn over operations to STM by 1 August 1986 was postponed. A compromise was reached on 25 September 1985 among the Ministry, representatives of the PM's Department, and NUTE officials. For five years, STM was not allowed to retrench any of the workers who opted to join it, except on disciplinary grounds. The workers were provided with two options. Option A granted those who chose to transfer to STM the same conditions of employment that they had as public servants. Option B consisted of a higher take-home pay and participation in bonuses and share offerings. All of NUTE's 22,000 workers, along with the 2,500 MTSU members, opted to work for STM.[41] Of the 28,724 personnel, 28,364 voluntarily joined STM, 102 were assigned to stay with JTM, and about 258 decided to retire early or resign.[42]

The Corporatisation of Telekom Malaysia

On 1 January 1987, the provision and operation of the telecommunications services was transferred to Syarikat Telekom Malaysia (STM), while

JTM took up the regulatory function. STM received land, buildings, and equipment valued at RM7 billion, and about RM5 billion in liabilities.[43]

Tan Sri Mohamad Rasdan Baba, Chief Executive of the Guthrie Group, was appointed STM's first Chairman. Daud Ishak, then Director-General of JTM, became its first Managing Director.[44] The company's management composition indicated how little involvement JTM had in its privatisation. Although JTM's Director-General and Deputy Director-General sat on the committee that managed the privatisation, JTM had very little input in the process. The EPU, which headed the privatisation committee, held actual decision-making authority. Reportedly, the Director-General of JTM set up an internal committee within the Department to work on personnel and legal questions. This internal committee prepared papers to submit to the Privatisation Committee, but their recommendations were ignored.[45]

The government issued a 20-year monopoly license to STM on 1 December 1986. The license spelled out 37 conditions for STM to fulfil, including responsibilities for universal rural service, limitations on rate setting, non-discrimination in providing services, the avoidance of cross-subsidisation in areas of service that were already liberalised, and keeping separate accounting books for different services. The license also specified the annual fees that STM had to pay to the new regulatory body, and required ministerial approval for any rate increases that STM levied for its basic or value added services. Arbitration procedures for disputes between the company and consumers were also laid out. Most importantly, the license contained provisions for a Golden Share that the government would hold, which was similar to the provision contained in the license granted to British Telecom when it was privatised.[46] The golden share gave government the power to veto company policies, even if it was no longer the majority shareholder, if these conflicted with the national interest. What constituted the national interest, however, was undefined.

Listing on the Kuala Lumpur Stock Exchange

Following corporatisation, the government planned to list STM on the stock exchange and divest its share holding to the general public. The Capital Issues Committee (CIC) rules required a company to have a record of at least three consecutive years of profitability before listing. Hence, increasing profitability and efficiency became the STM's new priority. Three controversies arose in attaining the goal of increasing company profitability. First, Minister Leo Moggie ordered STM employees to drop their civil service mentality and change their attitudes.[47] Senior managers who had been brought in from the private sector pointed to the staff's resistance to change as a major obstacle to improving company performance. Staff productivity was considered very low and the company was overstaffed, with a ratio of 22 employees to every 1000 lines. This was four times higher than the ratios of NTT and Bell.[48] However, the President of the Congress of Unions in the Public and Allied Services (CUEPACS), Ragunathan, called on the Minister to stop blaming the staff and to look at the overall system that did not encourage productivity.[49]

The delineation of STM's area of business was a second contentious issue. STM protested the Ministry's renewal of Uniphone's license to operate urban payphone services for 15 more years. STM argued that it was capable of handling urban payphone services given that it was ably providing payphones to the rural areas. It also pointed out that the urban payphone business was a means of improving company profitability. The Ministry ignored its objection and confirmed Uniphone's license.[50]

A third issue was the degree of STM's independence. The MEPT maintained control over the company's operating license, rates, and supposedly, the composition of the board of directors.[51] Government officials or politicians were usually appointed to the board, giving them great influence over decisions despite their limited (if not non-existent) experience in telecommunications. The basis for choosing these appointees, it seemed, was their political affiliations rather than their knowledge of the sector. A review of the STM board composition from 1988 to 2000 shows that politicians from the three major parties of the BN (UMNO, the MCA, and the MIC) held at least one seat each.[52]

When the financial records for its first year were released in late 1988, the net loss was RM96.63 million. This disqualified the company from listing before 1991. However, STM's financial performance for 1988 was an astonishing improvement. With the benefit of a three-year tax holiday, it reported a profit of RM180.12 million.[53] Pre-tax profit improved to RM366 million in 1989 and RM564 million in 1990. By December 1989, STM hired the Arab-Malaysian Merchant Bank to manage its public listing.[54]

In July 1990, the new Minister of Energy, Telecommunications and Post, Samy Vellu,[55] announced that Cabinet had approved STM's listing on the Kuala Lumpur Stock Exchange (KLSE), with 5 per cent of the shares allocated to employees and 15 per cent to Bumiputera institutions and individuals. The government decided to retain 80 per cent of the shares, and no foreign equity was allowed during the initial sale, although foreign investors could buy at market prices thereafter. JTM employees were also given the option to buy STM shares because they were also "part of the privatisation process." Mahathir decided that starting with Telekom's listing, privatised government agencies were required to set aside 5 per cent of the overall shares to be sold to their employees at a special price. Thus, STM and JTM employees were allocated RM100 million worth of shares.[56]

In August, a total of 470.5 million shares were offered for sale at RM5 each share. Of these shares, 100 million were reserved for Bumiputera institutions, 152.1 million for approved institutions, 70.5 million for eligible directors and employees, and 147.9 million for public sale, including 51.9 million to Bumiputera individuals. The Ministry of Finance made decisions about how to allocate the Telekom shares and which organisation or institution was allowed to buy the shares. However, Bank Negara, despite the fact that it was not an approved institution to buy shares, took up 88.25 million shares, which was equivalent to about 4.5 per cent of Telekom's equity. In addition, the entire 100 million shares for Bumiputera institutions were taken up by PNB (Permodalan Nasional Berhad).[57]

The listing of STM on the KLSE on 7 November 1990 marked its partial privatisation. The company changed its name to Telekom Malaysia Berhad (TM or Telekom), became the first public utility to

Table 5-1: Performance of Telekom Malaysia Before and After Privatisation

Indicators	Before Privatisation	After Privatisation	
	1990	1997	2000
Return on assets (%)	4.0	7.6	4.7
Average revenue per user (RM)	1,227.0	1,609.0	1,755.0
Production per employee (RM)	34,372.0	219,641.0	51,672.7
Number of direct exchange lines per employee	36.0	154.0	183.0
Percentage of responses to customer complaints within 24 hours	80.0	91.5	100.0
Operating revenue (RM millions)	1,644.2	7,165.7	—
Profit before tax (RM millions)	4.9	2,376.4	—
Debt equity ratio	2.3	0.7	—

Sources: Government of Malaysia, *Eighth Malaysia Plan* (Kuala Lumpur: Government Press, 2001), p. 186.

be corporatised and was the largest company listed on the KLSE, comprising eight per cent of the stock exchange's capitalisation. Its initial public offering was oversubscribed, valued at US$872 million.[58]

Following its listing, the company's revenue improved, as did its quality of service.[59] Moreover, new products and services were introduced. Then CEO Syed Hussein concluded that corporatisation and the partial privatisation of Telekom had been a win-win situation: customers received more and better services, the government's equity value increased fivefold, the employees would benefit from the upcoming stock option, and the company added prestige and investment to the Malaysian capital market.[60] This partly explains why the government still controlled about 82 per cent of Telekom in 2000: the company was making money and was a profitable investment.[61]

The Maika Controversy

Although Telekom's listing appeared beneficial to all, a controversy erupted with the distribution of company shares involving the then-Minister of Energy, Telecommunications, and Posts, and President of the Malaysian Indian Congress (MIC), Samy Vellu. The way in which the issue was handled, the resulting investigation by the Anti-Corruption Agency (ACA), and the resolution of the controversy tells much about the dynamics of politics in Malaysia, and the implementation of privatisation.

In April 1992, almost two years after Telekom's listing, then new Finance Minister Anwar Ibrahim told parliament that the Indian community was allotted 10 million Telekom shares during the company's listing. Anwar revealed that the Ministry of Finance, under his predecessor Daim Zainuddin, made a special allocation, usually given only to Bumiputera institutions, in recognition of the Indian community's relative poverty. The shares were originally to be allocated to Maika Holdings, the investment arm of the MIC, to hold them in trust for the Indian community.[62] However, Vellu informed then Finance Minister Daim Zainuddin that Maika could only afford to take up one million of the shares, and nominated three companies to take up three million each of the remaining nine million shares.[63] These three companies were Clearway Sdn. Bhd., S.B. Management Services Sdn. Bhd., and Advance Personal Computers Sdn. Bhd.[64] Anwar reasoned that the Finance Ministry acted in good faith when it awarded the shares to the three companies based on Maika's recommendations, thinking that they represented the interests of the Indian community and "would be able to pass on the shares to cooperatives and chambers of commerce with Indian interests."[65]

Speaking to parliament on 29 April 1992, Lim Kit Siang, the leader of the opposition Democratic Action Party (DAP), demanded explanation from Vellu how the diversion of nine out of ten million shares to three private companies benefited the Indian community. The shares, originally under-priced at RM5 supposedly to make them affordable to Bumiputera individuals and institutions,[66] were trading at RM13 at the time of the exposé. This meant that the three private companies would have already earned RM72 million from the purchase of the nine million shares. Lim argued that Vellu's decision to divest the shares was a "hijacking of the interest of the Indian community" that was tantamount to a criminal breach of trust and abuse of power.[67]

The situation grew more complicated when the Chairman of Maika board of directors and MIC Vice-President, G. Pasamanickam, claimed that the board did not reject the offer and that it was, in fact, able to secure a loan of RM50 million from the Arab-Malaysian Bank to buy the shares. According to Pasamanikam, the Ministry of Finance withdrew the offer and allotted only one million shares to Maika.[68] Former Maika Chairman and also a Vice-President of the MIC, Rama Iyer, who was then head of the company, concurred with Pasamanickam. Evidently, the Maika board was neither informed of nor involved in the decision to redirect the shares to the three private companies.

The day after the parliamentary exposé, Vellu left on an unplanned trip to the US purportedly for medical reasons. On 8 May 1992, the ACA, an office under the Prime Minister's Department, launched an investigation. It raided the MIC headquarters and the offices of the three companies that bought the Telekom shares.[69] When Vellu returned ten days later, he declared that he had personally requested then Finance Minister Daim to allocate Telekom shares to the Indian community. Contradicting the statements of two past Maika Chairmen, Vellu maintained that Maika could only afford one million shares.[70] Thus, he recommended that the three private companies take up the rest of the nine million shares. The directors of the three companies, Vellu said, were his close associates and he trusted them to channel all of the profits to benefit the Indian community.[71] Relations within the MIC and with the BN became strained; MIC members of parliament distanced themselves from their party leader. Finally, when asked to comment about DAP's statement that it was contemplating bringing the police into the matter, Vellu said, "why don't they ask about how much shares are allocated to other parties?"[72] Pregnant with meaning, the statement promoted the idea that nine million shares were insignificant compared those allocated to other "special groups."

While the issue was still under investigation, Vellu announced that he had already briefed the Prime Minister, and that the Prime Minister was satisfied with his explanation.[73] Mahathir contradicted the claim and asked Vellu to brief the Cabinet.[74] On 2 July 1992, Vellu explained the issue to Cabinet and afterwards told reporters that the matter was now closed.[75] Again, Mahathir contradicted him, announcing that the ACA investigation would push through because it was not a matter for the Cabinet to decide.[76] The Prime Minister's statement apparently was necessary to continue the investigation of a top leader of the BN. Soon thereafter, the ACA ordered Vellu and his son, Paari Vel, to declare all of their assets at home and abroad.[77]

Despite the investigations, Mahathir chose not to suspend Vellu or request him to take a leave of absence. There were calls for his suspension from his Ministerial position, but the Prime Minister, who had the sole power to elect, suspend, or expel a Cabinet member, stated that Samy was innocent until the ACA finds him guilty. Mahathir was thus sending mixed signals: on the one hand, he encouraged the investigations to go on. On the other he did not let go of Samy Vellu, who was the leader of the third major component party of the BN. The ACA investigations can be interpreted to mean that

Dr Mahathir did not want to be perceived as giving special treatment to his Cabinet members. However, he also supported Samy Vellu by not suspending him from his Cabinet post, despite pressure from within the ranks of the MIC and the opposition.[78] Sixteen months later in August 1993, the ACA cleared Vellu and the directors of Maika of any wrongful action or criminal offence.[79] Ironically, Lim Kit Siang, the opposition leader who exposed the matter, was suspended from parliament, due to "rowdy behaviour and initiating heated debate outside of the tabled business for the day."[80]

Despite the lack of convictions, the Maika controversy demonstrates that the implementation of privatisation benefited not only the Malay bourgeoisie, but also other political elites, albeit on a smaller scale. The episode shows how top politicians used their position in government to benefit themselves or people linked to them. Politicians justified their actions in the name of the racial communities that the political leaders represented, even though the real beneficiaries were selected individuals only. Vellu's statement hinted that a much larger number of shares were allocated to other parties, perhaps such as what UMNO received.

The Maika controversy highlighted how decisions about share allocations of privatised entities were decided at the discretion of the Prime Minister and then-Finance Minister Daim Zainuddin. The extent of their discretion was indicated by the way in which Vellu as President of the MIC felt no need to consult his party's leadership or the Maika board of directors. In a broader sense, the controversy provides a glimpse into the working of government in Malaysia, and how the power to decide ultimately resided in the Prime Minister and his senior allies.[81]

SUMMARY

This chapter has put forward three arguments. First, Prime Minister Mahathir was the main actor behind the adoption of the privatisation policy. Mahathir saw privatisation as a cornerstone of his administration's economic policy framework for the country's development, and as a key strategy to help create a Bumiputera business class. The worldwide economic slowdown and, in particular, the decline of Malaysia's terms of trade, were certainly reinforcing reasons. However, the Prime Minister's commitment to the policy was the most important factor in fully understanding the drive towards and commitment to privatisation.

Second, while the Malaysian state is capable of independently adopting its own economic policy, the privatisation of JTM was greatly influenced by the active lobbying of a Bumiputera businessman who commissioned a study on the matter. Through Shamsuddin's suggestion and lobbying, which corresponded with the direction of the Prime Minister's declared policy, JTM was identified as the first government department to be corporatised and slated for privatisation, even before the creation of a formal policy on privatisation. Despite opposition from parts of the bureaucracy, key unions, and the political opposition, privatisation occurred. In other words, a strong state that is able to formulate policies independently is not necessarily insulated from lobbying that shapes implementation.

Third, the Maika controversy over Telekom's shares, a seeming glitch in what was touted as a successful privatisation exercise, pointed to how the implementation of telecommunications privatisation redounded to the benefit of politicians, their political allies, and those within the BN patronage network.

NOTES

1. Mahathir Mohamad, "'New Government Policies', memo to senior government officials, 28 June 1983", in *The Sun Also Sets: Lessons in Looking East*, edited by K.S. Jomo (2nd edition, Kuala Lumpur: INSAN, 1985), pp. 305–7.
2. Mahathir (1985), p. 305. See also Khoo Boo Teik, *Paradoxes of Mahathirism: An Intellectual Biography of Mahathir Mohamad* (Kuala Lumpur: Oxford University Press, 1995), pp. 132–33.
3. Abdullah Ahmad Badawi, "Privatisation and the Implementation of the Malaysia Inc. Concept", Opening Address to the Malaysian Economic Association Seminar on Privatisation and the Implementation of the Malaysia Inc. Concept, 1983.

4. Gordon Means, *Malaysian Politics: Second Generation* (Singapore: Oxford University Press, 1991), p. 98.
5. Radin Soenarno Al Haj and Zainal Aznam Yusof, "The Experience of Malaysia", in *Privatisation: Policies, Methods and Procedure* (Manila: Asian Development Bank, 1985), p. 215.
6. Government of Malaysia, *Privatisation Masterplan* (Kuala Lumpur: National Printing Department, 1991).
7. Government of Malaysia, *Guidelines on Privatisation* (Kuala Lumpur: Economic Planning Unit, 1985), pp. 4–5.
8. Ali Abul Hassan, "The Role of the Government in a Privatisation Exercise: Pre- and Post-Privatisation", paper delivered at the National Conference on Privatisation: The Challenges Ahead, Kuala Lumpur, 7 October 1993, p. 6.
9. *Privatisation Masterplan*, pp. 22–24.
10. Ali Abul Hassan bin Sulaiman, "Privatisation — A Malaysian Success", in *Malaysia Inc.* (Kuala Lumpur: Lim Kok Wing Integrated S.B., 1995), p. 23; K.S. Jomo, Christopher Adam, and William Cavendish. "Policy", in *Privatizing Malaysia: Rents, Rhetorics and Realities,* edited by K.S. Jomo (Colorado: Westview Press, 1995), pp. 44, 83.
11. *Guidelines on Privatisation*, p. 11.
12. Ali Abul Hassan, 1993, p. 13.
13. *New Straits Times*, 19–24 June 1998.
14. *Business Times*, 27 December 1989.
15. Malaysia experienced a severe economic recession during the mid-1980s as its terms of trade declined due to the collapse of world prices for its major exports (petroleum, tin, rubber, cocoa, and palm oil), leading to a –1 per cent real growth rate in 1985, down from 8 per cent the previous year. In 1982, current account deficit reached 21.7 per cent of the GDP, from an average of 10 per cent of the GDP during 1971 to 1980. See Christopher Adam and William Cavendish, "Background", in *Privatizing Malaysia: Rents, Rhetorics and Realities,* edited by K.S Jomo (Colorado:Westview Press, 1995), p. 11.
16. K.S. Jomo and E.T. Gomez, "Rents and Development in Multiethnic Malaysia," in *The Role of Government in East Asian Economic Development: Comparative Institutional Analysis*, edited by Masahiko Aoki, Hyung-Ki Kim, and Masahiro Okuno-Fujiwara (Oxford: Clarendon Press, 1997), pp. 365–66.

17. Mahathir Mohamad, *The Malay Dilemma* (Singapore: Times Book International, 1981).
18. Edmund Terence Gomez and K.S. Jomo, *Malaysia's Political Economy: Politics, Patronage and Profits* (Cambridge: Cambridge University Press, 1999), pp. 79–81.
19. *Malaysian Business*, 1–15 September 1989.
20. Jomo, "Overview", in *Privatizing Malaysia: Rents, Rhetorics, and Realities*, edited by K.S. Jomo (Colorado: Westview Press, 1995), pp. 57–58.
21. G. Naidu and Cassey Lee, "The Transition to Privatization: Malaysia", in *Infrastructure Strategies in East Asia: The Untold Story*, edited by Ashoka Mody (Washington, DC: World Bank, 1997), p. 27. See also Table 4-3 in Chapter 4.
22. Laurel Beatrice Kennedy, *Privatization and Its Policy Antecedents in Malaysian Telecommunications*, Ph.D. dissertation, Ohio University (1990), pp. 33–34.
23. See various issues of the *Report of the Auditor-General on the Accounts of the Telecommunications Department* (Lapuran Juru Odit Negara, Jabatan Talikom), 1971, 1972, 1974–81, 1985–86.
24. Rita Hashim, *Privatisation of Telecommunications in Malaysia*, MPA thesis, University of Malaya (1986), pp. 59–62; Kennedy (1990), pp. 155–63.
25. Kennedy (1990), pp. 201–4.
26. Laurel Kennedy, "Telecommunications", in *Privatising Malaysia: Rents, Rhetoric and Realities*, edited by K.S. Jomo (Colorado: Westview Press, 1995), pp. 224–25.
27. Vincent Lowe, "Malaysia and Indonesia: Telecommunications Restructuring", in *Telecommunications in the Pacific Basin: An Evolutionary Approach*, edited by Eli Noam, Seisuke Komatsuzaki, and Douglas Conn (New York: Oxford University Press, 1994), p. 119.
28. Hashim (1986), pp. 81–82.
29. Syed Hussein Mohamed, "Corporatisation and Partial Privatisation of Telecommunications in Malaysia", in *Implementing Reforms in the Telecommunications Sector*, edited by Bjorn Wellenius and Peter Stern (Washington: World Bank, 1994), p. 268. See also Kennedy (1990), pp. 213–14.
30. *The Star*, 23 and 24 July 1985, and 12 November 1985; and *New Straits Times*, 24 July 1985.
31. *The Star*, 23 and 24 July 1985.

32. Interview with Francis Loh Kok Wah, Penang, 14 August 2001; and with Mary Asunta, Penang, 15 August 2001. See also *Aliran Monthly*, May 1984.

33. For instance, see Mavis Puthucheary, "Privatization — Proceed, But with Caution", in *The Sun Also Sets: Lessons in Looking East*, edited by Jomo (2nd edition, Kuala Lumpur: INSAN, 1985), pp. 360–69. See also Means (1991), pp. 98–99.

34. Means (1991), pp. 84–85.

35. The statistics vary. Vincent Lowe cites 22,000 members. Leong states that according to 1985 Ministry of Labour data, NUTE's membership was only 15,531. Hashim claims that NUTE has 18,000 members. See Lowe (1994), p. 121; Leong Choon Heng, *Late Industrialization Along with Democratic Politics in Malaysia*, Ph.D. dissertation, Harvard University, 1991, p. 92; Hashim (1986), p. 92.

36. Leong (1991), pp. 84–97.

37. Ahmad bin Nor, President of CUEPACS, "Privatisation and the Malaysia Inc. Concept — An Unionist View", paper presented at the MEA seminar of Privatisation and the Implementation of the Malaysia Inc. Concept, November 1983.

38. *Business Times*, 15 January 1986; *New Straits Times*, 5 January 1986; *The Star*, 3 January 1986, 19 May 1986, 2 June 1986, and 24 June 1986.

39. Hashim (1986), p. 93.

40. *The Star*, 21 May 1986.

41. *The Star*, 24 September 1986.

42. Lowe (1994), p. 121.

43. *Business Times*, 17 May 1989.

44. *Business Times*, 8 December 1987.

45. Kennedy (1990), pp. 21–18.

46. Ibid., p. 127.

47. *New Straits Times*, 7 January 85.

48. *Business Times*, 13 April 1989 and 17 May 1989.

49. *New Straits Times*, 8 January 1985.

50. *The Star*, 19 July 1988.

51. In various interviews, Telekom Malaysia staff members clarified that it was actually the Minister of Finance, not the Minister of Energy, Telecommunications, and Posts, who made the decisions about who should sit on the Telekom board. Interview with two Telekom Malaysia officials, Pantai Bahru, 17 August 2001 and 10 September 2001.

52. *Telekom Malaysia Annual Reports*, 1988–2000.

53. Although this turn-around was celebrated as a result of the management's cost-cutting measures, market analysts were doubtful and credited the turnaround to changes in the company's accounting practices. See *Business Times*, 10 July 1989.

54. *Business Times*, 12 December 1989.

55. Samy Vellu replaced Leo Moggie as Minister of Energy, Telecommunications and Post (METP) in a 1989 Cabinet reshuffle. Leo Moggie was relocated to the Ministry of Public Works. Moggie returned to the METP in 1995, while Samy Vellu was reappointed to the Ministry of Public Works.

56. *New Straits Times*, 27 July 1990.

57. *The Star*, 7 July 1990; *Business Times*, 23 November 1990.

58. Ali Abul Hassan bin Sulaiman (1995), pp. 22–27.

59. Syed Hussein (1994), p. 270.

60. Ibid.

61. The government's control of Telekom Malaysia was through the shares owned by Khazanah Nasional Berhad, the Ministry of Finance Inc., the Employees Provident Fund, and Permodalan Nasional Berhad. These four government agencies or government-controlled organizations owned 81.42 per cent of Telekom. With the company performing profitably, the traditional argument that it should be sold off because it was a drain on public resources did not stand. Moreover, Telekom continues to be one of the biggest corporate employers in Malaysia, with 28,000 employees. See *Telekom Malaysia Annual Report 2000.*

62. Maika was originally conceived as the Indian counterpart of UMNO's Fleet Group or MCA's Multi-Purpose Holdings. It was envisaged as a vehicle to increase the Indian share of national wealth, which was estimated to be below 1 per cent. Established in 1984, Maika Holdings Berhad's initial capitalization of RM100 million came from contributions by 66,000 Indian stakeholders, mostly estate workers, who took up loans or invested their savings to buy shares. Aside from Telekom Malaysia, Maika was also allocated portions of

public shares during the listing of companies such as TV3, MISC, MAS, EON, and the Bank of Commerce. Maika's other investments included plantations (Tumbuk Estate), book retailing (Anthonian Store Sdn Bhd), rubber-based manufacturing, computers, commodities trading, insurance, the manufacturing and marketing of soft drinks, and construction. See *New Sunday Times*, 31 May 1992.

63. *Singapore Straits Times*, 16 May 1992.

64. The first two companies were registered as having RM2 capitalization, while the third had RM250,000. Given the capitalization of these companies, it was curious how they could have afforded to buy shares costing RM15 million. Thus, the important question to ask is where did these companies get the money to buy the shares? The directors of the three companies were reportedly close to Samy Vellu. A director of Advanced Personal Computers, R. Selvendra was also a director of Maika. In addition, it was discovered that Paari Vel Samy Vellu, Samy's son, was the managing director of one of the three companies. See Lim Kit Siang, *Samy Vellu and Maika Scandal* (Kuala Lumpur: Democratic Action Party, 1992), pp. 91–98, 134.

65. Lim (1992), pp. 94–95.

66. *Singapore Straits Times*, 9 May 1992.

67. Lim's expose was the first high profile discussion on the matter, although the issue had been debated in the Tamil press for months. See Lim (1992), p. 120, and *Singapore Straits Times*, 21 May 1992.

68. *New Straits Times*, 14 May 1992.

69. *New Straits Times*, 9 May 1992.

70. *Singapore Straits Times*, 21 May 1992.

71. Ibid.

72. *Singapore Straits Times*, 14 May 1992.

73. *Singapore Straits Times*, 8 June 1992.

74. *Singapore Straits Times*, 9 June 1992.

75. *Business Times* and *Singapore Straits Times*, 2 July 1992.

76. *New Straits Times*, 3 July 1992.

77. *New Straits Times*, 5 July 1992.

78. *New Straits Times*, 16 July 1992, 31 July 1992.

79. *New Straits Times* and *Singapore Business Times*, 27 August 1993.

80. Aside from being suspended from Parliament for the rest of the year, Lim Kit Siang's comments on the issue were ignored by the BN-controlled media in Malaysia, prompting him to publish his own speeches to disseminate the information. See Lim (1992).

81. A couple of months after Mahathir's resignation as Prime Minister, there were calls for reopening of the ACA investigation on Vellu and the Maika share controversy. Rama Iyer, Maika's former chairperson and a former Treasury official, wrote to *Malay Mail* on 10 December 2003 calling for a reopening of the probe. On 12 January 2004, Parti Keadilan Rakyat (National Justice Party) lodged a police report against Vellu. Keadilan submitted copies of Vellu's bank statements between March and May 1992, at the height of the scandal, showing transfers amounting to millions of ringgit. Reportedly, this information was available in 1994 but the police and the ACA did not take any action. See *Malaysiakini*, 12 January 2004; and *The Star*, 19 and 21 November 2003.

6

THE LIBERALISATION OF TELECOMMUNICATIONS IN MALAYSIA

In this chapter we now turn our attention to the telecommunication industry's liberalisation and the ensuing outcomes. The narrative identifies the market entry beneficiaries, discusses how they obtained their licenses, and identifies their political linkages. The section also assesses the types of rent outcomes and their impact on the industry and economy.

Although the privatisation policy was well planned and deliberations with various stakeholders were held during a series of conferences and seminars,[1] the liberalisation of the sector was largely unplanned. Decisions to allow the entry of competition were made in an ad-hoc and secretive manner. The way in which licenses were issued suggests that the government favoured politically well-connected businessmen. Senior UMNO leaders presided over the distribution of the much sought after market entry licenses.

The opening of the industry commenced in the early 1980s, with the liberalisation of customer equipment terminals and the award of turnkey contracts. As was discussed in Chapter 4, these contracts were exclusively conferred to bumiputera-owned firms. Other methods used for liberalisation were public pay phones and radio paging licensing.[2] However, the most profitable sectors of the industry were basic network, mobile telephony, and international gateway services. The liberalisation of these segments marked the introduction of real competition in telecommunication services.

An examination of the introduction of competition in these three segments indicates the political nature of awarding licenses. The Minister of Energy, Telecommunications, and Posts had the authority to award licenses. Yet, Samy Vellu, who held the portfolio from 1989 to 1995, and under whom the proliferation of licensees took place, declared in 1993 that it was the Cabinet that decided on these matters.[3] This statement confirmed the widely held belief that the power to allocate licenses or contracts lay not with the Minister, but with someone higher than him.

From 1993 to 1995, the liberalisation of mobile telephony, international gateway, and basic network facilities took place rapidly. By 1995 there were seven basic network providers, seven cellular mobile phone systems, and five international gateway facilities. The lack of planning and the absence of a policy for liberalisation were made obvious when a National Telecommunications Plan (NTP) was released in 1994, after all but one of the licenses had been awarded. The NTP seemed to be an afterthought. One former senior Telekom official described it as an "act of trying to correct faults of what has been done, rather than a policy statement of the way forward."[4]

The ownership patterns of the firms that obtained the licenses indicate a great deal about the liberalisation process. First, licenses were given to companies owned by businessmen linked to the top three government officials at the time — Prime Minister Mahathir, Deputy Prime Minister and Finance Minister Anwar Ibrahim, and UMNO Treasurer Daim Zainuddin.[5] Second, all of the politically well-connected businessmen rose rapidly in the corporate world because they were also privy to state rents in other industries. The presence of an established link between these rent recipients and government leaders, indicating a patron-client relationship, suggests that they did not need to spend extra resources to gain market entry. In fact, as some scholars have argued, access to state-created market opportunities, in the context of executive dominance, is not a competitive process, because how the rents are allocated is already known beforehand.[6]

Third, apart from Shamsuddin Kadir of Sapura and Vijay Kumar of Syarikat Telefon Wireless, none of the recipients of telecommunications licenses had any background in the sector. Fourth, licenses were awarded in a non-transparent manner. No criteria were set and no public announcements were made. The liberalisation of the sector apparently followed the "first come first served" principle, whereby private individuals were simply awarded projects that they had proposed. Fifth, the political nature of the distribution of licenses was made more evident in the issuance of multiple mobile telephony licenses using the same technology to serve the highly populated Klang Valley area. The situation earned Malaysia the dubious distinction of having the most telecommunications operators per capita worldwide.[7] Finally, while most of the new players in other countries had foreign joint-venture partners because telecommunications is a capital- and technology-intensive industry, almost all of Malaysian companies had no foreign partners when they commenced business. They only took in partners to bolster their positions when the government called for a rationalisation of the industry in 1995.

THE MALAYSIAN TELECOMMUNICATIONS PLAYERS

Table 6-1 provides a list of the companies in the telecommunications sector in 1996, following the industry's liberalisation. The table lists the company's major shareholders, the services that they were authorised to provide, and the dates of the award of their licenses. The following discussion deals with the development of the companies that received licenses and the political connections, if any, of the main shareholders. The purpose of the discussion is to show three things: that the liberalisation of the telecommunications industry was unplanned, that market opening was due to the lobby of some businessmen, and that political connections were the basis of obtaining market entry in the liberalised telecommunications sector of Malaysia.

Telecommunications Company	Majority Owner	Licence and Issue Date	Foreign Partner
Telekom Malaysia	Government (MOF and Khazanah)	1. Basic network service — 1/1/1987 2. Cellular — 1/1/1987 3. International Gateway Facility — 1/1/1987	None
Celcom (Cellular Communications)	Technology Resources Inc. — Tajudin Ramli	1. Cellular — 1/9/1989 (ART 900) and 1/4/1994 (GSM) 2. Basic network — 5/5/1994 3. IGF — 1/7/1993	Deutsche Telekom AG (21%), October 1996
Maxis Communications (formerly Binariang)	Usaha Tegas — T. Ananda Krishnan	1. Cellular — 1/1/1993 2. Basic network — 1/3/1993 3. IGF — 1/3/1993	US West (20%), 1994; then BT (33.3%) and Media One (12%), 1996
Mobikom	Telekom, EON, PNB, Sapura	Cellular — 1/7/1993	None
Sapura Digital	Sapura Group — Shamsuddin Kadir	Cellular — 24/12/1993	None
MRCB Telecommunications	MRCB Group	Cellular — 1/6/1994	None
Time Telecom	Renong — Halim Saad	1. Basic network — 1/6/1994 2. IGF — 1/12/1994	None
Syarikat Telefon Wireless (STW, now Prismanet)	Shubila Holdings — Vijay Kumar	Pilot license — 1/12/1993 Basic network — 24/12/1994	International Wireless Communications
Digi.Com (formerly Mutiara Telecoms)	Berjaya Group — Vincent Tan	1. Cellular — 8/8/1994 2. Basic network — 18/1/1995	Swiss Telecoms, 1996

Celcom

Telekom Malaysia incorporated Celcom, originally named Syarikat Telekom Malaysia Cellular Communications, in January 1988. Celcom was 51 per cent owned by Telekom, while the Fleet Group held the remaining equity. The Fleet Group was UMNO's foremost investment holding company at that time, and as UMNO Treasurer, Daim Zainuddin, who was also then Finance Minister, was in charge of its business activities.[8]

Telekom successfully launched the country's first cellular telephone system in 1987. It established Celcom as a subsidiary to manage the expansion of its cellular capacity. The Fleet Group noticed the success of the business and wanted a 70 per cent stake in Celcom.[9] Telekom executives refused. After a few months, Telekom, which was under the control of the Ministry of Finance, eventually agreed to a 51–49 partnership with the Fleet Group. Yet, despite Telekom's majority stake in Celcom, the Fleet Group secured management control and made decisions about technology and equipment issues. At Celcom's official launch in August 1989, Tajudin Ramli, representing the Fleet Group, gave the opening speech. Telekom staff boycotted the ceremony. Furthermore, Celcom had apparently launched its mobile telephone service before it was granted formal authority to operate. However, through its links to Daim, Celcom received a backdated technical licence from JTM.[10]

In December 1989, Telekom sold its 51 per cent stake in Celcom to an obscure company, Alpine Resources, for only RM4 million. The reason was "a hitherto unknown political facet to the entire issue." As an industry analyst put it: "someone out there wanted a piece of the action."[11] Telekom executives "resented and opposed the sale of Celcom for a song."[12] Nevertheless, it was forced to sell its shares to a predetermined buyer at a predetermined selling price, on the promise that it would be awarded a new mobile telephony license.

Three months later, in February 1990, the government reneged on that promise. Then Minister Samy Vellu announced, "the Cabinet has decided to postpone the implementation of a third cellular network so as to give Celcom sufficient lead time to get off the ground and eventually go nationwide."[13] The Cabinet gave Celcom a five-year monopoly before new players, including Telekom, were allowed into the mobile market. Moreover, the Cabinet barred Telekom from offering paging services, which was popular during the early 1990s when cellular phones were still expensive. From a regulatory point of view, these restrictions made sense because Telekom was the dominant company. Yet it was evident that the main consideration was not regulatory but to benefit Celcom.

Alpine Resources, the buyer of Telekom's share in Celcom, was an investment holding company owned by Tajudin Ramli, a close associate of Daim Zainuddin. Tajudin was one of the few handpicked Malays that the state nurtured to become major corporate players. Aside from Celcom, Tajudin benefited from other government largesse, such as the privatisation of Malaysian Airlines (MAS). Tajudin had been a director of property developer Peremba, the private arm of the government's Urban Development Authority (UDA), which Daim set up and headed in 1979.[14] He also sat on the board of Daim's family companies, Seri Iras and Taman Maluri. He was also once a director of the Fleet Group and one of its major private subsidiaries, Fleet Trading and Manufacturing. It was while he was a director of Fleet Group that Tajudin became involved in Celcom.

Tajudin owned 99.2 per cent of Alpine Resources, and his brother Bistamam held the remaining stake. Its directors included UMNO-linked personalities, such as Abdul Rashid Abdul Manaff and Zaki Azmi.[15] According to Tajudin, his big break in telecommunications came when Telekom "offered" to sell to Alpine Resources its stake in Celcom.[16]

Through a complicated series of corporate calisthenics, Tajudin secured nearly total ownership of Celcom. Four months after Alpine Resources had purchased Telekom's stake in Celcom, the Fleet Group sold its 49 per cent stake in the firm to Time Engineering, a company controlled by another Daim associate, Halim Saad. The price was RM81.5 million, 20 times more than what Telekom had received for its shares. In November 1991, Alpine Resources offered to sell its 49 per cent stake in Celcom to Technology Resources Industries (TRI) for RM259.44 million. Tajudin owned over a quarter of

Figure 6-1: Celcom's Takeover by TRI

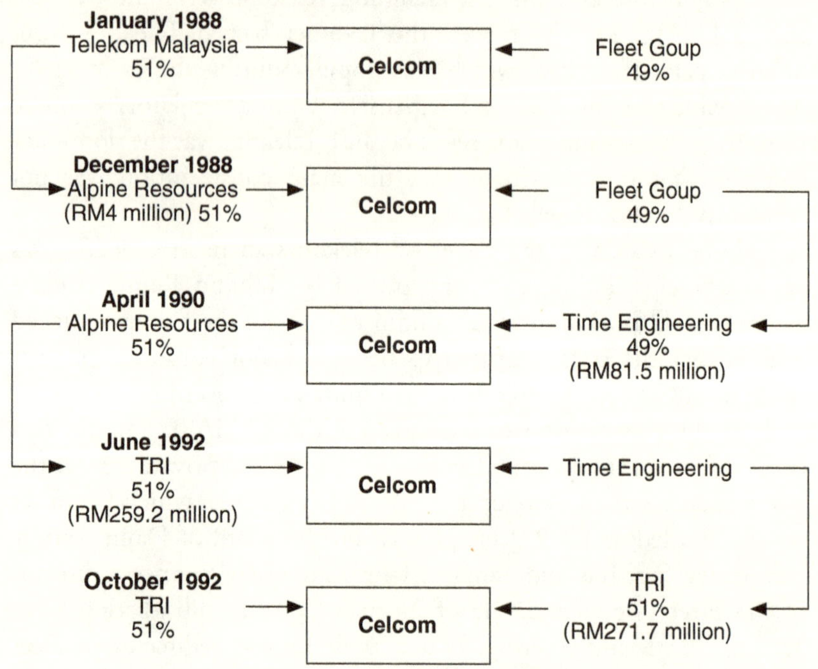

TRI.[17] However, Celcom's other major shareholder, Time Engineering, opposed the sale, insisting that it had the right of first refusal if the shares were for sale.[18] Time Engineering's objection stopped the deal. Because Time Engineering was under the control of Halim Saad, this suggested that two of Daim's protégés were now in competition with each other.

In June 1992, Tajudin and his brother Bistamam reportedly offered to sell 100 per cent of Alpine Resources to TRI for RM259.2 million. In this instance, Time Engineering's consent was not needed because what was being sold was the holding company itself rather than Alpine's stake in Celcom. By then Time Engineering had withdrawn its earlier protest because the government promised the company a new license.[19] The license enabled Time Engineering's sister company,

United Engineers Malaysia (UEM), to construct and operate a fibre-optic network along the North-South Highway. In October 1992, Time Engineering relinquished its 49 per cent stake in Celcom to TRI for RM271.7 million.[20] Meanwhile, Alpine Resources sold its shares of Celcom to TRI through a share-swapping exercise. Tajudin's interest in TRI increased from 27.4 per cent to 48.5 per cent.[21] Tajudin thus emerged as the main shareholder in two companies, TRI and Celcom.

How Celcom evolved reveals how state patronage was crucial in Malay enterprise development. First, in 1989, Telekom, then wholly-owned by the government through the Ministry of Finance, was "convinced" to sell its stake in Celcom to a firm owned by Tajudin, a close associate of then Finance Minister Daim. No public bidding occurred. The sale was announced only after the deal was consummated. Telekom executives declined to sell a profitable business, but its resistance proved futile because allies of powerful politicians wanted entry into this potentially lucrative market. Second, in the next stage of the sale involving Celcom equity, Tajudin controlled both the seller (Alpine Resources) and the buyer (TRI). Third, in the final stage, two of Daim's close associates, Tajudin and Halim, were in competition. Clearly, the issue of "who gets what" was not decided in the market. Rather, key figures in the ruling party made the decision. In the case of two individuals who were both clients of the same patron but competing for a controlling stake in one company, the public display of thorny conflicts was avoided through negotiated deals that led to each getting their own section of the telecommunications market. This indicates that competition also existed among "Daim's boys" but was not publicly played out.

With a cellular telephony monopoly for five years, Celcom rapidly gained subscribers, even though it offered poor service and its network was congested. Celcom also received licenses for an international gateway in July 1993 and a domestic network operation in May 1994, bringing it into direct competition with Telekom.[22] Following its total acquisition of Celcom, TRI spread its focus from telecommunications to the transportation, tourism, and manufacturing sectors.[23]

Maxis Communications

The second licensee in the mobile sector was Binariang Communications, presently known as Maxis Communications. T. Ananda Krishnan owned a majority stake in the company through his holding company, Usaha Tegas. Binariang's entry into the telecommunications and cable TV business started in January 1993 when Prime Minister Mahathir handpicked Ananda to operate Malaysia's first satellite. In addition, Binariang received a cellular license to use GSM technology. In March 1993, Binariang also acquired licenses to operate a basic network and international gateway facility. These licenses were apparently secured through Ananda's connections to Mahathir.[24]

Ananda, a Tamil born in Kuala Lumpur, graduated from the University of Melbourne and has an MBA from Harvard University. Ananda first made his corporate mark abroad in oil dealing.[25] From his oil ventures, Ananda invested his profits in horse stud farms in Australia, a gas field in Oklahoma, and cartoon animation studios in California and Manila. In Malaysia, Ananda's main holding companies are Usaha Tegas and MAI Holdings.[26]

Ananda served as a director of Petronas and of Bank Negara from 1982 to 1987. He is reputed to have been close to Mahathir, and to have played a key role in the reconciliation between Mahathir and Razaleigh Hamzah, a former Finance Minister who had threatened the Prime Minister's position.[27] Ananda was also involved in the development of the KLCC Twin Towers and secured the rights to operate a privatised lottery operation. UMNO members openly objected to Ananda's involvement in these projects at the 1992 UMNO General Assembly.[28]

Although Ananda, like other beneficiaries of state patronage, had "built bridges with powerful politicians, who are the traditional sources of lucrative licenses and contracts," he also possessed entrepreneurial skills.[29] During the 1997 currency crisis when his companies ran into financial problems, he negotiated to sell a 33.3 per cent stake in Maxis to British Telecoms (BT). BT wanted a clause in the contract that gave it the right to buy additional Maxis shares if Ananda sold more of his stock. Ananda agreed, but insisted that he also be given the right of first refusal if BT sold its Maxis equity. Thus, when BT decided to unload its Maxis shares in May 2001, Ananda's Usaha Tegas acquired BT's Maxis equity for RM1.9 billion. This price had a premium of only RM100 million from the amount that BT had paid when acquiring the equity in 1998. This transaction raised Ananda's stake in Maxis to 70 per cent.[30]

As of September 2006, Maxis was the leader in cellular telephony, with 41 per cent of Malaysia's market and an estimated 8.9 million subscribers. Maxis supplanted Celcom from the top position, despite the lead-time given the latter to establish itself in this sector.

Mobikom

Mobikom, the third mobile licensee in Malaysia's liberalised industry, was incorporated by Telekom Malaysia on 11 December 1989 in anticipation of the government promise of a third license. Yet, the company only received its license on 1 July 1993. By then, many other parties were also interested in the business as demand for cellular telephony increased. At least three international firms with local partners were vying for the AMPS 800 license that Mobikom secured. They included Hutchinson of England, Bell South of the US in joint venture with the Berjaya Group, and the US' GT&E in joint venture with Mirzan Mahathir, the Prime Minister's son. Numerous discussions were held between the various interested parties, and intensive lobbying transpired. Eventually, the government decided that the best solution was to award the license to a consortium.[31] The government, therefore, persuaded Telekom to form a consortium with Edaran Otomobil Nasional (EON), Permodalan Nasional Berhad (PNB), and Sapura.[32]

PNB was the government's leading trust agency, while EON was the distributor of Malaysian-made Proton cars. Sapura, owned by Shamsuddin Kadir, had still not secured a telecommunications license, even though it had presented the proposal to privatise Telekom. Telekom, EON, and PNB each held 30 per cent of Mobikom, and Sapura held the remaining 10 per cent.[33] From the outset, the four shareholders squabbled over control of the company's management, hindering Mobikom's development. Its official launch was delayed by around six months because of a dispute surrounding the award to Perwira Ericsson of a contract to build Mobikom's network.[34] Then Finance Minister Anwar Ibrahim was brought in to settle the matter, and he directed Mobikom to re-call the tender. Yet, following a review by an "independent" committee, the original tender was deemed legitimate.[35] The case of Mobikom's development demonstrates that politicians decided the settlement of contentious corporate issues involving competing interests of companies that were owned by influential businessmen or trust agencies.

Another aspect of the launch delay was JTM's inadvertent awarding of too broad a frequency spectrum to Celcom, which left no room for other entrants. The need to allocate the spectrum rationally and divide it among competing companies was a scenario not foreseen by the regulators when Celcom was awarded the license in 1989.[36] Although this situation indicates the rapid pace of development in the telecommunications sector, it also reveals that the awarding of licenses was a political process, at the end of which JTM was expected to sort out the ensuing technical problems.

Sapura Digital

On 24 December 1993, Sapura Digital obtained the first of three licenses to operate a Personal Communication Service (PCS) network.[37] The owner of Sapura Digital was the Sapura Group, controlled by Shamsuddin Kadir. Shamsuddin has acknowledged his close relationship with Mahathir. He has previously served as a director of UMNO's cooperative, Koperasi Usaha Bersatu. In addition, Sapura Group's then Vice Chairman Rameli Musa was a confidant of Deputy Prime Minister Anwar.[38] These political connections seem to have been useful in Sapura Group's securing a telecommunications license.

Shamsuddin, who earned an engineering degree from a British university, had been in the telecommunications equipment supply business since the mid-1970s. From 1959 to 1971, he worked for the JTM, which gave him useful knowledge about the sector, as well as valuable contacts. In 1971, he joined Eric Chia's United Motor Works (UMW) as its Executive Chairman.[39] Under Shamsuddin's management, UMW secured a contract to operate a payphone business. In 1975, Shamsuddin bought the payphone business from Chia and set up Sapura Holdings.[40] He also formed Uniphone as a subsidiary of Sapura to manage the payphone business. In the same year, Shamsuddin won three big contracts from JTM: a 10-year contract to provide payphones in urban areas, a contract to supply telephone sets to JTM, and turnkey projects to lay down cables. In 1983, Sapura was one of four bumiputera firms that secured the then biggest build-operate-transfer (BOT) contract, worth RM2.5 billion, to modernise the telecommunications infrastructure.[41] Clearly, the Sapura Group's rapid development was due to the government contracts that it cornered.

Shamsuddin's telecommunications training and background set him apart from Tajudin and Halim, and other bumiputra entrepreneurs in the liberalised telecommunications industry. Shamsuddin was probably one of the few Malay businessmen who productively utilised the government-allocated rents that he received, building up productive capacity through investment in research and development, and eventually becoming competitive internationally. Furthermore, Shamsuddin, unlike Tajudin and Halim, entered into joint ventures with foreign firms, such as Sumimoto and Marubeni of Japan, and insisted on technology transfers, thus giving Sapura the capacity to manufacture cables and telephone sets and systems. Sapura's success in manufacturing was evident when it began exporting its telephone products. Apart from manufacturing telecommunications equipment, Sapura expanded its business into manufacturing IT products and vehicle component parts, and the distribution of Apple computers.[42]

Time Telekom

Time Telekom, a wholly-owned subsidiary of Time Engineering of the Renong Group was the second basic network provider in Malaysia. The company constructed a fibre-optic network covering the stretch of the North-South Highway that spans the entire peninsula, from Thailand to Singapore. In contrast to the other new telecommunications companies, Time Telekom was not originally involved in cellular telephony, but in basic network facilities. Time Engineering's license to operate the fibre-optic network was transferred to Time Telekom on 1 June 1994. The awarding of this license was apparently in connection with a promise made to Time Engineering in 1992 as an incentive for it to sell its stake in Celcom to TRI. The company, which was part of Renong Group's stable of companies, was able to obtain a telecommunications license largely through Halim Saad and his connections to Daim.

A New Zealand-trained accountant, Halim Saad, was a leading bumiputera businessman in the 1990s. He served as Peremba's Corporate Services Manager and was one of the original directors of Hatibudi, an UMNO investment holding company.[43] Starting in the 1980s, Halim sat on the board of a few publicly-listed firms in which Daim, who was the Finance Minister at the time, had an interest in, including Cold Storage, D&C Bank, and Roxy (which became TRI). In 1988, the Mahathir government awarded United Engineers Malaysia (UEM), a subsidiary of Renong, the contract to build and operate the North-South Highway project. Earnings from this asset helped Renong to build a conglomerate that was involved in engineering, construction, power generation, financial services, hotels and properties, oil and gas prospecting, investments and trading, and telecommunications. Halim became the Executive Vice Chairman of UEM, and Chief Executive of the controversial Projek Lebuhraya Utara-Selatan (PLUS), the company formed to build and operate the privatised North-South highway project.[44] In 1990, Halim became Chairman and Chief Executive Officer of the Renong Group. He later admitted in a court affidavit that his shares in Renong were held "in trust for the ultimate beneficial owner, UMNO."[45]

Like Tajudin, Halim quickly emerged as a leading bumiputra businessman through the receipt of numerous privatisation contracts and licenses. Halim went on to lead Renong to become one of Malaysia's biggest and most indebted conglomerates. In its heyday, Renong embodied UMNO's corporate face, as it took over most of Fleet Group and Hatibudi's assets in 1990 when UMNO ostensibly divested its interests in the corporate sector.[46] Halim lost ownership and control of the Renong Group in October 2001, not long after Daim resigned as Finance Minister in June of that year.

MRCB Telecommunications

MRCB Telecommunications, a subsidiary of Malaysian Resources Corporation (MRCB), was awarded the sixth mobile telephony license on 1 June 1994. Its mobile telephony service called Emartel was Malaysia's second PCS network.[47]

Publicly-listed MRCB gained prominence in 1993 when it was used as the vehicle for the management buy-out of the New Straits Times Press (NSTP) and Sistem Televisyen Malaysia, the operator of TV3, from Renong.[48] Subsequently, MRCB started to further diversify its interests, moving into construction, property development, and power generation.[49]

At the time of the awarding of the cellular license, Nazri Abdullah, Mohd Nor Mutalib, and Khalid Ahmad, businessmen linked to then Finance Minister Anwar, were managing MRCB. The license was just one concession in a long list of government awards and contracts that MRCB obtained. During Anwar's tenure as Finance Minister, the company gained the most desirable contracts and licenses, which was reminiscent of Renong's fortune under Daim as Finance Minister. The awarding of a cellular license to MRCB, among other things, suggested that Anwar was in the process of building his own set of favoured businessmen while he was at the helm of the Finance Ministry. At its height, MRCB owned four quoted firms: NSTP, TV3, power generator Malakoff, and the banking group Commerce Asset Holdings.

Syarikat Telefon Wireless

Vijay Kumar, a telecommunications engineer, and Conny Dolonius, a Swedish banker, established Syarikat Telefon Wireless, the only company that obtained a license because it offered a technological solution to the lack of telephones in rural areas. Kumar proposed the use of wireless local loop (WLL) technology to provide telephones to areas where Telekom did not provide fixed line services. In December 1993, STW obtained the Prime Minister's approval for a one-year pilot project to serve Langkawi, then identified as a free port and tourist development area, where telephone services were poor.[50] Of all the companies that had been awarded licenses, only STW went through the process of providing EPU a business plan that detailed its technological solution and proved the viability of its proposal. Moreover, only STW was first given a provisional license to pilot test its proposal. The pilot test catering to 300 subscribers was so successful that STW immediately had a waiting list of 5,000 in Butterworth, Alor Setar, Perlis, and Penang.[51]

Kumar pointed out that STW was complementary and not competitive to Telekom's services because STW aimed to go into the unserved areas. Many rural areas were without telephone services due to the high capital cost involved. However, the cost of WLL technology was less than a third of the cost of copper lines. In November 1994, STW was finally given a national network license. Furthermore, in contrast to the other new telecommunications companies, STW immediately took in International Wireless Corporation (IWC), an American company, as a joint venture partner.

Even though STW was clearly a competent entrant, Kumar and Dolonius were non-bumiputeras, and had to bring in Shuaib Lazim and Khalil Akasah as bumiputera partners. Shuaib is a cousin and reportedly a close associate of Daim. Khalil Akasah, also linked to Daim, was a shareholder of Daim-controlled companies Raleigh and Berjaya.[52] In 1995, following the government's call for a rationalisation of the telecommunications sector, Telekom offered to buy STW for RM250 million. When STW's shareholders refused the offer, they found that the release of their imported equipment was being delayed at Customs, setting-back construction. They eventually gave up the fight and sold their stakes in STW to Rosli Man[53] and Shuaib Lazim in August 1996, just after the government called off the rationalisation exercise. STW was renamed Prismanet, and is presently under receivership. STW's shareholders were allegedly not yet paid for the sale of the company, and are still pursuing court action to regain ownership.[54]

Digi.Com

Formerly known as Punca Mutiara Communications, Digi was the last firm to obtain a license on 8 August 1994. It was the seventh mobile telephone licensee and the third company to build a PCS network. Digi also secured a basic telecommunications network license on 18 January 1995. In fact, Digi's licenses were issued after the government had announced that it would not issue any more licenses. Vincent Tan of the Berjaya Group controls Digi and was closely associated with both Daim and Mahathir.[55]

Tan's first claim to corporate fame was in 1982, when he secured the exclusive right to McDonald's franchises in Malaysia. In 1984, he linked up with Azman Hashim of the Arab Malaysian group to acquire a 38 per cent stake in Berjaya Corporation, then involved in the manufacturing of steel wire products. In 1985, Tan's private firm, B&B Enterprise, won the contract to buy 70 per cent of Sports Toto, a lottery operator owned by the Ministry of Finance Inc (MOF). Tan reportedly won the contract due to his "unique proposal" to the Prime Minister.[56] In 1989, in another controversial deal, MOF's remaining 30 per cent interest in Sports Toto was sold to Raleigh, which was then controlled by Tan's Berjaya Corp.[57]

Tan's Berjaya Group has been described as one of Malaysia's most active corporate raiders, having bought and sold stakes in numerous quoted companies.[58] Tan also held a stake in the news media firm, *The Star*, and owned the *Watan* and *The Sun* newspapers. Tan's business interests include cosmetics distribution, hotel and resort management, property development, telecommunications, and media services. The rapid corporate expansion of Tan's Berjaya Group is another example of a non-bumiputra gaining access to state concessions through personal links with senior politicians.

THE CALL FOR RATIONALISATION

In 1989, the Minister of Energy, Telecommunications, and Posts (METP), Leo Moggie, asserted that the Ministry was keen on maintaining Telekom's monopoly status while regulatory policies were being developed. Only after regulatory procedures had been put in place would market be opened to competition.[59]

Following a cabinet reshuffle in 1989, Moggie relinquished this portfolio. The new minister, Samy Vellu, immediately reversed this policy. Moggie believed that Vellu might have thought that Telekom would be forced to perform better if other firms were allowed market entry. Yet, Moggie acknowledges that in retrospect, "we should probably not have given so many licenses at that time."[60] As the regulations were not laid down before competition was introduced, undoing the repercussions of the decisions made in the ministry under Vellu was precisely what Moggie embarked upon when he was re-appointed to the portfolio in 1995.

Moggie announced in September 1995, just eight months after the last license had been issued, that the government would encourage rationalisation of the industry to avoid wastage of resources and the duplication of infrastructure. Telekom, Celcom, and Binariang were believed to be the three telecommunications companies chosen to remain in the industry as full service companies.[61] Tan and then Chief Executive Officer of Time Telekom, Jaafar Ismail, issued statements expressing dissatisfaction with the new policy. Tan also sought a foreign partner, and sold 30 per cent of Digi to Swiss Communications.[62]

Although the telecommunications operators never openly acknowledged it, negotiations for mergers took place due to the Ministry's pressure. However, the possibilities of mergers were scuttled because of disagreement on valuation. Companies allegedly favoured using future earnings valuation while Telekom Malaysia's preferred real asset valuation. STW, the only company without strong political backing, declared that the government's consolidation exercise amounted to expropriation. STW's foreign partner, IWC, apparently sought US President Clinton's intervention to stop the consolidation exercise.[63]

In July 1996, Moggie retracted the policy. "To avoid further uncertainties in the market," he said, "the government has decided that it will not get involved in trying to persuade the companies to merge … and let market forces decide." It is difficult to determine the real reason for this decision. It is most probably because consolidation negotiations among companies owned by the politically influential had proven to be a Herculean task.[64]

Nevertheless, the call for rationalisation was not totally unsuccessful. Two mergers occurred — the sale of MRCB Telecommunications to

Telekom, and of Sapura Digital to Time Engineering, Time Telekom's parent company. In July 1996, barely a year into its operation, Sapura decided to sell its mobile business, and Uniphone, its public payphone business, to Time Engineering. Time Engineering wanted the Sapura Group's mobile and payphone licenses to complement its fledging landline services. Analysts saw these acquisitions as a last ditch effort by Time Telekom to stay in the business.[65] Meanwhile, a senior Sapura executive disclosed an additional reason for the company divesting its interests in the sector. He argued, "it was not worthwhile to be a telecommunications service operator when your biggest customer, Telekom Malaysia, becomes your competitor and stops buying from you."[66] The Sapura Group was fortunate to have left the telecommunications service provision industry a year before the currency crisis occurred.

The second consolidation involved Telekom's purchase of MRCB Telecommunication for RM640 million in August 1996.[67] Emartel had been in operation for just over a year when MRCB realised that it would need to invest at least another RM1 billion to complete its network. MRCB was then pressed for funds for its other major projects, most notably the KL Sentral project and a rail link to the new KL International Airport.[68] Telekom took over Emartel and renamed it TM Touch, which became its first wholly-owned digital mobile service. The deal, according to market analysts, was a bailout of MRCB, as the price for an incomplete network with a low volume of subscribers and faltering service was too expensive.[69] At the time of the sale, business associates of then Finance Minister Anwar were managing MRCB, while Telekom was under the Ministry of Finance's control.

THE 1997 FINANCIAL CRISIS

When the currency crisis struck in 1997, telecommunications companies, which had incurred huge debts while developing their networks, were also affected by a slowdown in demand. Although the government had decided not to insist on the consolidation of the sector in 1996, suggestions about how to deal with the debt-burdened companies were widely discussed in the media after the onset of the crisis. The government did not force consolidation, as it did in the banking and financial sector, because the telecommunications industry was seen to be in a less problematic state of affairs.[70] Unlike in banking some prominent politicians were even willing to allow foreign participation in the sector. Thus, the government appeared divided over how to deal with telecommunications. Its response can be categorised into two parts. First, the government bailed out troubled telecommunications companies owned by well-connected businessmen rather than allowing them to relinquish their interest in the sector. Second, the government increased the allowable foreign equity and was actively involved in deciding which foreign company could enter.

Two notable bailouts occurred. In May 1998, Telekom bought out the other shareholders of Mobikom for RM232.7 million.[71] Market analysts politely termed this a "national service," and argued that it made no sense for Telekom to own three mobile phone systems using different technologies — Atur, Emartel, and Mobikom.[72]

A second, more controversial bailout involved Time Telekom. When the company undertook public listing on the KLSE in March 2001, it changed its name to Timedotcom. Despite the hype and publicity that surrounded the listing, there was little market confidence that Timedotcom had the capacity to register profits.[73] This was most evident when the company's equity was massively under-subscribed by 75 per cent. Public agencies, such as the Employees' Provident Fund (EPF) and the pension fund Kumpulan Wang Amanah Pencen (KWAP), were among the institutions that lapped up the unsubscribed shares at the initial offer price of RM3.30 per share. On the first day of trading, Timedotcom's share price fell to RM2.43. Analysts argued that the offer price of RM3.30 per share was exorbitantly high and valued the shares at only RM1 each.[74] Public criticism arose over the use of EPF and KWAP funds to buy at inflated prices the unsubscribed shares of this ailing but well-connected firm.

Subsequently, the government took a very active role in the consolidation of the industry. In 1998, the Cabinet decided to increase the allowable foreign equity in telecommunications companies from

30 to 49 per cent, before increasing it further to 61 per cent, with a caveat that foreign equity is brought down to 49 per cent within five years of acquisition.[75]

While allowing more foreign involvement in telecommunications, the government was particular about the choice of foreign partners. Three cases can be cited to indicate government intervention in this matter. First, at the height of the crisis in 1998, Maxis started negotiations with Singapore Telecommunications (SingTel). Just as Maxis' majority owner, Ananda Krishnan, was about to close a deal, he was "prevailed upon" to sell 33.3 per cent of Maxis' equity to BT for RM1.8 billion, even though BT's bid was lower than that offered by the Singaporean firm.[76] Second, in May 2000, Halim Saad initiated negotiations, again with SingTel, to inject capital and technology into Timedotcom. An agreement was signed for SingTel to buy 14.48 per cent of Time Engineering and 20 per cent of Timedotcom for RM3 billion. The deal collapsed at the last minute after UMNO members criticised it as "selling national assets to a company controlled by the Singaporean government."[77] With the collapse of the deal, Khazanah Nasional, the government's investment arm controlled by the Finance Ministry, infused RM2.3 billion into Time Telecom.[78] Khazanah helped financially but did not add as much value compared with the industry expertise that SingTel could have brought in. Evidently, political manoeuvring and other considerations contributed to Khazanah's acquisition of Time Telekom. Finally, in early 2000, Japan's Nippon Telephone and Telegraph (NTT), the world's second largest telecommunications company, was reportedly poised to buy Khazanah's share in Telekom, but a final agreement could not be reached. Telekom claims that this occurred because NTT was asking for more control than it was prepared to give.[79] It must be noted that the government is still the largest shareholder of Telekom.

The government's involvement in the industry's rationalisation was justified on the grounds that the crisis had brought Malaysia into "unusual times where the ailments of some companies are contributing, though indirectly and in a small way, to the financial problems of the country."[80] This observation, although containing a grain of truth, glossed over how the sector required rationalisation because non-market factors had determined the allocation of market entry licenses. The "unusual times" that the 1997 crisis brought about only highlighted the political root of the industry's problems. This suggests that the government's priority was to rescue companies without revamping their management, not pushing through corporate reforms.[81] As will be detailed next, however, corporate restructuring took place from 2001 because of a political fall-out between the top politicians in the government.

IMPACT OF THE MAHATHIR-DAIM FALLING OUT

Daim Zainuddin returned to government as Finance Minister when Deputy Prime Minister and Finance Minister Anwar Ibrahim was sacked from his post and expelled from UMNO in September 1998. Daim resigned from this post in June 2001. Neither he nor the Prime Minister offered an explanation. The political and business elites, particularly UMNO members, were rumoured to be very displeased that only businessmen linked to Daim were taking over assets from Anwar's business associates. The bailout of Halim Saad's Time Telekom, the government buy-back of MAS from Tajudin Ramli at an exceedingly high price, and Daim's attempt to secure overwhelming influence in the consolidation of the banking industry were surmised to have contributed to the rift between Daim and Mahathir.[82]

Mahathir became acting Finance Minister, and the government took over the companies of Daim's associates ostensibly in the name of improving transparency and corporate governance. Although these takeovers suggest the Prime Minister wanted to restructure corporate debt, their timing and targets imply that political factors were more important.

Three examples are relevant. First, in late June 2001, Halim Saad resigned without explanation from the vice-chairmanship of UEM, Renong's most profitable listed company, which controlled PLUS, the owner of the North-South Highway concession and Timedotcom. A week later, the government announced its takeover of UEM to help restructure Renong's RM20 billion debt.[83] Although these were positive moves, some observers were surprised at the swift pace of the takeover because Renong had not previously been subject to any government discipline. In fact, the speed of restructuring seems to have breached the Securities Commission's rules and procedures for general offers.[84]

Second, in August 2001, the government announced the appointment of two professionals to MRCB's board to oversee the conglomerate's debt restructuring. MRCB's chairman, Abdul Rahman Maidin, who had a large, indirect interest in the company and was a close Daim associate, resigned quietly.[85]

Third November 2001, the Securities Commission rejected Tajudin's attempt to rescue TRI (the controlling stakeholder in Celcom), from debt using the cash that he made from selling MAS back to government. In April 2002, Tajudin lost control of his interests in TRI and Naluri (which formerly controlled MAS) to Danaharta, the government's debt restructuring body.[86] His 13.2 per cent stake in Celcom was sold to Telekom for RM717 million the day after Tajudin defaulted payment of RM130 million to Danaharta. On 6 May 2002, Danaharta also called for an auction of Tajudin's shares in cash-rich Naluri. Mahathir publicly gave his approval for the takeover. He argued that the collapse of Tajudin's corporate empire was not a reflection of the failure of Malay businessmen, but that people like Tajudin, who had borrowed heavily from banks, rightfully lost control of their businesses when they failed to pay up.[87] Tajudin filed a case in the High Court to stop Danaharta from selling his shares, but the court ruled against him. He resigned from TRI on 3 July 2002, clearing the way for Telekom's take-over of Celcom.[88]

The restructuring of UEM, the management changes in MRCB, and the sale of TRI's shares to Telekom transpired with remarkable haste, and in the case of the former, without public bidding. The speed with which Halim, Tajudin, and Abdul Rahman were displaced from controlling supposedly private companies points to Mahathir's direct involvement, overseeing the restructuring campaign as acting Finance Minister.[89] This raises doubts about the legitimacy and the real motive of the corporate restructuring taking place in Malaysia. More importantly, it provides evidence that the state is still actively involved in shaping the corporate sector, and has the power to intervene and decide who gets state patronage and who goes bankrupt.

ASSESSMENT OF THE MALAYSIAN TELECOMMUNICATIONS LIBERALISATION

The form of development and rationalisation of the telecommunications industry demonstrates the interactions of the state and politically influential interests in policy planning and implementation. Privatisation and liberalisation neither removed the structures of patronage and rent-seeking nor lessened state involvement in the industry. Market opening, by virtue of altering the existing property rights regime, actually created further venues for rent-seeking.

While privatisation was introduced to promote growth, encourage efficiency, and reduce state involvement in the economy, its main focus was fostering the rise of a bumiputra business class through the patronage of a chosen few. Through the "first come first served" policy, well-connected businessmen had access to potentially lucrative licenses from the government. This situation was primarily due to the centralisation of power in the hands of the executive, which controls the allocation of government-created rents. State intervention in the private sector continued, despite supposed market-opening exercises. The recipients of these rents were well-connected businessmen, hence political ties inevitably shaped the industry.

The recipients of telecommunications licenses can be categorised based on their political linkages to the top three politicians at the time (see Figure 6-2). Four politically well-connected bumiputeras, one Chinese, and one Indian received telecommunications licenses. Two of the bumiputeras, Tajudin and Halim, were Daim's protégés, while Shamsuddin reputedly had close ties with Mahathir. The MRCB group was controlled by Anwar's associates. Vincent Tan reportedly was close to both Daim and Mahathir, while Ananda was apparently a personal friend of Mahathir. Vijay Kumar was associated with Daim, but only because he took in Shuaib Lazim as a partner. The government awarded licenses to these favoured individuals in a surreptitious manner. The Ministry or Cabinet did not publicise the process or announce the criteria for giving licenses. The liberalisation of telecommunications allowed private individuals to secure a license without needing to go through an open tender system or a transparent selection process.

Figure 6-2: Types of Rent Recipients and Their Political Patrons

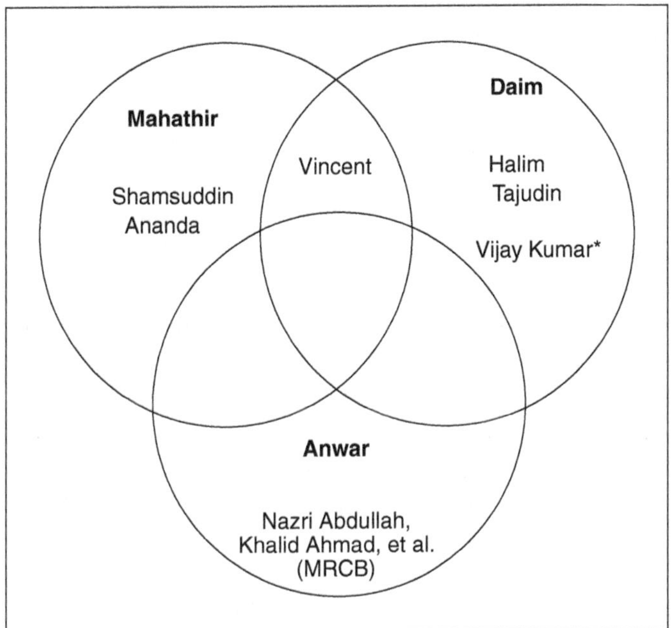

* Vijay Kumar was indirect linked to Daim through Shuaib Lazim, his bumiputera partner in the company.

Privatisation and liberalisation helped strengthen rentiers rather than promote entrepreneurship, although these categories are not mutually exclusive. Some of the licenses were relatively well-managed, as was the case with Ananda, who showed entrepreneurial capacity. Others, such as Halim and Tajudin, did not develop their licenses prudently and even required bailouts after the currency crisis. Tan, who had little experience in the telecommunications sector, was shrewd enough to allow his foreign partners to run Digi. However, the fall-out between political elites affected ownership patterns in the telecommunications industry. Tajudin and Halim lost control of Celcom and Timedotcom following a rift between Mahathir and Daim. Similarly, Anwar's associates lost control of MRCB after Mahathir removed him from all of his party and government posts.

Table 6-2 indicates the outcome of the reforms and consolidation of the telecommunications sector. As of 2002, the industry had six telecommunications players, one of which, Prismanet (formerly STW), was under receivership. Of the remaining five companies, the government controlled three. Only Maxis and Digi are fully privately owned. The government, through Khazanah Nasional, has majority control of Telekom and Timedotcom.[90] Meanwhile, Telekom took control of Celcom after buying a 32 per cent stake in TRI in September 2002.[91] A further round of consolidation took place when Maxis bought Timedotcom's cellular business in May 2003.[92] This effectively left the Malaysian telecommunications industry with three players: Telekom Malaysia, Maxis, and Digi.

Competition only occurs in the mobile telephony sector, where it has led to lower prices and the introduction of various services. As of 2005, there were 19.5 million mobile subscribers, who comprised 81.7 per cent of the total telephone subscribers. (See Tables A3-1 and A3-2 in Appendix Three)

Yet, the construction of parallel mobile telephony networks in the Klang Valley by five companies has led to a considerable waste of resources and duplication of networks. A market analyst estimated that the companies collectively spent around RM20 billion, which could have been better spent on constructing a national network.[93] Although the government tried to undo the mistake of issuing too many licenses through a rationalisation exercise, five companies remain. This is primarily because ruling politicians still determine the consolidation process, as they did the liberalisation of the sector. On the positive side, however, the presence of stiff competition in the mobile telephone sector means the availability of choice, lower prices, and more value-added services.

Tables A3-3 to A3-6 in Appendix Three document the improvements in fixed line telephony in Malaysia in comparison to the Philippines, Singapore, Indonesia, and Thailand. Table A3-3 shows that as of 2005, 99.1 per cent of telecommunications demand had been satisfied. Tables A3-4 and A3-5 trace the rise in fixed line teledensity in urban and rural areas, and the availability of payphones. Table A3-6 provides comparative information on the cost of fixed

Table 6-2: Malaysian Telecommunications Market After Consolidation, 1996–2002

Telecommunications Company	Majority Owner	Foreign Partner	Acquisition/Status
Telekom Malaysia	Government (Khazanah and MOF)	None	1. MRCB's Emartel — 1996 2. Mobikom's Mobifon — 1998 3. Celcom — July 2002
Celcom (Cellular Communications)	Telekom	Deutsche Telekom (21%)	Telekom as of July 2002 controls 31.25% of TRI; Telekom bought Tajudin's 13.2% stake in April 02
Maxis Communications (formerly Binariang)	Usaha Tegas — T. Ananda Krisnan	US West (sold to BT) Media One (12%) British Telecom (33.3%, until 2001, sold to Usaha Tegas)	Private company and over 70% controlled by Usaha Tegas
Time Telecom	Renong Berhad — Khazanah	None	1. Sapura's Adam and Uniphone — 1996; 2. Khazanah bought 30% — July 2000 3. Danasaham acquired UEM, parent company of Time Engineering — July 2001
Digi.Com (formerly Mutiara Telecoms)	Berjaya Group — Vincent Tan	Swiss Telecoms, then Telenor (61% as of 14 Aug 2001)	
Prismanet (formerly STW) under receivership	Shubila Holdings — Shuiab Lazim	International Wireless Communications	

Table 6-3: Malaysian Telecommunications Market After Consolidation, 2003

Telecommunications Company	Majority Owner	Foreign Partner	Acquisition/Status
Telekom Malaysia	Government (Khazanah and MOF)	None	1. MRCB's Emartel — 1996 2. Mobikom's Mobifon — 1998 3. Celcom — July 2002
Maxis Communications (formerly Binariang)	Usaha Tegas — T. Ananda Khrisnan	US West (sold to BT) Media One (12%) British Telecom (33.3%, till 2001, sold to Usaha Tegas)	1. Acquired Time Cellular from Timedotcom in May 2003
Digi.Com (formerly Mutiara Telecoms)	Berjaya Group — Vincent Tan	Swiss Telecoms, then Telenor (61% as of 14 August 2001)	

line telephone services for residential and business users, showing a decline in the cost of service after the introduction of competition to the industry.

Despite liberalisation, Telekom still controlled about 95.5 per cent of fixed line telephony in 2002. The rest was distributed among Time Telecom, Celcom, Maxis, and Digi, with an estimated 50,000 subscribers. Although these companies had licenses to build fixed line networks, they were slow to enter this market segment, preferring to concentrate on the more lucrative mobile telephony. This is in direct contrast to the Philippines, which required cellular and international gateway service providers to provide a certain number of fixed lines in exchange for market entry. Telekom, among all operators, was given the responsibility to provide "universal service." Because of this social responsibility, Telekom continues to receive government support according to Minister Leo Moggie.[94]

As will be discussed in the next chapter, the government's majority ownership of Telekom contributed to a great deal of contradiction for regulation and policy planning. Yet, the government is unlikely to divest ownership of Telekom. No local company is considered big enough to buy Telekom, and selling the firm to a foreign enterprise is unlikely, given the strategic role that it plays in terms of the provision of rural telephony, the building of capital intensive projects,[95] and its role as an investment arm of the government that "carries the flag" in Asian and African countries. Telekom has also performed other "national service" duties, such as bailing out ailing companies like MRCB Telecoms and Mobikom. A former top Telekom official pointed out that there is no record of the government categorically stating that it would divest its stake in Telekom when it was being prepared for privatisation. He further contends that all of the decisions by Telekom to invest overseas — in line with Mahathir's promotion of South-South relations — were made independently, and not at the behest of the government.[96] However, it is still the Ministry of Finance that appoints members of the Telekom board, which makes company decisions.

Being in a favoured position, however, seems to have been coupled with high trade-off costs. In the case of Telekom, although it was given preference in the protection of its fixed line segment, it lacked management independence. Favoured Malay businessmen also lacked autonomy. A former top executive of Time Telekom argued that Malays were given preferential contracts but were expected to undertake "national service" duties. In effect, they could not make independent decisions once they had committed to a special relationship with the state. They could not sell shares or give up corporate control at will, and had to wait for capable bumiputras to step in. Thus, the popular criticism that these people were mere rentiers[106] without the capacity for productive investment becomes a self-fulfilling prophecy.

Since 2001, the Malaysian government has been moving towards corporate restructuring of the highly leveraged conglomerates that gained economic strength in the 1990s through their owners' political connections. This exercise points to the pragmatism and realism of government in reversing the policies that were announced with so much fanfare a decade ago. It appears, however, that the decisive impetus towards these restructurings is factionalism among the political elite, particularly the falling-out between Mahathir and Daim. These reforms are being implemented with little transparency. Such a strategy in the past was costly for Malaysia, which is now paying for huge bailouts using public funds. However, Mahathir apparently continued to believe in the attainment of his dream of creating entrepreneurial and globally competitive Malay enterprises.[97] Up until he left office, Mahathir believed that he could select winners and appoint them as trustees to run enterprises central to the country's economy. The market reforms that Mahathir introduced failed to sever the nexus between politics and business that characterised his administration.

With Mahathir's retirement in October 2003, the policy of privatisation and liberalisation was unlikely to be reversed. Prime Minister Abdullah Ahmad Badawi had already stated that he will continue Mahathir's economic policies, including privatisation, with the important caveat that he supports open bidding in future privatisation programs. He has also stated that the time has come for "know-how rather than know-who" to be the basis of business success in Malaysia.[98]

SUMMARY

Table 6-4: Malaysian Actors and Their Positions During the Reform of the Telecommunications Sector

Actors	Role and position in the issue
1. Prime Minister Mahathir	Main push for privatisation, announced as policy in 1983. Personally oversaw corporate restructuring in 2001 after political falling-out with Daim
2. Daim Zainuddin	Saw privatisation as strategy for creating Bumiputera business class, a strategy which Mahathir agreed with. Chose his allies as recipients of licenses.
3. Leo Moggie	Minister during privatisation; called for rationalisation of the industry in 1995, Minister up to 2005
4. Samy Vellu	Minister during liberalisation, 1989-1995, when licenses proliferated
3. Shamsuddin Abdul Kadir	Commissioned a study on privatisation of JTM in 1984, influenced decision on privatising the department
4. Privatisation Committee	Headed by EPU, official body tasked to oversee privatisation process
5. JTM	Opposed entry of Shamsuddin's company to the industry became regulator, and opposed early liberalisation of the industry but had no say in it
6. JTM employees' union (NUTE)	Opposed privatisation of JTM, but most members agreed to work for new company after compromise agreement
7. DAP — Lim Kit Siang	Criticised privatisation at Parliament and tried to delay passage of law unsuccessfully
8. Telekom Malaysia Officials	Opposed early liberalisation of the industry, "convinced" to sell Celcom to Tajudin
9. Politically well-connected businessmen	Got market entry licenses to the sector through their political connections

The lack of planning and the opaque manner of awarding licenses to the politically well-connected in the liberalisation of the telecommunications industry shows that the strong, developmental state of Malaysia was not insulated from rent-seeking. Mahathir allowed rent-seeking businessmen to influence the liberalisation process because their activity corresponded with his goal of creating a bumiputra business class. This chapter detailed how licenses were distributed and linked the new licensees to particular politicians. The top three politicians acted as political patrons to favoured businessmen. Malaysian telecommunications liberalisation proves that rents are not always eliminated during market opening. Rent-seekers supported liberalisation, and market entry became a type of rent that was awarded based on political connections.

The state's bailout of companies after the financial crisis was costly. Politically well-connected businessmen were not allowed to fail. After the relations between Mahathir and Daim soured, Daim's Malay protégés lost control of conglomerates that included telecommunications companies. These state actions point to Mahathir's pragmatism and ultimate control of power in the Malaysian state, as he personally led Malaysia's corporate restructuring as acting Finance Minister. This indicates the centralisation of power in the Prime Minister's hands and his pivotal role in deciding which businessmen were to be saved. The rent-seeking activities of politically well-connected businessmen certainly shaped the liberalisation of the telecommunications industry, but these businessmen rose and fell with their political patrons. The events that took place after 1997 and 2001, demonstrate Mahathir's firm control over the Malaysian state. Having a long time frame of office made such actions possible.

NOTES

1. Interview with Ismail Salleh, Kuala Lumpur, 3 August 2001. See also Means (1991), p. 98.

2. At its height, 38 companies were licensed to provide radio paging services. Earlier Studies by Hashim (1986) and Kennedy (1990) cover this period of liberalization of the sector.

3. *New Straits Times*, 25 May 1993.

4. E-mail correspondence with former senior official of Telekom, Kuala Lumpur, 19 September 2001.

5. Gomez makes a distinction between the rationale behind Mahathir, Daim, and Anwar's patronage of businessmen. He argues that Anwar used his influence in government to develop his political base in UMNO. Mahathir was more selective in his distribution of rents, genuinely believing that he could "pick winners". Daim appeared to have used close associates as proxies for his own business interests. See Edmund Terence Gomez, "Political Business in Malaysia: Party Factionalism, Corporate Development and Economic Crisis", In *Political Business in East Asia* edited by Edmund Terence Gomez (London: Routledge, 2002), p. 108.

6. Gomez and Jomo (1999), p. 7.

7. *Business Times*, 25 June 1996.

8. Fleet Holdings, established in 1972 by Tengku Razaleigh Hamzah, marked the commencement of UMNO's corporate involvement. Razaleigh ran Fleet Holding until 1982 when Mahathir appointed Daim Zainuddin to head the company. Fleet Holding's assets were later consolidated under Renong after UMNO was deregistered in 1987 when Razaleigh and former Deputy Prime Minister Musa Hitam challenged Mahathir's leadership. See Edmund Terence Gomez's two books: *Politics in Business: UMNO's Corporate Investments* (Kuala Lumpur: Forum, 1990) and *Political Business: Corporate Involvement of Malaysian Political Parties* (Townsville: James Cook University, 1994).

9. Interview with a former telecommunications executive, Kuala Lumpur, 13 August 2001.

10. Celcom was authorized to operate a mobile cellular system on the 900-megahertz spectrum using analogue technology (ART 900). A former JTM executive disclosed that his department had to give Celcom a backdated licence because it was already launching its mobile service but had not yet received JTM's approval. He stated that JTM did this because Celcom was linked to Daim. Interview with a former JTM official, Kuala Lumpur, 17 August 2001.

11. Both quotations came from *Malaysian Business*, 16–31 March 1990.

12. Interview with a senior executive of Telekom Malaysia, Makati City, 18 April 2001.

13. *Malaysian Business*, 16–31 March 1990.

14. Peremba became the training ground for a group of young Malay professionals, most of whom were qualified accountants. These Malays, including Tajudin, Halim Saad, Wan Azmi Wan Hamzah, and Samsuddin Abu Hassan, benefited from state patronage as part of Mahathir's strategy to create Malay billionaires. They became active in the management of Daim's firms and in the development and management of UMNO's business empire. See Gomez (1990).

15. *Business Times*, 6 June 1992 and 11 June 1992.

16. *Business Times*, 24 February 1993.

17. TRI was originally known as Roxy Electric industries, a quoted firm in the business of locally manufacturing Sharp products. Roxy had been under the control of Alex Lee, a former Deputy Minister in the government and a close Daim ally. In June 1990, Tajudin bought a 25 per cent interest in TRI from Lee. See Sally Cheong, *Bumiputera Entreprenuers in the KLSE*, vol. 1 and 2 (Kuala Lumpur: Corporate Research Services, 1997), p. 277; Gomez and Jomo (1999), p. 149. At the time of TRI's buy-in, Tajudin owned 27.4 per cent of the company through his holding company, Arah Murni. See *Business Times*, 6 June 1992; *New Straits Times*, 8 June 1992.

18. *Malaysian Business*, 1 July 1992.

19. Interview with a former senior executive of Time Engineering, Kuala Lumpur, 5 July 2001.

20. *New Straits Times*, 28 October 1999; *Computimes*, 5 November 1992.

21. *Business Times*, 6 June 1992 and 11 June 1992; *New Straits Times*, 8 June 1992.

22. *Business Times*, 5 July 1996.

23. *Business Times*, 28 October and 12 November 1992; *New Straits Times*, 20 and 27 November 1992.

24. Interview with a former telecommunications executive, Kuala Lumpur, 13 August 2001.

25. *Asian Wall Street Journal*, 10 September 2001.

26. *Far Eastern Economic Review*, 5 December 1991.

27. Razaleigh challenged Mahathir for the presidency of UMNO in 1987, leading to a split in the party. Mahathir narrowly won, garnering 761 votes against Razaleigh's 718. Razaleigh resigned from his government position and later established Semangat 46 and forged a united opposition front against the BN

during the 1990 general elections. Interview with a political aide to Tengku Razaleigh Hamzah, Kuala Lumpur, 15 June 2001. See also *Asiaweek*, 18 May 2001; and Means (1991), pp. 199–206.

28. Gomez and Jomo (1999), pp. 159–65.

29. *Asian Wall Street Journal*, 10 September 2001.

30. *Asiaweek*, 18 May 2001.

31. *Business Times*, 1 July 1994.

32. *New Straits Times*, 16 October 1992 and 28 November 1992.

33. *Business Times*, 27 January 1993.

34. *Business Times*, 29 March 1993 and 15 March 1993.

35. *New Straits Times*, 24 January 1994.

36. *New Straits Times*, 24 January 1994.

37. Formerly known as Electronic & Telematique, Sapura Digital launched its mobile telephone service called Adam on 8 August 1995. Its digital network covers major cities in peninsular Malaysia. See *Computimes*, 18 May 1995; *New Straits Times*, 15 May 1995 and *Business Times*, 2 August 1995. A Personal Communications Service (PCS) refers to digital mobile networks using the 1900 Mhz frequency in the United States. In other countries, it refers to a digital mobile network using the 1,800 frequency. See International Telecommunications Union, *World Telecommunication Development Report 1999: Mobile Cellular*. (Geneva: ITU, 1999), p. 99.

38. Gomez and Jomo (1999), p. 73.

39. Eric Chia, a successful Chinese businessman, is said to have "long-standing ties with Mahathir". See Searle (1999), pp. 161, 170 and Gomez and Jomo (1999), p. 102.

40. Cheong (1997), p. 55.

41. Hashim (1986); Kennedy (1990); Lowe (1994).

42. Searle (1999), pp. 169–75.

43. Gomez (1990), p. 107.

44. *Business Times*, 4 October 2001.

45. Gomez (1990), p. 110.

46. Hatibudi, one of UMNO's company vehicles, was incorporated in 1984. Its major shareholders were Halim Saad and Mohamad Razali Abdul Rahman, both close associates of Daim Zainuddin. See (1990, 1994).

47. *New Straits Times*, 14 June 1994.

48. *Far Eastern Economic Review*, 9 August 2001.

49. *Malaysian Business*, 1 February 1996.

50. Daim Zainuddin was head of the Langkawi Development Authority after he resigned as Finance Minister.

51. *Business Times*, 31 October 1994.

52. Harold Crouch, "From Alliance to Barisan Nasional", in *Malaysian Politics and the 1978 Elections*, edited by Harold Crouch, Lee Kam Hing, and Michael Ong (Kuala Lumpur: Oxford University Press, 1980), pp. 21–22. A third shareholder in STW was VXL Holdings of Lim Chee Wah, the son of Genting's Lim Goh Tong, also reputedly a close Daim ally.

53. Rosli Man was reputedly responsible for the development of Telekom's Atur 450 and Celcom's ART 900 systems. He left Celcom in 1996, apparently after a falling out with Tajudin.

54. Interview with Vijay Kumar, Kuala Lumpur, 13 August 2001.

55. Searle (1999), p. 192.

56. Gomez and Jomo (1999), p. 153.

57. Searle (1999), p. 194.

58. Gomez and Jomo (1999), p. 154.

59. This position was supported by JTM. Interview with Dato Hod Parman, Kuala Lumpur, 10 October 2001.

60. Interview with Minister Leo Moggie, Kuala Lumpur, 5 October 2001.

61. *Business Times*, 29 September 1995 and 27 February 1996.

62. The company was valued at RM1 billion. See *Business Times*, 1 January 1997.

63. Interview with Joseph Vijay Kumar, 6 and 13 August 2001. See also *Business Times*, 1 January 1997.

64. In an interview, Moggie emphasized that his ministry did not retract or scrap the call for rationalization, as was reported in the media. He maintained that

the players negotiated the consolidation among themselves. For a contrasting report, see *Business Times*, 5 July 1996. An alternative explanation lies in Leo Moggie's position in the government. Moggie is the leader of a small Dayak party, the Parti Bangsa Dayak Sarawak (PBDS or Dayak People's Party of Sarawak), a component party of the Barisan Nasional. Because he represents a minor BN party, he was not considered a powerful figure in the government and thus did not have enough clout to maintain his rationalization policy.

65. The deal involved the acquisition by Time Engineering of a 75 per cent stake in both Sapura Digital and Uniphone for RM1.2 billion. See *Business Times*, 26 June 1996; *New Straits Times*, 8 July 1996.

66. Interview with a Sapura executive, Kuala Lumpur, 6 July 2001.

67. *New Straits Times*, 11 and 12 June 1996.

68. *Malaysian Business*, 1 February 1996.

69. Interview with telecommunications market analysts, Kuala Lumpur, 19 July 2001 and 2 October 2001. See also *Business Times*, 12 June 1996.

70. Interview with Dr Zainal Aznam, Kuala Lumpur, 13 July 2001.

71. *New Straits Times*, 15 May 1998, *Singapore Straits Times*, 15 May 1998.

72. Interview with a telecommunications market analyst, Kuala Lumpur, 19 July 2001. ATUR uses radio technology, Emartel is a PCN network operating at 1800 megahertz, and Mobikom uses AMPS 800. As of 2001, Telekom has yet to rationalize and merge the operations of TM Touch (formerly Emartel) and Mobikom. See also *Far Eastern Economic Review*, 26 March 1992 and Kennedy (1995), p. 231.

73. *The Edge*, 10 March 2001 and 4 March 2002.

74. *The Edge*, 13 March 2001; *Singapore Business Times*, 13 March 2001. Ram, "The Cost of Time to the Nation", *Aliran Monthly*, no. 2, 2001.

75. *Business Times*, 27 February 1998, 4 and 5 May 1998; *Singapore Business Times*, 1 May 1998.

76. *Singapore Business Times*, 17 February 1998; *New Straits Times*, 30 May 1998; *Business Times*, 25 July 1998 and 3 August 2001.

77. *Singapore Business Times*, 18 May 2001, 21 May 2001, and 4 July 2001.

78. *Business Times*, 16 March 2000, 17 March 2000, 18 March 2000, and 3 August 2001; *International Herald Tribune*, 18 May 2000.

79. *Business Times*, 17 June 2000, 19 July 2000; *New Straits Times*, 16 August 2000; *The Edge*, 19 July 2000.

80. E-mail correspondence with a former Telekom senior executive, 19 September 2001.

81. *Asian Wall Street Journal*, 16 May 2000.

82. *Far Eastern Economic Review*, 3 May 2001, 5 July 2001, 19 July 2001, and 23 August 2001, Gomez (2002).

83. *Singapore Straits Times*, 9 July 2001.

84. *New Straits Times*, 10 September 2001.

85. Abdul Rahman Maidin took over control of MRCB from Anwar's associates after Anwar was removed from his post. See *Far Eastern Economic Review*, 9 August 2001 and 23 August 2001.

86. In March 2002, Tajudin filed a case at the High Court to stop Danaharta from selling his shares, but the court ruled in favour of Danaharta. See *Asian Wall Street Journal*, 25 and 26 March 2002.

87. *New Straits Times*, 3 May 2002; *Singapore Business Times*, 3 May 2002.

88. *The Edge*, 2 September 2002.

89. *Far Eastern Economic Review*, 23 August 2001.

90. *The Edge*, 17 July 2002 and 23 September 2002; *The Star*, 24 September 2002.

91. *The Star*, 9 September 2002.

92. *The Edge*, 20 November 2003, *The Star*, 21 November 2003.

93. Interview with a market analyst, Kuala Lumpur, 2 October 2001.

94. Interview with Minister Leo Moggie, Kuala Lumpur, 5 October 2001. The Universal Service Obligation is the obligation of a particular licensed network operator to ensure that basic telecommunications and related services are available and reasonably accessible to all people wherever they reside at prices that do not reflect the cost of provision. The targets for the availability and penetration of services are set by the government, and can differ for rural and urban areas. http://www.mcmc.gov.my/mcmc/facts_figures/glossary/u.asp (accessed 23 June 2003).

95. For instance, Telekom invested RM1 billion in constructing the backbone for the Multimedia Super Corridor (MSC).

96. Telekom has investments in Cambodia, Bangladesh, Thailand, Sri Lanka, and India in Asia, and Ghana, South Africa, and Malawi in Africa. E-mail correspondence with a former top Telekom official, 26 September 2001.

97. Towards the end of his term, Mahathir favoured, Ananda Krishnan and Syed Mokhtar Albuhary allegedly because of their entrepreneurial capacities.

98. *The Edge*, 16 April 2003.

7

REGULATORY REFORMS IN MALAYSIA

The need for regulatory reform became evident after the adoption of privatisation and liberalisation policies in the telecommunications industry, as was the experience of countries worldwide. Although market opening greatly improved the availability of telephone services, regulations were needed to ensure the development of infrastructure in line with the country's needs, the presence of competition, and the protection of the public interest.

The chapter describes how Malaysia legislated regulatory policies and established regulatory institutions after privatisation and liberalisation. Interactions among state actors, the former monopolist, new private players, and international consultants shaped Malaysia's regulatory responses. International actors played an important role, but domestic political considerations were the decisive factors. In particular, the Prime Minister's strong commitment to the use of information and communications technology (ICT) and the development of information technology (IT) infrastructure as important for the country's next stage of development led to a forward-looking regulatory regime.

This chapter reviews the regulatory regime under JTM before looking into the creation of the Communications and Multimedia Commission (CMC), the new regulatory body that operates based on the principles of convergence. A final section summarises the experience of Malaysian telecommunications reforms.

REGULATION UNDER THE JABATAN TELEKOM MALAYSIA

In contrast to the Philippines, where telecommunications have always been in the hands of the private sector, a government department in Malaysia provided the services and regulated the sector before privatisation and liberalisation. Only after the corporatisation of Telekom was Jabatan Telekom Malaysia (JTM) directed to act as a separate regulator.

In 1985, the Legislature amended the Telecommunications Act 1950 authorising the transfer of JTM's operational responsibilities to its successor company, Telekom Malaysia (Telekom). Meanwhile, JTM was restructured to take on a new regulatory role while remaining an office under the Ministry of Energy, Telecommunications, and Posts. Section 3B of the Telecommunications Act spelled out the duties and responsibilities of the Director-General of the JTM:

> to exercise regulatory functions in respect of the conduct of telecommunications and the running of telecommunications services and their enforcement, and to promote the interests of the consumers, purchasers and the use of telecommunications apparatus in Malaysia in respect of the quality, the prices charged for, the variety of services available, and the apparatus supplied.[1]

JTM was responsible for setting technical standards and allocate spectrum in the operation of telecommunications networks. JTM's central regulatory task was to ensure that telecommunications systems ran smoothly for the benefit of consumers, and that quality, price, availability, and access were reasonable. The law provided that the Minister had the authority to issue licenses to new market entrants while JTM's Director-General was mandated to advise the Minister, though such advice was not usually sought. JTM's regulatory participation was typically limited to ensuring that the new players were fulfilling the technical aspects of their licenses.

Nearly all of the almost 30,000 JTM employees joined Telekom. A few opted to retire early while 108 were chosen to remain with JTM.[2] Most of the staff who remained with JTM were engineers knowledgeable of the technical and operational side of telecommunications. With its new job description, JTM researched existing regulatory models and considered the examples of Britain, Australia, Japan, and the US.[3]

Building regulatory capacity was JTM's focus in the first two years after corporatisation. Alongside this was the liberalisation of value-added services such as paging, trunk radio, and user-equipment supply. As was pointed out in Chapter 4, the liberalisation of these segments had actually been a state response to the lobbying of Malay businessmen. Nonetheless, the dominant persuasion within the Ministry and JTM was to allow sub-sector competition in order to understand its nature while full-scale competition would be introduced later.

Leo Moggie maintained that until 1989, when he was relocated to the portfolio of Public Works and Highways,[4] the thinking in the Ministry had been to maintain Telekom's monopoly while preparing the regulatory procedures for market opening.[5] However, the newly appointed Minister, Samy Vellu, immediately reversed that decision. Moggie surmised that the decision to liberalise early must have been brought about by the desire to push Telekom Malaysia to improve its services, and at the same time encourage new companies to invest.

Indeed, communication technologies were rapidly progressing during the early 1990s as various technological standards were being developed in the US and Europe.[6] As anywhere worldwide, equipment vendors and businessmen were exuberant about the prospects of introducing new services to the Malaysian market.

Because the authority to grant licenses was in the hands of the Minister, JTM had no real influence in the process, although it was empowered to provide advice to the Minister before any decision was made. Whether such advice was taken into consideration, however, is unknown. Publicly, JTM was perceived to be too close to Telekom and lacked impartiality. Telekom — which remained majority-owned by government even after its listing in 1990 — was very critical of the rapid and unplanned liberalisation of the sector. It actively and publicly pressured the government, arguing that complete liberalisation of the industry at an early stage could lead to its breakdown.[7] As discussed earlier, Telekom reluctantly agreed to sell its stake in Celcom, the first analogue mobile telephony system. In paging, 38 companies were licensed to offer geographically divided services, but Telekom was barred from entering the sector.

Telekom officials argued that customer welfare did not improve with the entry of new players offering paging and cellular services because service quality was very low. Congestion was a major problem since network capacity did not grow as fast as the number of subscribers. Telekom officials pointed out that although the market was big enough for others to enter, too many licenses were issued to operators who had no technical experience and limited financial resources crucial for network expansion.

Telekom executives complained that regulation was biased against them. Its regulatory obligations were enshrined in its license, but the rights to its present business and access to new businesses were not guaranteed. This was further heightened by the fact that only Telekom was obliged to build up the basic national network and provide rural telephony, which was not required of new entrants.

From a regulatory perspective, the decision to prevent Telekom from entering the cellular market and mandating interconnection with new operators made perfect sense. These decisions were justified as ways to contain the dominant player's monopolistic and anti-competitive tendencies. However, no clear regulatory guidelines were drafted. Decisions on which market segment to open and how many new licensees to allow were made on an ad-hoc basis. Moreover, the decision to allow competition was made in Cabinet. JTM was neither privy to any of these decisions nor was its advice sought. One former official lamented that JTM only learnt about the licensing of new operators through media reports.[8]

There was neither a clear liberalisation policy nor any regulatory rules in place. JTM was still in the process of organising itself and setting up regulatory procedures when the paging and cellular markets were opened to competition. During public forums, Telekom executives raised several suggestions that must be in place before opening the industry to competition. First, an organisationally sound regulatory body, overseeing policy-making and the regulation of the sector, had to be established along with flexible, transparent, and objective regulatory procedures. Second, the regulator had to be empowered to issue licenses to establish and operate network facilities and services. Third, an equitable tariff regulation had to be instituted to balance national interests and the survivability of operators. Finally, a framework in which all network operators would share the burden of social and national obligations was needed.[9]

These appeals fell on deaf ears. With the active lobbying of various influential personalities, new licenses were awarded in the telecommunications sector. Between 1993 and 1995, the speedy liberalisation of mobile telephony, international gateway, and basic network facilities took place. By 1995, one satellite, seven cellular, seven basic network, and four international gateway facilities licenses had been issued. All licensees, except one, were politically well-connected individuals.

The National Telecommunications Plan

The lack of planning and the absence of policy for liberalisation were made apparent by the issuance of a National Telecommunications Plan (NTP) in 1994, after all but one of the licenses had been issued. JTM opposed the opening of the market while it was still adjusting to its new regulatory role, and advised the government to act with caution in liberalising. Yet, the Mahathir government allowed the entry of many new players in the industry.

JTM drafted the NTP in an attempt to establish order in the industry and guide its future direction.[10] Even though it was issued belatedly, the NTP was an important document because it was Malaysia's first formal policy statement outlining a broad framework for the industry to ensure that it fit in with the goal of attaining developed nation status by the year 2020.[11] The document, which contained both general and specific policy recommendations, aimed to act as a catalyst for the growth and development of the sector, and to coordinate the emergence of a competitive atmosphere. In its declaration of objectives, the NTP specifically mentioned the goal of "encouraging active participation of Bumiputera entrepreneurs in the development of all the sectors of telecommunications, in line with the government's policy to create a Bumiputera Business and Industrial society."[12] The NTP recognised the central role of the private sector in the construction of a telecommunications network that would support the country's economic growth and overall development. One of the NTP's measurable targets was the goal of having 50 telephones per 100 people by 2020, and a teledensity of 25 telephones per 100 people for the rural areas.

The NTP contained four important points that foreshadowed the regulatory reforms introduced in 1999. First, it recognised the central role of competition policy in the development of the telecommunications industry.[13] Second, the NTP proposed a revision of the sector's structure into network infrastructure, telecommunications services, corporate telecommunications network, and IT superhighway network infrastructure.[14] The first two categories were adopted in the 1999 licensing regime. Third, the NTP specifically stated that all telecommunication providers would be required to contribute to upgrading rural telecommunications facilities. Finally, the NTP provided for the creation of a Consultative Forum, where the government, operators, and users of telecommunications services can discuss regulatory issues.

These four key points indicated that JTM was capable of drawing up an unambiguous policy and regulatory direction for the industry. Yet, JTM's capacity to regulate was often publicly perceived to be lacking, and even government and Ministry officials dismissed JTM as an institution that was only knowledgeable of the technical side of regulation. This view forwarded as the main reason why JTM was not invited to participate in drafting the Communications and Multimedia Act (CMA).[15] However, the NTP, and JTM's efforts to draw up regulatory directives, which will be discussed below, proved that the regulatory body knew what was happening in the industry. The biggest problem it faced was that it was not privy to decisions were being made elsewhere.[16]

Despite having no involvement in the issuance of licenses, JTM had to untangle problems arising from the indiscriminate entry of so many players. One major problem was the allocation of frequencies for cellular telephony. Because the original intention had been to maintain Telekom's monopoly in basic services and establish regulations first, almost the entire frequency band was allocated to Celcom. Thus, when new licenses were issued without JTM's involvement, frequency allocation turned into a quagmire.[17]

A second thorny issue was interconnection. Interconnection allows the subscribers of one company to be able to call the subscribers of another. Without interconnection, subscribers would only be able to call people who are connected to the same network. The licenses that were issued to companies were not public documents, thus the scope, terms, and conditions of each franchise were not known. Although interconnection between operators was mandatory, it was very difficult for companies to reach agreements to interconnect their systems on their own. For instance, Telekom and Celcom had no interconnection agreement for six years (1989–95). Thus, Celcom's cellular subscribers had difficulty calling Telekom Malaysia fixed-line subscribers, and vice versa.[18] Other new players took between two to three years arriving at an interconnection pact with Telekom.[19]

Interconnection in reality is an issue involving two aspects, technical and commercial. Technical interconnection is the provision of comparable technical and operational interconnection facilities between differing switching and transmission facilities. Technology solutions and international standards for the interconnection of various technological systems had made it possible for different telecommunications networks to be able to connect with each other. The more difficult issue to settle, however, was commercial interconnection, which dealt with the question of the structure and level of rates and charges. In particular, the new companies that were connecting to the incumbent operator argued that the cost of interconnection, the access charge, should be computed based on the direct cost incurred. The incumbent, however, argued that setting up the national backbone had historical costs that should be considered in computing the access charge.

To settle interconnection and other regulatory problems, JTM established various steering committees in 1994 composed of all the telecommunications operators. These committees became venues for discussion to resolve regulatory issues and eventually draw up frameworks and guidelines. In some issues, foreign consultants were hired to provide analysis, technical input, and comparative perspectives on best practice. From these working groups came significant regulatory documents, the most important of which was the General Framework for Interconnection and Access (GFIA), launched in May 1996.[20]

The General Framework for Interconnection and Access (GFIA)

The GFIA established the policy objectives and regulatory principles for the interconnection of licensed telecommunications operators based on fair, equitable, and non-discriminatory principles. All licensed operators had the right to interconnect with each other's networks.[21] The GFIA stated that a dominant network operator in any market segment should not abuse its market power to limit access to essential facilities, and that no carrier should be asked to carry an inequitable share of the universal service obligations out of proportion to its market share.[22] The framework also provided for the establishment of bilateral interconnection agreements between network operators. The companies were given 90 days to reach agreements, after which the Director-General was empowered to arbitrate and issue a decision that would legally bind all parties. In addition, the Director-General had powers to settle disputes, impose penalties for non-compliance, and enforce any decisions or determinations.

The long-term objective of the GFIA was to arrive at an access charge based on cost. However, the real cost of interconnection was in dispute until the costs of operations were unbundled or broken down into their component costs. Thus, the GFIA provided for the establishment of an industry-wide Charter of Accounts and Cost Allocation Manual as the basis for an access charge, a Universal Service Obligation (USO), and the investigation of anti-competitive practices. However, until a Chart of Accounts and Costs was ascertained, interconnection access charges were made on the basis of revenue sharing between interconnecting parties. In the event that the network operators could not agree to a method of calculation for the access charges, the Director-General would determine them.[23]

The GFIA also had sections providing for the implementation of equal access from 1 January 1999. The Equal Access program allows consumers to choose a network different from that of their usual provider when making long-distance or international calls.[24] The implementation of Equal Access was delayed from 1997 to 1 January 1999 because of stiff opposition from Telekom, which considered it unfair that new companies would have immediate access to its 3.2 million subscribers. Telekom argued that new players should develop services to tap new subscribers rather than poaching existing subscribers.[25]

Aside from the GFIA, JTM issued various regulatory determinations that covered technical, quality, and other issues in 1998 and 1999. In 1998 alone it issued nine Regulatory Determinations, the most significant of which was Telecommunications Regulatory Determination (TRD) 6/98, which provided for cost-based interconnection and universal service obligation. TRD 6/98 was a result of consultation between JTM, industry players, and Analysys Limited, a UK based consultancy firm jointly hired by JTM and the companies.

TRD 6/98 provided for the creation of a Local Access Fund by 2001 for the provision of universal service. All telecommunications operators were required to contribute based on their weighted network revenues. In the interim, the government assigned Telekom to be the sole provider of Universal Service, after which any interested operator would be able to claim from the fund to provide service comprised of basic telephony (including emergency and directory assistance), public payphones in rural areas, and services to the disabled.

TRD 6/98 also provided for the implementation of the first stage of Equal Access by 1 January 1999, with the second stage to commence two years later. A 20 per cent floor on retail price reduction for long distance and IDD retail rates was introduced to ensure an orderly transition to cost-based interconnection changes.[26] Fully allocated cost, or the total cost of providing trunk, local, and IDD services, would be the basis of the interconnection rate during the two-year interim period (1999–2001). It was expected that by 2001, the government would be able to determine cost-based interconnection pricing. Given the difficulty of arriving at a cost-based interconnection rate, the Director-General of JTM was mandated to set a "fair price," by taking into consideration the interests of the users, companies, and the national objective of ensuring the attainment of telephone penetration targets.

The introduction of a floor on discount rates basically placed a cap on how much operators could reduce their prices. New companies saw the 20 per cent floor and the postponement of the implementation of Equal Access as a way of protecting Telekom Malaysia's control over its fixed line subscribers. New players argued that there would be no incentive for subscribers to move to other service providers when the cost of a call was the same as with their current provider. The players decried this sort of decision as a proof of JTM's bias towards Telekom Malaysia. Although JTM publicly denied such claims, its regulatory decisions are best seen in the light of Telekom's Universal Service obligations.

To summarise, the above discussion points to the fact that JTM was not an incompetent regulator. It tried to do its job, but it was operating in a difficult environment in which top politicians made major decisions without involving JTM. That the agency issued the NTP, the GFIA, and other regulatory determinations is evidence that JTM attempted to face up to its regulatory role. It consulted with industry players and sought experts' opinions. Indeed, JTM attempted to balance the various interests of players, most of whom were linked to top politicians. The difficulties of managing an industry in which the major players were owned by politically well-connected businessmen was also confirmed by the Ministry's call for industry rationalisation. Minister Leo Moggie's rationalisation call immediately upon his return to the MEPT, was a very clear admission of the mistake of over-issuing licenses. Yet, even if the Minister presided over meetings to encourage mergers, the licensed operators refused to budge. At the end of the day, even the Minister had to withdraw his call for rationalisation and decided to allow the telecommunications companies to decide mergers on their own.

Given that it was not involved in liberalisation, JTM had difficulty regulating an industry that was dominated by politically influential players. Nonetheless, its regulatory role was crucial in making sure that competition existed. The new players needed JTM to ensure that

Telekom did not exercise its unfair advantages as the dominant player. Given that Telekom shouldered the Universal Service Obligation for the country and that JTM and Ministry officials sat on the board of Telekom, new players criticised the regulator for its bias towards Telekom. Thus, regulation of the liberalised telecommunications industry was faced with contradictions and difficulties.

THE COMMUNICATIONS AND MULTIMEDIA COMMISSION

At the macroeconomic level, regulatory reforms lagged behind privatisation and liberalisation. In fact, the regulatory reforms that led to major changes in the institutional and regulatory frameworks of the telecommunications industry were brought about by a confluence of events in which the Prime Minister's personal involvement and commitment were crucial.

In 1996, the launch of Mahathir's pet project, the Multimedia Super Corridor (MSC), signalled Malaysia's desire to shift its developmental emphasis from the export of commodities and low-end manufacturing to the creation of a knowledge-driven economy. The MSC is a 15 by 50 km zone that extends from the Kuala Lumpur City Centre, the site of the Petronas Twin Towers that symbolises Malaysia's economic growth, to the new Kuala Lumpur International Airport in Sepang. The MSC was launched as a test bed for the development of information technology in the country, copying the model of Silicon Valley in California. It was an experiment to catalyse the development of technology parks and innovation clusters by providing physical and legal infrastructure, R&D support, tax holidays, and other incentives intended to attract international attention and investment.[27] This shift in development strategy took place at the height of international exhilaration over the rise of a new global network economy made possible by the communications and microelectronic technology revolution.[28] Malaysia became one of the trailblazers among developing countries as it swiftly enacted laws in 1997 such as the Digital Signature, Telemedicine, and Computer

Crimes and Copyrights (Amendment) Acts as part of the MSC's legal infrastructure. All of these laws were fast-tracked because of Mahathir's personal commitment. It also helped, of course, that the Barisan Nasional controlled the majority of seats in Parliament.

Technological changes, specifically the convergence of voice, data, and computing technologies, had altered the character of vertically segmented communication, information, and broadcasting industries. In 1997, Mahathir directed the Ministry of Energy, Telecommunications, and Posts (METP) to look into how the country could be prepared for convergence.

On 8 August 1997, the National Telecommunications Council (NTC) met for the first and only time, to discuss the proposal for reform of the telecommunications industry. The NTC was an exclusive group, set up to act as an advisory body in 1996. Its chair was the Prime Minister and was composed of top government officials, high profile private sector personalities, and CEOs of major telecommunications companies. Mahathir opened the meeting by stressing the need to move fast in bringing about reform. Then Secretary-General of the METP Nuraizah Abdul Hamid presented a plan for action. The proposals were based on a 1996 study that Telekom commissioned. The report, conducted by Cutler and Company, a Melbourne-based IT consultancy group, urged a reorganisation of JTM and suggested upgrading its capacity to handle convergence. Because they had Mahathir's enthusiastic endorsement, no objections were raised against the reform proposals. The NTC agreed on the need to formulate a comprehensive national policy framework for convergence, develop a new regulatory framework and institutional arrangements, and to draft a Multimedia Convergence Act and table it in parliament.[29]

In September 1997, Cutler and Co. was hired to assist the Ministry draft the reform package. With the Prime Minister's mandate, the consultants worked with an inter-agency committee, and within three months issued a five-volume report.[30] Volumes Two and Three of the reports contained draft legislation for the regulatory and institutional framework for convergence, which eventually became the Communications and Multimedia Act (CMA 1998) and the Malaysian Communications and Multimedia Commission Act

(MCMC 1998). These two acts presented a new policy and regulatory framework for the communications and multimedia sector, recognising that the convergence of technologies dissolved boundaries between telecommunications, broadcasting, and information industries. A new independent and partially self-financed regulatory body fashioned after the Securities Commission was to replace JTM. The report also identified necessary subsidiary instruments to replace current legislation, and provided a detailed analysis of the existing licensing arrangements, how to go about the transition to convergence, and a schedule of implementation.

According to some Ministry officials, various roadshows informed industry players and interested parties of the changes that were about to take place.[31] The Communications and Multimedia Bill and the Communications and Multimedia Commission Bill were tabled in Parliament in July 1998. Minister Leo Moggie explained that the bills contained a detailed policy structure encompassing the convergence of the telecommunications, broadcasting, and information technology industries. The complex and technical nature of the legislation precluded detailed discussion. Yet, both government and opposition members of parliament unanimously supported the bills. Thus, the speed of drafting was matched by the speed of the bills' passage through Parliament. The Parliament approved the two acts in the same month, and received the King's assent on 23 September 1998.[32]

On 1 November 1998, the Ministry of Energy, Telecommunications, and Post was renamed the Ministry of Energy, Communications, and Multimedia (MECM). Along with this, the Communications and Multimedia Commission (CMC) was created to replace JTM. Dr Syed Hussein Mohamed, Telekom's Executive Director from 1987 to 1993, and a non-executive director of Telekom's Board until 1998, was appointed the first CMC Chairman.[33]

The CMC officially began operation on 1 April 1999, when the Communications and Multimedia Act (CMA) came into effect, repealing the Telecommunications Act 1950 and the Broadcasting Act 1988. The new act gave policy-making power to the MECM while mandating the CMC to supervise and regulate the communications and multimedia industry guided by four principles: national interest, transparency, less regulation, and flexibility.[34]

The major goal of the CMA was to break down the vertical market structure of the telecommunications, broadcasting, and IT sectors.[35] The new regulatory framework introduced six notable changes. First, it radically restructured the licensing regime to reflect a foreseen convergent market. In the past, companies had been licensed based on the type of service they offer (fixed line, cellular, or international gateway) or the technology they use (copper lines, fibre optic, GSM, PCN, satellite, etc.). The new licensing regime created four new categories: network facility, network service, network application, and network content application. In each category, individual and class licenses would be issued.[36] Second, transparency and consultation were incorporated as principles of the legal and regulatory process. The CMA provided for a formal process of public inquiry and feedback before the adoption of new regulations. To facilitate transparency and institutionalise a process of feedback from industry players and interested parties, a public registry would be created to contain regulatory documents and the licenses of all players.[37] This change is significant in contrast to the past practise whereby licenses were confidential documents. In addition, the CMA provided for the establishment of an Appeals Tribunal to review CMC decisions and its direction.

Third, consumer, content, technical, and access forums were established to devise rules and regulations, and to establish voluntary codes of conduct for their members. The aim was to introduce self-regulation, with the CMC playing less the role of regulator and more the role of promoter of the industry and consumer interests. Fourth, a Universal Service Fund was created to finance the rollout of services to underserved areas. All major players, including network and spectrum users, were required to contribute six per cent of their gross income to the fund. Before this, universal service was solely the obligation of Telekom. Fifth, the CMA provided for Malaysia's first formal competition policy. Finally, the Act provided for CMC's financial autonomy, with an initial grant of RM122 million from the government.[38]

Table 7-1: New Malaysian Licensing Regime, 2001

Provider Categories	Individual License	Class License	Exempted/Not licensed
Network Facilities Providers (NFP) — Owners of network infrastructure and facilities used to provide content and application services	Satellite earth stations, broadband fibre optic cables, switching equipment, radio communications, including transmission stations, and public pay phone facilities	Niche or limited purpose facilities such as mobile, trunk, and paging	Incidental facilities, private facilities, broadcasting and production studios, customer premises equipment, and Internet cross-connection equipment
Network Service Providers (NSP) — Provide basic connectivity and bandwidth support for a variety of application services	Access and connectivity services, bandwidth services, mobile satellite communications services, mobile services, and broadcasting distribution services	Niche access and connectivity services (e.g. trunk mobile)	Private network services, incidental network services, LANs, and router internetworking
Applications Service Providers (ASP) — Provide specific and particular functions of capabilities to end users	PSTN, public IP, public switched data services, and public payphone	Internet access providers, SMS, paging, directory assistance, telegram services, private payphones, card services, and audiotext hosting	E-transaction services, call centres/interactive voice response, interactive information services, webhosting/client servers, and network advertising boards
Content Application Service Providers (CASP) — Provide application services with integrated content	Satellite broadcasting, terrestrial free to air TV, radio broadcasting, and subscription TV		Internet content services, Internet webcasting, and limited content services

Source: Communications and Multimedia Commission, 2001

The changes brought about by the CMA were both monumental and radical. Compared to the first 10 years of liberalisation, where non-transparent and politicised decisions were the rule and JTM had no say in what was happening, the attempt to create an independent regulator and the institutionalisation of consultation and transparency were giant steps towards reform. There is little evidence that suggest politically well-connected businessmen running the new companies played any significant role in pushing for these reforms. The international consultants who drafted the new regulatory framework developed the ideas behind the CMA. These ideas were adopted as policy because of Mahathir's personal commitment to the development of information technology in the country. With its focus on convergence and competition, the new regulatory framework promised to create a level playing field among telecommunications market players.

Despite the potential of the new regulatory set-up, a major source of apprehension was the CMC's capacity to independently regulate. Market analysts pointed to the government's continued majority ownership of Telekom. They were also worried that the first two Chairs of the CMC were formerly connected with Telekom, the first as a top ranking Telekom executive, and the second Chair as a member of the Telekom Board. One industry analyst called this an "incestuous relationship," in which the regulator sat on the board of the regulated company.[39] Thus, when the CMC Chair stated in 2001 that "the delays in the introduction of measures to boost competition were to safeguard national interest,"[40] observers interpreted this to mean that the dominant player, Telekom, would continue to receive protection from the government. Despite the policy of encouraging competition and the issuance of three competition guidelines, the new CMC was hesitant to fully enforce competition rules among players. A good example of this is the delay in the implementation of Equal Access from 1996 to January 1999, and the indefinite suspension of the second stage of the program, which was supposed to be enforced by 1 January 2001. Furthermore, the 20 per cent floor on discounts that companies could offer was an obvious protection of Telekom and not in the interests of the consuming public. These issues however are expected to be resolved when the Universal Service Fund becomes operational.

A second issue related to licensing. Despite the new convergent licensing regime, no real crossover between the telecommunications, broadcasting, and IT industries took place, except perhaps in the licensing of Voice over the Internet service providers. This is because, at the launch of the new regulatory framework in April 1999, Moggie announced a one-year moratorium on the provision of convergent services to prepare for the migration of companies to the new licensing regime.[41] Some analysts viewed this as defeating the purpose should convergence be indefinitely put on hold.[42] Moreover, migration to the new licensing regime was fraught with problems. The CMC issued the new licenses months after the deadline for registration. This situation prompted some operators to declare that they were "in limbo"[43] operating without a valid license. The CMA provided that no existing network licensee would be forced to migrate to a new license. Yet, those who did not do so were not eligible for any of the benefits of the Act. Existing license holders were given 12 months to register with the CMC and to notify it whether they wanted to migrate to the new licensing system. New licenses were finally issued in July 2001, three months later than the expected release.

Third, although the CMA was unique in providing a mechanism for consultation and transparency in the issuance of policy determinations, consultations so far were mere formality. Some telecommunications operators (now known as network facilities providers and network services providers) expressed their unhappiness with the way in which their views were not considered because their submissions were received at the CMC office "6 minutes after the 12 noon deadline."[44] The companies argued that if the CMC was really serious about consultation, such submissions would not be invalidated. The CMC officials put the blame on the companies and contended that if they were serious about being heard, they would have submitted their material before the published deadlines.[45] Moreover, in place of formal consultative networks, real decisions were made at monthly lunches that were attended by CEOs of the telecommunications companies and the Chairman of the CMC. Some felt that this practice defeated the purpose of consultation enshrined in the law. Finally, although the CMA had extensive provisions about public consultation and transparency, the decisions were ultimately to be made by the Minister of Energy, Communications, and Multimedia, and the extent to which views expressed in consultations would be considered was unclear.

Fourth, while the CMA's provision for self-regulation via the establishment of industry forums was a novel idea, the readiness of the industry to embrace the concept was suspect. In 2001, a Consumer Forum and a Content Forum were established, as well as temporary committees on technical and access forums. Although self-regulation might work in an ideal world, the unbalanced market positions of the various players meant that arriving at unanimously agreed upon regulatory principles was difficult. In fact, one of the reasons why the Access Forum was not yet established was because interconnection charges were a contentious issue. Telekom believed that historical costs should be used while new operators favoured a cost-based rate. The CMC was inclined to adopt a (LRIC) Long Run Incremental Cost rate, but until a Chart of Costs is established the cost of access will still be based on revenue sharing, agreed upon via JTM's TRD 006/98.[46]

What was not publicised was the fact that except for the new licensing regime, almost all of JTM's regulatory determinations were still in operation. The continued relevance of TRD 6/98 and other JTM regulatory determinations point to the fact that migration to a convergent environment is a long process. More importantly, it also indicates that JTM was not merely an incapable regulator, caught up with the technical side of regulation, as it was portrayed. Rather, it was actively moving towards establishing clear regulatory principles before it was suddenly replaced by the CMC. Indeed, although it is commendable that the government had the capacity to move swiftly in making decisions and introducing changes when the crucial support of the Prime Minister was present, it was alarming that JTM staff members were not included in the drafting of the new regulatory framework that eventually led to JTM's demise. Moreover, of JTM's 190 employees, only ten were chosen to join the CMC. The rest were given the option to join other government offices or retire. This created much resentment from JTM's former staff, especially in the way that CMC clearly demarcated the start of the new regime, and its attempt to replace the regulatory rules that JTM created.

To be fair, the CMC is still a new body trying to grapple with migration from an old regime to a new regime. Many issues are still being studied, and the CMC has hired international consultants to consider competition, access, tariff rebalancing and accounting rates, and spectrum mapping. Given the early mishaps and complaints, the regulator should find its feet and address them. Creating a convergent regulatory environment is a huge step to make, and the transition cannot take place overnight. The challenge, however, is for the CMC to demonstrate its capacity to regulate an industry that developed largely through past political decisions.

SUMMARY

Table 7-2 summarises the main actors involved during regulatory reform process in Malaysia. Meanwhile, Tables A3-7 and A3-8 in Appendix 3 summarise the existing regulatory regime, the functions of the regulatory agencies, and the accountability, autonomy, and competency of the regulatory process in Malaysia and the Philippines.

This chapter discussed the regulatory responses of the Malaysian state, and how telecommunications, in line with the experience in other countries, had to be actively regulated to resolve various issues after liberalisation. The section detailed how JTM was not involved in the liberalisation of the telecommunications sector and was left to regulate the industry after the politicised manner of the issuance of market entry licences. JTM attempted to institute order by the issuance of the NTP, GFIA, and various regulatory determinations. The Ministry, meanwhile, had a difficult time forcing mergers to take place among the politically well-connected, and had to abandon its call for rationalisation.

Ironically, all of the new entrants needed JTM's regulation to make sure that Telekom played fair and interconnected with them.

Table 7-2: Malaysian Actors and Their Positions During Regulatory Reforms

Actors	Role and Position in the Issue
Mahathir	Main push for regulatory reform, in line with launch of MSC project and the focus on ICT as new strategy for country's development
MECM	Formerly METP, oversaw the drafting of new regulatory framework
NTC	Advisory body to the Prime Minister composed of high profile government and business personalities, supported shift to convergence and establishment of a new regulatory body
JTM	Top level officials consulted "but not consistently"; abolished after the creation of CMC; the rest of the employees caught unaware of the change
Cutler and Co.	Melbourne-based consultant which drafted the new regulatory framework
Telekom Malaysia	No active role, Cutler and Company was originally its consultant on regulatory issues
New companies	No active role, but were pushing for clearer competition rules

As all incumbents had done worldwide, Telekom delayed if not refused interconnection. The goal of introducing competition in the fixed line business through Equal Access was postponed twice, and eventually abandoned. New entrants saw these actions as protecting Telekom. Unwittingly, the politically well-connected needed JTM to enforce competition and interconnection regulation, but both JTM and the Ministry held seats on the Telekom board, which led to conflicts of interest.

Meanwhile, Mahathir's MSC dream made evident the importance of a communications infrastructure and a forward-looking regulatory regime. This resulted in the abolition of JTM and the establishment of the CMC. The creation of the CMC shows once again the central and crucial role of Mahathir and his vision of modernisation for Malaysia, bulldozing an existing department to create a new, forward-looking entity designed by foreign consultants. Such changes were possible only through Mahathir's personal commitment to the development of ICT. This can be compared to the case of the Philippines that will be discussed in the next chapter, where governmental power is diffused among its three branches and where there is a fixed six-year term of office for the executive, thus making it difficult to introduce long-term institutional reforms. In contrast, the Prime Minister effectively controls state power in Malaysia. Thus, it was feasible in Malaysia to introduce radical reform such as the abolition of a government department and the establishment of an entirely new entity. Nonetheless, the independence and capacity of the CMC to regulate the industry is yet to be tested, although the policies announced so far hold much promise.

This chapter also provided evidence for the argument that the consistency and credibility of reform requires the active involvement of a state leadership (usually the executive) that is committed to furthering and deepening market oriented reform by creating the legal and institutional framework for regulation. The Philippines' case in the next chapter shows that such a commitment was missing. In contrast, the adoption of regulatory reform in Malaysia was fast-tracked due to the central role and personal commitment of Mahathir, who launched the MSC project in 1996, providing top-level commitment to reform, and exercising the political will necessary to establish the appropriate legal and physical infrastructure.

A SUMMARY OF MALAYSIA'S TELECOMMUNICATIONS REFORMS

Table 7-3 provides a summarizes the highlights of telecommunications reform in Malaysia.

The Malaysian state is considered to be strong and capable of independently adopting policies and implementing them consistently to achieve developmental goals. Under Mahathir, state power was increasingly centralised in the office of the Prime Minister. Authoritarian laws and institutions were utilised to demolish criticism of, or opposition to, governmental policies. The strong developmental state, with Mahathir at its helm, adopted privatisation as a cornerstone of the country's economic program not because of the standard rationale for the policy but more due to the aim of creating a Bumiputera business class. This goal has been the central rationale in implementation of privatisation, and in the ensuing bailout of companies after the 1997 financial crisis. In addition, the adoption of a new regulatory regime points to the key role of the Prime Minister. The strength of the Malaysian state and Mahathir's pivotal role is also confirmed by his decisive implementation of corporate restructuring after the financial crisis, to the point of bailing out or renationalising privatised firms.

Although a strong state that is capable of independently adopting general economic policy, Malaysia is nevertheless highly permeated by rent-seekers with regards to microeconomic policies and their implementation, as JTM's privatisation and the subsequent liberalisation of the telecommunications industry show. In fact, rent-seeking and patronage were key factors in JTM's privatisation and in the liberalisation of the industry as politically well-connected businessmen lobbied for market opening. The lobbying of rent-seekers not merely facilitated the adoption of liberalisation, despite the lack of readiness of the regulatory body and its advice to the contrary, but also shaped the way in which liberalisation was implemented by the state. The most enthusiastic support for liberalisation came from politically well-connected businessmen who wanted market entry, knowing full well the benefits they would reap in the telecommunications sector. Thus, they lobbied senior politicians and were granted licenses to enter the market.

Because the main support for the opening of the telecommunications market came from businessmen linked to top politicians, a license to enter the market was a type of rent that was reaped by a chosen few. The state deliberately adopted market liberalisation to create rents with which to reward people within the patronage network of UMNO, as well as to achieve the goal of creating individual Bumiputera businessmen. The state used privatisation and liberalisation policies in general, and of the telecommunications sector in particular, to create business opportunities for a selected group of people. The recipients of rents were all politically well-connected, reaping market entry licenses through existing patronage networks and linkages with the top three politicians.

Gaining market entry, nonetheless, does not necessarily redound to profitability in an industry with a dominant player. The introduction of a new regulatory regime with the creation of the CMC augurs well for competition. However, whether the CMC can act consistently and independently remains to be seen.

Table 7-3: Highlights of Telecommunications Reforms in Malaysia

	Malaysia
Year of Reform	1987, 1990, 1993, 1999
Type of Reform	Corporatisation (1987) Privatisation (1990) Liberalisation(1993) Regulatory Reforms (1999)
Regulatory body	JTM (1987–99) CMC (from April 1999)
Pre-reform market structure	Government monopoly (JTM)
Post-reform market structure (2002)	Competitive, but existence of dominant carrier (TM); 3 major players (TM, Maxis, Digi)
Major telecommunications laws	Telecommunications Act 1950 CMA, 1998 MCMC, 1998
Interconnection	Mandatory
Universal service obligation	USO fund established by 2000, Telekom sole provider before 2000
Convergence	Allowed by law

NOTES

1. 1985 Amendment to the Telecommunications Act 1950.
2. Interview with a former JTM staff member, Kuala Lumpur, 17 August 2001.
3. Interview with Hod Parman, Kuala Lumpur, 10 October 2001.
4. In Malaysia, the power to appoint Cabinet members is vested in the Prime Minister. Leo Moggie served as Minister of Energy, Telecommunications and Post from the late 1970s to 1989. Samy Vellu took over the portfolio from 1989 to 1995. Moggie went back to the Ministry in 1995 until 2004.
5. Interview with Leo Moggie, Kuala Lumpur, 5 October 2001.
6. Different technological standards, mostly in mobile telephony, such as Nordic Mobile Telecommunications (NMT), Total Access Communications System (TACS), and Global System for Mobile (GSM) in Europe, and Advanced Mobile Phone System (AMPS) and Digital Advanced Mobile Phone System (DAMPS) in the United States, were being launched.
7. Syed Hussein Mohamed, "Asian Telecom Market: What Are the Options?" paper delivered at the Mobile Communications 1992 Conference, 25–26 August 1992, Kuala Lumpur, p. 5.
8. Interview with a former JTM staff member, Kuala Lumpur, 17 August 2001.
9. Mohamed Said Mohamed Ali, "Privatisation — Lessons to Be Learnt from Local Experience", Paper presented at the National Conference on Privatisation — The Challenges Ahead, 7–8 October 1993, Kuala Lumpur, pp. 14–15.
10. In January 1994, JTM was reorganized to prepare for its new regulatory role. Haji Hod bin Parman was appointed Director-General of JTM in April. See *Jabatan Telekom Malaysia Annual Report 1994/1995.*
11. Vision 2020 is a policy document that outlines Malaysia's goal of attaining the status of a developed country by the year 2020. Dr Mahathir launched Vision 2020 in his speech to the Malaysian Business Council Meeting in Kuala Lumpur on 28 February 1991.
12. Ministry of Energy Telecommunications and Post, *National Telecommunications Policy 1994–2020*, Section 14.2.12, 17 May 1994, (hereinafter referred to as *NTP*).
13. *NTP*, Section 15.3.1.
14. *NTP*, Sections 16.1.1 to 16.1.4.

15. A former senior official said that JTM was "consulted in a way but not consistently" in the drafting of the Communications and Multimedia Act. Rank and file staff members however had no idea of the reform changes that were about to happen. Interview with a former senior JTM official, Kuala Lumpur, 10 October 2001.

16. One former senior official declared in an interview that JTM's hands were tied. Interview with a former senior JTM official, Kuala Lumpur, 28 July 2001.

17. *New Straits Times*, 24 January 1994.

18. *Business Times*, 2 August 1995.

19. *Business Times*, 20 February 1995, 21 June 1995, 6 July 1995, 3 August 1995, and 2 December 1995.

20. The length of time that it took for an agreement to be arrived at by all of the companies points to the fact that the issue of interconnection was very difficult to settle. Before the development of the GFIA, interconnection was based on interim agreements (*New Straits Times*, 3 August 1995).

21. Director General of Telecommunications, Malaysia, *General Framework for Interconnection and Access*, Section 5.1, paragraph 46, 17 May 1996 (Hereinafter referred to as *GFIA*).

22. *GFIA*, Section 4, paragraphs, 37–39, 44.

23. *GFIA*, Section 5.2, paragraph 53.

24. The first step was the dialling of access codes or prefixes before the number being called to identify which network the consumer wished to use for a call. The second stage was pre-selection, whereby a consumer would be able to use the services of another company each time a long-distance or international call was made.

25. *New Sunday Times*, 25 February 1996.

26. JTM, Telecommunications Regulatory Determination 006/98: Determination of Cost-Based Interconnect Prices and The Cost of Universal Service Obligation, Section 3.1.

27. Terry Cutler, "The Development of Malaysia's Multimedia Super-Corridor", paper read at the Third Biennial Australia-Malaysia Conference, Canberra, 15 March 2002.

28. Manuel Castells, *The Rise of the Network Society, the Information Age: Economy, Society and Culture*, vol. 1 (Mass: Blackwell Press, 1997).

25. Interview with Tan Sri Nuraizah Abdul Hamid, Kuala Lumpur, 11 October 2001.

30. The inter-agency committee was composed of the Ministries of (1) Energy, Telecommunications and Post, (2) Information, (3) Science, Technology and Environment, (4) Home Affairs, (5) Domestic Trade and Consumer Affairs, (6) Finance, and (7) Housing and Local Government, (8) the Attorney General's Chambers, (9) the Economic Planning Unit, (10) the Public Services Department, and (11) the Modernisation of Administration Unit (MAMPU).

31. Interview with Ministry of Energy, Communications and Multimedia officials (Suriah binte Abdul Rahman, Abu Hassan Ismail, and Teo Yen Hua), Kuala Lumpur, 24 July 2001.

32. Laws of Malaysia. *Communications and Multimedia Act, Act 588* (Kuala Lumpur: Percetakan Nasional Malaysia Berhad, 1998); and *Malaysian Communications and Multimedia Commission Act, Act 589* (Kuala Lumpur: Percetakan Nasional Malaysia Berhad, 1998).

33. *Business Times*, 31 October 1998; *Computimes*, 28 December 1998.

34. Syed Hussein Mohamed, *New Regulatory Framework for Communications and Multimedia Sector: Subsidiary Legislation.* Communications and Multimedia Communications Notes, no date. Document found at the CMC library.

35. Nomura Asian Equity Research, "Wireless Operators", 18 April 2001, p. 11. Convergence is defined as the progressive integration of the value chains of traditional communications and content industries within a single value chain based on the use of distributed digital technology, creating economic markets for network application services, network services providing the connectivity for applications, networking facilities that create the infrastructure platforms and goods and services used in conjunction with the above.

36. Statement by the Minister of Energy, Communications and Multimedia, 31 March 2000.

37. Any CMC regulatory decision had to be posted in the registry. The public is given 45 days to submit feedback before the finalisation of a regulatory determination.

38. Laws of Malaysia, Malaysian Communications and Multimedia Commission Act, 1998, Sections 38–48.

39. Interview with a telecommunications market analyst, Kuala Lumpur, 19 July 2001. In contrast, former Telekom CEO Mohamed Said Mohamed Wira

defended the government's choice of appointing Syed Hussein and Nuraizah. He argues that the two were professionals appointed based on their merit and past performance. Telephone interview with Mohamed Said Mohamed Wira, Kuala Lumpur, 26 September 2001.

40. *The Star*, 19 February 2001.

41. Statement by the Minister of Energy, Communications and Multimedia, 31 March 2000.

42. Interviews with two telecommunications market analysts, Kuala Lumpur, 19 July 2001 and 2 October 2001.

43. Interview with regulatory officers of two telecommunications companies, Kuala Lumpur, 9 August 2001 and 23 August 2001

44. Interview with an executive from a telecommunications company, Kuala Lumpur, 9 August 2001.

45. Interview with a CMC official, Kuala Lumpur, 3 September 2001.

46. The CMC is contemplating the long-run incremental cost (LRIC) as the interconnection costing method. LRIC is a forward-looking method of costing that uses incremental cost or the cost that arise as a result of providing an increment rather than the use of historic cost accounting. In contrast to other costing methods, LRIC only includes the cost that is directly caused by the provision of an increment. See http://www.oftel.gov.uk/ind_info/international/lric498.htm viewed on 25 November 2003.

8

REFORMING THE TELECOMMUNICATIONS SECTOR OF THE PHILIPPINES

At the end of 2005, the Philippines had 34.8 million mobile phone subscribers and 3.4 million fixed line subscribers, leading to a total teledensity of 45.1 per 100 persons. According to the National Telecommunications Commission, the industry's regulator, Filipino mobile phone users sent an average of 250 million text messages per day or an average of 6 messages per person each day in 2005.[1] Because of this, the Philippines has earned the moniker "Text Capital of the World." This situation is a far cry from the condition in 1990 when then Singapore Prime Minister Lee Kuan Yew quipped that 99 per cent of Filipinos were on queue for a phone while the remaining one per cent were waiting for a dial tone. How did this huge change come about.

This chapter examines the first stage of reforms in Philippine telecommunications. The discussion is divided into five sections. The first looks into the adoption of liberalisation as part of the government's economic reform agenda and emphasises the role of President Fidel Ramos in the process. The second focuses on how the "coalition for reform" manoeuvred to liberalise the industry and the responses of the monopolist. The third looks into the implementation of liberalisation while the fourth analyses the passage of a law ostensibly to safeguard liberalisation gains. A final section summarises the chapter's discussion and arguments.

LIBERALISATION IN THE PHILIPPINES

As Chapter 4 detailed, PLDT successfully obstructed attempts to open the telecommunications market under the Aquino administration. This chapter discusses how an historically weak and penetrated state was able to liberalise the telecommunications industry in the face of an influential vested interest. In particular, liberalisation came about through decisive executive action and support from a coalition for reform, which identified the oligarchic control of the economy as the main reason for economic underdevelopment in the Philippines. A crucial factor in the success of liberalisation was the rise to the presidency of Fidel V. Ramos, who personally sustained efforts to open the economy and made such reforms a key aspect of his administration's agenda.

Ramos, the 12th president of the Philippines, proudly asserts that his government stabilised the economy and created avenues for growth by dismantling monopolies and cartels. In his inaugural address in June 1992, Ramos identified the breaking up of oligarchic cartels and monopolies that dominate the Philippine economy as one of his main tasks. He argued that the Philippine state had long been preyed upon by "oligarchies that used their privileged access to the bureaucracy to accumulate great fortunes and tremendous political power."[2] In an economy where people with political influence extracted and transferred wealth instead of creating it, the chief task was to "restructure the entire regime of regulation and control that rewards people who do not produce at the expense of those who do."[3] Ramos' aims were "levelling the playing field" by "dismantling oligarchic cartels and monopolies" and introducing competition to a closed and protected economy.[4] A celebrated example of this reform effort was the break-up of the PLDT telecommunications monopoly.

The role of Ramos, and in particular the exercise of political will that he demonstrated, was crucial to the economic liberalisation process. This was absent in the earlier administration of Corazon Aquino, where economic and political reforms were stalled as powerful vested interests influenced policy and decision making in Congress and the bureaucracy. Although the Aquino administration initiated economic liberalisation by reforming sectors controlled by cronies of former President Marcos, Aquino's efforts fell short due to her own family's lobbying.

Ramos, the first Filipino president from a professional military background, did not have the traditional family interests of most past presidents. He did not have a personal stake in monopolies and cartels nor was he beholden to the oligarchs for winning the election, and hence, was able to make decisions regardless of how they affected the country's economic and political elites.[5] The presence of political will, the lack of traditional family interests, and Ramos' being genuinely convinced of the need to dismantle "oligarchic control of the economy" were potent factors in the eventual implementation of liberalisation.

Fidel Ramos's victory in the 1992 presidential election marked the first peaceful transition in post-Marcos Philippines. While he was linked to the Marcos regime because he served as Chief of the Philippine Constabulary during Martial Law, Ramos' withdrawal of support from Marcos contributed to the peaceful "Edsa Revolution" in 1986. Under the Aquino Administration, Ramos served as Chief of Staff and Defence Secretary. He played a key role in quelling eight military coups of the Rebolusyonaryong Alyansang Makabayan (RAM or the Revolutionary Nationalist Alliance, formerly called Reform the Armed Forces Movement), thus assisting Aquino retain the presidency. Such pivotal role of Ramos was not lost on Aquino, who endorsed him for the presidency in the 1992 elections. Ramos ran under a relatively unknown political party for his presidential campaign, with Aquino's endorsement and the support of urban-based groups. He narrowly won the 1992 presidential election, garnering a plurality of 23.6 per cent of the votes in a seven-way contest. He defeated the more organised and well-funded campaigns of Ramon Mitra and Eduardo Cojuangco, as well as the popular Miriam Santiago.[6]

The victory of Ramos at the polls represented a departure from the traditional pattern of former presidents who usually came from or were supported by landed elites. Ramos' most organised support came from the urban academic, professional, and business communities, and from military officers known for their opposition to the old landed oligarchy and Marcos' crony capitalism.[7] Rigoberto Tiglao classified into three groups Ramos' support base, as reflected in his choice of top officials. The first group consisted of military men, led by Jose Almonte, believed to be the unofficial power centre at that time. This group included Customs Commissioner Guillermo Parayno, Press Undersecretary Honesto Isleta, and Government Service Insurance System chairman Jose Magno.

The second group was composed of businessmen and technocrats, several of whom were members of the Makati Business Club, which was at the forefront of the business opposition against the Marcos regime. This group included the Secretaries of Finance (Ramon del Rosario), Foreign Affairs (Roberto Romulo), Trade and Industry (Rizalino Navarro), Agrarian Reform (Ernesto Garilao), and Economic Planning (Cielito Habito). A third group consisted of professional politicians, represented by House Speaker Jose de Venecia and congressman Edelmiro Amante. De Venecia played an important role in getting Congress to cooperate with the administration, by building the "Rainbow Coalition" and recruiting members of the opposition to Ramos' party after the election, creating a national political base for the President. De Venecia also made sure that Ramos's legislative platform was given priority in the Lower House, although most bills got stuck at the Senate.[8] This base of support made it possible for Ramos to undertake his liberalisation agenda.

In contrast to Aquino, Ramos was able to establish political stability during his term by initiating peace talks with the Communist Party of the Philippines (CPP), the Moro National Liberation Front (MNLF), and rightist military group RAM.[9] Political stability provided a crucial context for the occurrence of reform.

At the regional and international level, competitive liberalisation pressure mostly from its Southeast Asian neighbours was increasingly felt with discussions on the establishment of an ASEAN Free Trade Area (AFTA). According to Ramos' national security adviser and closest confidant, Jose Almonte, "this is our last chance. If we do not make it, we will go down to the dogs." The business supporters of Ramos also shared this feeling of urgency and the need to catch up. Trade Secretary Navarro encapsulated this sentiment when he stated: "We're sick and tired [of the fact] that our neighbours see us as hopeless. We know how competitive we can be."[10]

Given the international climate and his support base, Ramos delivered reforms in areas where political will can make a difference. While Ramos exercised political will and commitment, his most trusted adviser and National Security Council (NSC) head from 1992 to 1998, retired Brigadier General Jose T. Almonte provided the strategy.[11] Almonte formulated the analysis that underpinned the Ramos reform agenda: the capture of the weak state by a powerful oligarchy was the root cause of the country's economic problems. Almonte believed economic power had to be dispersed and the state strengthened and freed from the influence of powerful families for the country to develop. This thesis was not new; academic writings since the 1960s had put forward similar arguments. However, it became a powerful idea when translated into policy with the backing and personal commitment of no less than the President himself.

Almonte assisted the President in strategising the phases of reform. To do this, he organised an informal group of like-minded Filipinos to flesh-out what would later become the Ramos administration's national agenda, the Philippines 2000 strategy. Within the Cabinet, Almonte worked in tandem with Antonio Carpio,[12] who served as Ramos' Chief Presidential Legal Adviser. Almonte and Carpio were described as the top two hands in the first years of the Ramos government: Almonte as the chief ideologue and Carpio as the chief operating officer.[13]

THE COALITION FOR REFORM

In the middle of 1992, Almonte, through Serafin Talisayon, a Professor at the University of the Philippines, organised a series of workshops with reform-minded members of government, business, the academe, and non-government organisations (NGOs). The purpose was to generate ideas, coordinate action, and facilitate advocacy for the government's reform agenda. The informal network met several times. The role of this network was never publicised, and its work was never formally recognised. Talisayon described the group members as having "passion for anonymity." For lack of a name, the group was called "People's 2000," echoing the "Philippines 2000" reform agenda that they helped craft.[14]

During the workshops, a point of overwhelming consensus was the need to deal with monopolies and cartels. Subgroups were established to focus on advocacy work in reforming various sectors of the economy. Topping the list of economic sectors identified for reform was the telecommunications industry, as well as banking, airline, shipping, insurance, and retail trade.

The "People's 2000" network tasked a subgroup to campaign for telecommunications reform by heightening people's awareness and mobilising popular support for liberalisation of the sector. Almonte tapped the members of this subgroup and put together a core group to strategise the phasing of telecommunications reform. Besides Almonte, Carpio, and Talisayon, this group included Antonio Abaya, an early supporter of the Ramos presidential bid, Pancho Villaraza, a senior partner at Carpio's law firm, and Anthony Abad, a young lawyer at the same firm.

The battle to liberalise the telecommunications industry and break PLDT's monopoly was fought on various fronts: popular, legal, at various branches of government, and within the PLDT boardroom.

At the grassroots level, the "People's 2000" subgroup became involved in various activities, which Anthony Abad described as "softening the ground" in preparation for a difficult reform that could create a huge political backlash from vested interests.[15] Despite the unpopularity of PLDT and the countless complaints against its perennially inefficient service, collective action demanding change was disorganised. Thus, the People's 2000 subgroup helped make visible a popular constituency for telecommunications reform. To do this, the group touched base with people and groups already involved in the issue and were calling for the dismantling of the PLDT monopoly.

First, the group met with consumer and business groups. In January 1993, a meeting at the AIT Hotel in Quezon City led to the formation of a broad coalition calling for the liberalisation of the telecommunications industry. The coalition was named Movement for Reliable and Efficient Phone System (MORE Phones). MORE Phones became the crucial public face of the clamour against PLDT's inefficient monopoly during the campaign to liberalise the industry.[16] From its inception in 1993 until the passage of the Telecommunications Act in 1995, the coalition organised public rallies, issued press statements, and attended public hearings in both Houses of Congress.

Anthony Abad of the People's 2000 subgroup convened the first meetings of MORE Phones. They also paid for coalition expenses such as meeting venues, snacks and lunches, and printing statements. Abad, however, avoided drawing attention to himself. Retired University of the Philippines Professor Helen Mendoza became the coalition's spokesperson. Mendoza launched a signature campaign against PLDT's monopoly in 1988 due to a bad personal experience. She also sued PLDT's management for alleged misuse of SIP funds.[17]

The fact that members of "People's 2000" convened the initial meetings was not lost on some NGOs who were suspicious of their actions. A few organisations worried about being used by politicians did not join MORE Phones. Nevertheless, support for dismantling the PLDT monopoly snowballed once the coalition started its campaign. The "People's 2000" group's direct involvement with MORE Phones ended when the government arrived at a settlement with PLDT in April 1993, which will be discussed in detail below. However, the coalition had gained a life of its own. Until mid-1990s, MORE Phones continued to pressure PLDT to be more transparent and accountable.[18]

Second, the People's 2000 group contacted Ricardo Manapat, the author of *Some are Smarter than Others*, which documented the illegitimate activities of the Marcoses and their cronies, and requested him to expound on what he wrote about PLDT. Manapat penned a follow-up report entitled *Wrong Number: the PLDT Telephone Monopoly*. The report discussed who really owned PLDT, described how the Cojuangcos controlled the company even though they owned only 1.6 per cent of the total stock, and provided a detailed list of questionable management practices. The People's 2000 group published and widely distributed the Manapat report to intensify public awareness on PLDT's excesses.

Third, the group also undertook some surreptitious activities to publicise PLDT's practices. On 28 January 1993, three of the major Philippine newspapers carried a report by the Philippine Center for Investigative Journalism (PCIJ) exposing corruption in the judiciary. The report claimed that PLDT's lawyer, and not a Supreme Court justice, wrote the decision on a case between PLDT and its competitor company, ETPI. This decision, discussed in Chapter 4, favoured PLDT in its dispute with ETPI by barring the latter from operating an international gateway facility, which would have competed with PLDT's most lucrative business segment.[19] The claim of faked authorship was based on the testimony of Professor David Yerkes of Columbia University in New York. Yerkes was an expert in identifying authorship of English language work. He executed an affidavit on behalf of ETPI, supporting its motion for reconsideration of the Court's decision, arguing that his "expert evidence establishes the possibility of bias, partiality, and the lack of independence of the judicial decision making process"[20] in the case at hand.

Yerkes analysed the writing style of the author of the court decision and compared it to earlier decisions written by Justice Hugo Gutierrez, the decision's supposed author, and the writings of Attorney Eliseo Alampay, PLDT's lawyer. Yerkes examined the "syntax, grammar and vocabulary, as well as tone or flavor and overall point of view ... the structure and form of the documents including their overall organization and layout, spelling, capitalization and punctuation."[21] He concluded that the PLDT vs. ETPI decision contained partisan and rhetorical language that was inconsistent with the dispassionate and formal writing style of Justice Gutierrez. The decision, he claimed, "looks, reads and sounds like the writings of PLDT's Counsel."[22]

The Yerkes testimony was one of the first concrete evidence of corruption in the judiciary. It demonstrated how influential companies like PLDT could obtain favourable decisions from a corruptible judiciary. Worse, it demonstrated that interested parties themselves could write court decisions. David Yerkes was supposedly hired by ETPI to assess the authorship of the Court decision. He executed the affidavit on 14 September 1992. Yet, the story of his testimony only became public knowledge in January 1993, perfectly coinciding with the build-up of the campaign against the PLDT monopoly. Both Justice Gutierrez and Attorney Alampay denied Yerkes' claims. However, four days after the release of the report, Justice Gutierrez resigned from the Supreme Court.

At the executive branch of government, the actions of the Ramos administration signalled that it wanted to break up PLDT's monopoly. First, in early January 1993, Magtanggol Gunigundo, the newly appointed PCGG Commissioner who was given the task of recovering the Marcoses' ill-gotten wealth, announced the revival of the government's claim of ownership of the sequestered PLDT shares.[23] The Aquino government set aside pursuit of these shares, believed to be Marcos-owned through various dummies. Gunigundo revived the contention that the government, not the Cojuangcos, should be in control of PLDT through the government's legal claim to the sequestered Marcos properties. On 26 January 1993, Ramos appointed Gunigundo and businessman Mario Jalandoni to replace the two non-performing Aquino appointees to the PLDT board. In addition, Ramos appointed Luis Sison, the author of the 1986 PCGG report that exposed the ownership structure of PLDT and its questionable management practices, as Vice Chairman of the ETPI board. This placed Sison at the helm of PLDT's number one competitor.[24]

Second, at the Department of Transportation and Communications (DOTC), the newly appointed Secretary Jesus Garcia reversed the controversial decision of his predecessor that favoured PLDT. On 15 February 1993, Garcia awarded to Digitel a 30-year lease/purchase contract to operate DOTC's telecommunications facilities in Regions 1–5.[25] As detailed in chapter 4, Digitel's bid of P40 billion was the clear winner, compared to PLDT's bid of only P7 billion, and was the unanimous choice of the Bidding Committee.[26] However, then DOTC Secretary Prado refused to award the contract, agreeing with PLDT's claims that the bid was too high and that Digitel did not have the proper franchise to offer telecommunications services.

Antonio Carpio, Ramos' Presidential Legal Counsel, provided the legal basis for Garcia's decision. He argued that Digitel, a joint venture between Cable and Wireless and local businessmen, did not require a legislative franchise for the contract because it would operate the telecommunications facilities as a lessee acting for the DOTC and not as a telephone company in itself.[27]

Third, recalling that PLDT delayed or refused interconnection with its competitors, President Ramos signed Executive Order (EO) 59 on 24 February 1993. EO 59 required interconnection among all authorised telecommunications companies, to enable subscribers of one operator to reach the subscribers of another. The EO empowered the National Telecommunications Commission (NTC), the industry's regulator, to set the terms of interconnection in case parties could not agree and to establish penalties for violations. The EO also enumerated severe penalties for refusal to interconnect. Before this, the only law governing any failure to comply with NTC regulations was Section 21 of Commonwealth Act 146 (Public Service Law), which provided a maximum penalty of P200 per day. The President also considered holding company executives criminally liable for maliciously denying interconnection with other companies. In February 1993, Congressman Renato Yap, a party-mate of President Ramos, filed House Bill No. 6939, proposing to penalise the officials of telecommunications companies that refused mandatory interconnection.[28] Meanwhile, in the Senate, Teofisto Guingona, filed a bill that sought to declare monopolies illegal and called for the establishment of an anti-trust commission. The bill had the support of nine other senators.[29] Similar anti-monopoly bills were filed in the House of Representatives. It was widely reported that PLDT was the main target of these bills.

At the NTC, Commissioner Simeon Kintanar announced that the NTC would award licenses to several cellular companies to meet the huge demand for telephone services. He disclosed that telecommunications applicants were willing to invest around P3–4 billion each for network expansion.[30]

Even before EO 59, Congress had already approved nine telecommunications franchises (see Table 8-1). These companies had applications pending at the NTC for provisional authority to operate international, cellular, or value-added services. Compared to Malaysia, where entry to the telecommunications industry was through a license that the Minister issued, entry into the Philippine market was a heavily regulated affair that involved a two-step process. A new company intending to operate a public utility such as telecommunications needed to first secure a congressional franchise. This meant that Congress controlled who could obtain a franchise.[31] To secure a legislative franchise, a bill had to be filed in both houses of Congress, undergoing three readings in each house. Public hearings are held during the second reading, where opponents raising objections normally delay the process.[32]

Secondly, the company had to apply for a Certificate of Public Convenience and Necessity (CPCN) at the NTC. Through the CPCN, the NTC assigned the type of service, its area of operation, the allowable rate for the service, and managed radio spectrum allocation. On top of the time involved in securing a franchise, acquiring a CPCN was also a lengthy process entailing public hearings before the NTC issued a decision. Delays usually took place when a competitor was bent on opposing a new entrant. Thus, this two-step process was time consuming, cumbersome, and open to abuse. PLDT used it to successfully bar entry of competitors as it constantly filed objections to new applications.

If its attempt to stop or delay its competitors in Congress and the NTC failed, there were the courts where PLDT normally sued a new entrant to block its operation. Until 1993, PLDT successfully obstructed new players by filing various legal challenges. If the lawsuits still failed, PLDT refused to interconnect. Thus, with EO 59 mandating interconnection, the Ramos administration hoped to break PLDT's monopoly.

Antonio Carpio wrote EO 59 after realising that President Ramos did not need to ask Congress for a law to mandate interconnection. Going to Congress was expected to take a long time as legislators allied with PLDT could oppose or stall the bill. Through an executive order, the President only needed to direct the NTC, which was already empowered by law to regulate the industry.[33] Carpio explained that the

Table 8-1: Philippine Telecommunications Franchises

Company	Republic Act	Date of Issuance	Comments	CPCN (Certificate of Public Convenience and Necessity)
PLDT	7082	3 August 1991	Extending the term of the original franchise granted under RA 3259	Basic network, CMTS, IGF
Globe Telecom (Globe McKay Cable and Radio Corporation)	7229	19 March 1992	Approving merger between Globe McKay Cable and Radio and Clavecilla Radio System, term of franchise granted under BP 95	IGF and CMTS
Smart (Smart Information Technologies Inc.)	7294	27 March 1992	*New company*	*IGF and CMTS*
Piltel (Pilipino Telephone Corporation)	7293	27 March 1992	Extending the original franchise granted under RA 6030	CMTS and IGF
Islacom (Isla Communications Company)	7372	10 April 1992	*New company*	CMTS and IGF

Bayantel (International Communications Corporation)	7633	20 July 1992	Extending the original franchise granted under RA 3259	IGF and CMTS in 2000
ETPI (Eastern Telecoms Philippines Incorporated)	5002 (and PD 484)	2 May 1967	Amending RA 808 (PD 484 authorizing transfer to ETPI of Eastern Extension Australasia franchise)	IGF
Philcom (Philippine Global Communications Company, Inc.)	4617	19 June 1965		IGF
Digitel (Digital Telecommunications Philippines)	7678	17 February 1994	*New company, landline and cellular*, offshoot of ETPI	*IGF and CMTS in 2000*
PT&T	6970			IGF
Bell Tel	7692	25 March 1995		IGF
Extelcom (Express Communications Company Inc.)	2090	22 June 1958	Cellular company	CMTS

Ramos administration was convinced that new companies were willing to invest once interconnection with PLDT was guaranteed. Thus, EO 59 was a central aspect of the plan to liberalise the industry.[34]

Nonetheless, Carpio knew that it was not sufficient to mandate interconnection. It was also necessary to make sure that PLDT would obey the rules. A crucial strategy to this end was for the government to exercise its right of ownership to the sequestered shares in PLDT. Carpio argued that as early as 1986 under the Aquino Administration, the government had enough votes on the PLDT board to name a majority. Yet Aquino did not act on this. Carpio and the PCGG team proposed to President Ramos to use the government's legal right in electing company directors to influence PLDT's behaviour. The appointment of Gunigundo and Jalandoni to the PLDT board was the first step in this direction.

The PCGG argued that, based on its records, the government was actually the single biggest stockholder of PLDT, through its control of PHI's 46 per cent stake in PTIC, the company that had majority control of PLDT's voting shares. Jose Campos surrendered to the government these PHI shares in PTIC as part of the wealth that he claimed he held in trust for former President Marcos. Aside from these, the government held an additional 10.6 per cent in PLDT via the Social Security System.[35] Carpio believed that asserting control over PLDT itself would be a central means of implementing liberalisation.

The reformers found an ally in Alfonso Yuchengco, a long time PLDT investor and board member. Yuchengco would have bought majority shares in the company in 1967 had Marcos not prevented him from doing so. In 1988, Yuchengco filed an appeal with the PCGG and the Sandiganbayan (graft court), claiming that he was the rightful owner of some of the PHI shares that the PCGG sequestered. The case was still pending in the Sandiganbayan in 1993. A team-up between the PCGG, which was pushing for a third seat on the 11-seat board, and Yuchengco, who controlled three board seats, would have meant that Cojuangco's five seats were outnumbered.[36] This way, the government, in tandem with Yuchengco, could have taken over PLDT by virtue of its majority of shares in the company. Yet, the Ramos government treaded carefully for economic and political reasons. Ramos, whose presidential bid was significantly buoyed by Cory Aquino's endorsement, was sensitive to the fact that Tonyboy Cojuangco was reputedly the former president's favourite nephew. In addition, PLDT was a key company to the economy. PLDT comprised almost 30 per cent of the Philippine Stock Exchange's capitalisation. Its shares were among the most actively traded, both locally and on the New York Stock Exchange. The company was thus a gauge of foreign investor confidence in the country.

Thus, the campaign to demolish PLDT's monopoly led by Almonte and Carpio pressed for liberalisation on various fronts using legal and clandestine means — from increasing public awareness, mobilising popular support, using the executive's powers, and introducing legislations.

PLDT STRIKES BACK

PLDT did not take these attacks lightly. It hired six publicists to work with the company's media machinery, and used the Cojuangco-owned newspaper, the *Manila Chronicle*, to counter the assault.[37] A massive blitz on television, radio, and print media was launched to create an image of PLDT keeping Filipinos "in touch" with their loved-ones anywhere in the world. Also, the usually inaccessible CEO, Antonio (Tonyboy) Cojuangco Jr., appeared on television and radio shows to promote a positive company image. To articulate his case, Cojuangco also tailed President Ramos on his international and domestic trips. The intensity of Cojuangco's lobbying efforts reportedly astonished Ramos himself.[38]

PLDT also went on the offensive. It attacked Almonte and Carpio, the identifiable leaders of the anti-monopoly campaign, depicting them as conspirators out to take over companies to put in their cronies. A columnist in the Cojuangco-controlled *Manila Chronicle* wrote, "without any effort or investment of ideas, the power grabbers of President Ramos want to take over big enterprises on the pretext of people empowerment."[39] Negative reports against Carpio's former law office also proliferated, with one columnist describing the firm as "gaining ill-repute as the country's biggest protection racket."[40] PLDT spread the idea that the anti-monopoly rhetoric was a veiled power-grab by Almonte and Carpio, reminiscent of the Marcos-style rhetoric in the 1970s that led to corporate takeovers. Big businessmen who were worried they might be the next targets sympathised with PLDT and expressed support for the Cojuangcos. Jaime Zobel de Ayala, then Chairman of Ayala Corporation, the country's biggest conglomerate, uttered such support and articulated the growing fears of the business community.[41]

Within the Cabinet, Cojuangco found sympathetic ears in Ramos' Defence Secretary, Renato De Villa, who told President Ramos of Cojuangco's complaint of being victimised.[42] A so-called White Paper was submitted to President Ramos alleging an Almonte-Carpio conspiracy to take over businesses.

With the personal attacks on Almonte and Carpio and big business' support for the Cojuangcos, the President's ultimate decision was uncertain. PLDT was the first monopoly to be dealt with, and was thus a crucial test of the administration's resolve to break monopolies and cartels. PLDT represented the archetype of a powerful, politically well-connected company with allies in almost all sectors of government. Some observers argued that if Ramos could break PLDT's monopoly, then he was really committed to his liberalisation agenda. However, if the government failed with PLDT, then the economic reform agenda would be in shambles. Carpio recalled that it was propitious that Almonte was there and that the President had full confidence in him. Almonte assured Ramos that reforming PLDT had to be done. PLDT was the tough test case for Ramos' resolve to deliver on his rhetoric about breaking cartels and monopolies. Almonte described the PLDT case in military parlance as the "breakthrough point to shatter the enemy's trenches so that the main force could drive through."[43] Then Press Secretary Jess Sison rationalised the government's actions in this way: "the Ramos administration's drive against monopolies and cartels — initially in telecommunications — is part of our effort to level the playing field of competition and dismantle the mercantilist regime left over from the colonial periods. The drive is not against bigness itself. It is against the inequity and resulting inefficiency inherent in crony capitalism."

A few days before a scheduled annual stockholders' meeting on 21 April 1993, the Ramos government and PLDT reached a compromise. PLDT agreed to three things: drop opposition to the entry of new players, comply with EO 59 and allow competitors to interconnect with PLDT's network, and grant voting rights to the shares held by PLDT subscribers through the Subscribers Investment Plan (SIP).[44] In exchange, Antonio Cojuangco Jr. would retain the presidency of the company, despite the fact that the government in alliance with Yuchengco controlled six out of the eleven board seats. The compromise effectively removed PLDT's opposition to the entry of new players but left Cojuangco at the helm of the company. According to a member of the People's 2000 core group, "Cory probably intervened to let her nephew stay."[45]

PLDT officials would neither confirm nor deny the existence of this compromise. Yet, PLDT records documented it as follows: "In April 1993, Antonio Cojuangco, President of PLDT and acting as representative of the heirs of the late Ramon Cojuangco, entered into an agreement with the PCGG to avoid the postponement of the Company's annual meeting. The agreement, among other things, authorised then President Fidel V. Ramos to nominate five of the eleven directors to the company's board that year."[46]

With this, liberalisation could finally take off. PLDT's management backed down and settled for a compromise. This outcome can be attributed to the President's exercise of political will and the success of the actions of the People's 2000 group. Those activities ranged from mass mobilisations, the exposé on the Court's decision, the decisions of the executive branch, and the support from the President's allies in Congress. A key factor in the hands of the government was the disputed ownership of the PTIC shares in PLDT. Almonte and Carpio knew that this was a significant pressure point that the Aquino administration failed to use. Ultimately, President Ramos used this pressure point to push through liberalisation. The Cojuangcos agreed to it, in exchange for their continued control of

PLDT. In the meantime, the question of ownership was pending in the Sandiganbayan.

EO 109 AND THE IMPLEMENTATION OF LIBERALISATION

Following the negotiated settlement, the core group of reformers led by Almonte bowed out of the telecommunications liberalisation process.[47] The NTC then took the lead in implementing liberalisation by drawing up a plan called the Service Area Scheme.

On 12 July 1993, President Ramos signed EO 109, entitled "Policy to Improve the Provision of Local Exchange Carrier Service." The executive order's main objective was to improve the provision of local telephones service in unserved and underserved areas. The policy laid down a strategy of expanding the national infrastructure based on the cross-subsidisation of non-profitable services by profitable international services.[48]

Two months later, on 12 September 1993, the NTC issued Memorandum Circular No. 11-9-93, which provided implementation guidelines for EO 109. All authorised International Gateway Facility (IGF) operators were required, within three years of the issuance of their authorisation, to install and maintain a minimum of 300,000 lines. Cellular mobile telephone system (CMTS) operators were required to install at least 400,000 telephones lines within five years. In addition, at least one rural exchange line had to be provided for every ten urban lines installed. The entire telecommunications network had to be interconnected in accordance with EO 59. Market opening was thus used to attain a social goal, with market entry to the industry incurring a service obligation cost to new entrants.

The Service Area Scheme (SAS) concretised EO 109. It emerged in early 1994 through consultations involving the NTC, DOTC, and industry players. According to then NTC Commissioner Simeon Kintanar, the SAS was a response to something that had to be done fast while also ensuring that investments were distributed nationally. When Kintanar and his staff examined the new companies' deployment plans, they found that all the companies wanted to go to Metro

Manila, Cebu, Davao, Baguio, and Subic, where telephone demand was high. None were keen to provide service to the rural areas. The SAS attempted to correct this by allowing a company to earn profits but at the same time, ensured that part of those profits would be channelled to serve less profitable areas.[49]

The SAS divided the country into 11 areas to be served by eight international gateway and cellular telephone operators. New entrants were allocated a profitable and an unprofitable area to ensure both operational viability and the provision of rural telephony. For instance, Smart Communications, which was granted a license to operate cellular and international gateway networks, was required by law to install 700,000 landlines. Smart was assigned the profitable areas of Pasay City, Las Pinas, Paranaque, Pateros, Taguig, and Muntinlupa in Metro Manila along with the unprofitable provinces of Abra, Ilocos Norte, Ilocos Sur, La Union, Pangasinan, Mountain Province, and Benguet.

In a seminar attended by CEOs and their foreign partners, PLDT, Smart, Globe, ETPI, PT&T, Bayantel, Islacom, Piltel, and Philcom agreed with the NTC on the division of service areas. Only a couple of deadlocks took place, which were arbitrated by the NTC in less than a week. All companies, along with their foreign partners, agreed to the SAS subdivision. All were very optimistic about the prospects of the industry, with an expressed backlog of a million lines, not to mention the unexpressed demand for telephone service. While PLDT participated in the planning process, it was not assigned a service area, though the company committed itself to launching an expansion program of its own.[50] Thus, eight companies were given licenses to operate regional fixed line networks, in addition to their international gateway or cellular networks.[51]

Yet, the SAS was not without its critics. Former NTC Commissioner and then Undersecretary of the DOTC, Josefina Lichauco, believed that too many players were given licenses to enter the market. She further argued that there were other ways of binding players to provide rural service.[53] Lichauco's criticism of Kintanar's scheme was depicted in the media as a personality and turf conflict between the NTC and the DOTC. By law, the DOTC was supposed to provide policy directions while the NTC was the industry regulator. In this case, the policy directive for the SAS came from the President via EO 109. A more plausible explanation of the bitter conflict related to how the SAS overtook the National Telecommunications Development Plan (NTDP) that the DOTC prepared under Lichauco's supervision.

Both Houses of Congress were also critical of the SAS and saw it as an invalid exercise of the principle of delegation of powers and a usurpation of Congress' legislative power. On 19 April 1994, Congressman Jose Zubiri, head of the House Committee on Legislative Franchises, denounced the SAS as a "grossly irregular, patently illegal and highly unconstitutional plan" that circumvented the intent of EO 109 to demonopolise and deregulate the telecommunications industry. Zubiri argued that instead of dividing the country into 11 service areas and assigning each to the telecommunications companies, the NTC and DOTC created smaller monopolies, which would not have competition in their service areas. He also argued that the SAS had the effect of amending the legislative franchises that Congress issued, which was something that only Congress can do as the sole franchise granting body.[53]

Zubiri called for a House public hearing to investigate the constitutionality of the SAS and invited position papers from the industry. All major telecommunications companies that participated in the scheme, including PLDT, supported the SAS and affirmed its legality. PLDT, in contrast to how it had challenged the NTC's authority in the past, argued that, "in compliance with the mandate to regulate and develop the industry, the NTC and DOTC enjoy plenary power to undertake the project and such other measure they may deem proper and essential to meet the telecommunications requirements of our people. As long as telecommunications companies concerned possess the requisite legislative franchise to operate telecommunications services in the assigned areas, we do not perceive any legal infirmity on the authority of the DOTC and NTC to allocate various services areas to qualified telecommunications operators."[54]

Only the Philippine Association of Private Telephone Companies Inc. (Paptelco), the organisation of over sixty provincial telephone companies, and Bell Telecommunications registered opposition to the SAS. PAPTELCO argued that the SAS was illegal because the NTC was not empowered by any law to formulate and implement any national policy on telecommunications. It contended that the authority to prescribe a national policy on telecommunications lay with Congress.[55] Not surprisingly, the members of PAPTELCO were threatened by the influx of bigger companies into their areas. Bell Telecommunications, a company owned by the Ortigas, Puyat, Maramba, and Madrigal families, was still waiting for the issuance of its franchise, and was keen to launch a national service. The nascent company criticised Kintanar's supposed awarding of service areas through personal letters and not circulars or memoranda, hinting that his actions were not binding or legal.[56]

Zubiri proposed to trim down the franchises of the companies that acceded to the plan to reflect the areas to which they were assigned and "legitimise" the SAS.[57] Yet, these companies firmly opposed Zubiri's proposal, stating that it would adversely affect the industry, because they were already in the process of obtaining letters of credit and signing commitment agreements with equipment suppliers. With all of the big companies in the industry unanimously supporting it, the scheme pushed through.

THE PASSAGE OF THE 1995 TELECOMMUNICATIONS ACT OF THE PHILIPPINES

From the NTC, the next stage shifted to Congress. Some legislators felt that the executive had pre-empted and bypassed Congress in liberalising the industry. A number of legislators endeavoured to put their imprint on market reform, in some ways reversing what the executive had done.

The legislation of a national telecommunications policy was long overdue. Until 1995, the colonial Commonwealth Act 146, otherwise known as the Public Services Act of 1935, governed the operations of telecommunications and other public utilities. The first efforts to introduce a new telecommunications law were made in the 1988 post-Marcos Congress. However, telecommunications development was not considered a priority in the face of issues such as the US military bases and the decentralisation of power to local government units.[58]

Before a bill can become law in the Philippines, it must gain approval from the House of Representatives and the Senate. The House and Senate versions of a bill are combined and conflicting provisions reconciled in a Bicameral Conference Committee composed of the members of the pertinent committees of both Houses. The Chairman of the House Committee on Transportation and Telecommunications, Jerome Paras sponsored the Lower House version of House Bill 14028, entitled "An Act to Promote and Govern the Development of Philippine Telecommunications and the Delivery of Public Telecommunications Services." The House version was passed without much fuss on 16 December 1994.

However, it was a different affair in the Senate. John Osmeña, Chairman of the Senate Committee on Public Services and an avowed supported of PLDT, sponsored the Senate version of the bill. In his sponsorship speech, Osmeña denounced Ramos' EOs 59 and 109, describing them as "usurpation of legislative authority."[59] He further argued for the urgency of the bill to set parameters and boundaries, otherwise "the NTC would go wild and they could do whatever they want,"[60] referring to the SAS.

During the second reading of the bill at the Senate, opposition senators lambasted Ramos and the SAS. Senator Ernesto Maceda described the NTC and the DOTC as "flagship centers of corruption" and alleged that President Ramos used liberalisation as a means of raising funds for his political party. Maceda claimed that his informed source was an undersecretary of the DOTC who revealed that the bagman who collected money from all applicants to the SAS was a nephew of the NTC Commissioner.[61] Maceda insinuated that Senator Osmeña, who was from Cebu, should be happy that part of the alleged kickbacks were flowing into his province, in reference to NTC Commissioner Kintanar and DOTC Secretary Garcia who also came from Cebu. Kintanar and Garcia openly acknowledged that their appointments were through the recommendation of John Osmeña's brother Lito, who ran as Ramos' vice-presidential partner. Lito Osmeña, who was not in good terms with John, lost the election but continued to serve as the Treasurer of Ramos' political party, Lakas-NUCD.

While Maceda criticised the SAS because it was allegedly hatched to raise money for Ramos, John Osmeña argued that it would have been more preferable for the government to follow the case of England or Germany, where only one or two entities were allowed entry. He declared that the SAS' creation of 11 dwarfs to fight one giant defeated the goal of "destroying the so-called PLDT monopoly." Osmeña described the new entrants as "midgets who are biting more than they can chew."[67]

Osmeña and Maceda wanted to repeal EO 109 but their technical staff advised that this would be in breach of vested rights of companies that had already invested money in the plan. Osmeña thus proposed to weed out the weak companies that "were only given those areas because they were strong with certain parties" by inserting a provision into the law reducing the period to accomplish the rollout of landlines from five to three years.

On 20 February 1995, the Bicameral Conference Committee met to reconcile the provisions of Senate Bill 11 and House Bill 14028. Only two of the 16 members were present in the meeting — Senator Osmeña and Congressman Paras. In a seven-minute meeting, Osmeña and Paras agreed on the provisions of the bill, with Osmeña reiterating that "the intention of the law is that there should be minimum discretion on the part of the NTC."[63]

Curiously, only Senator Osmeña's personal staff was present in the meeting and the entire House Committee staff was absent. In contrast to the Senate committee staff, which was basically composed of the personal staff of the Senator who heads the committee, House Committee staff comprised of permanent employees who remain in the same committee even if the Chairperson changes. Thus, the House structure leads to better documentation and institutional memory. With this particular bill, the head of the House Committee Secretariat for Transportation and Communication recalled that the House staff was not informed of the date of the Bicameral Conference Committee meeting. Instead, it was just given copies of the final draft of the law, which was unusual and contrary to practice.[64] It seemed, however, that this was not a simple oversight. Senator Osmeña stated on record that he was responsible for two important points in the telecommunications law that were inserted at the Bicameral Conference Committee level: first, changing the timeframe for SAS compliance from five to three years, which amended EO 109, and second, removing the role of the NTC in interconnection agreements, which amended EO 59. Osmeña explained that the two insertions were a result of rivalry between himself and his brother, who controlled the DOTC and the NTC. The shortened period for rollout obligations was intended to pressure companies who were raising capital to comply with their commitments, thus spoiling his brother's money-making plans. With regard to the law being silent on the role of the NTC in interconnection agreements, Osmeña admitted that PLDT lawyers suggested this idea to him. PLDT wanted interconnection to be negotiated between parties as opposed to being mandated by the NTC. Without a clear role for the NTC, PLDT could again choose to delay interconnection with its competitors.[65]

The Telecommunications Act of the Philippines (Republic Act 7925) became law on 1 March 1995. It institutionalised liberalisation and competition, emphasised the role of private enterprises in the provision of telecommunication services, affirmed the policy of cross-subsidisation, and provided for the privatisation of all existing government communications facilities. The NTC was identified as the principal administrator of the law and gave the DOTC the responsibility for long-term strategic national development planning. RA 7925 also altered the basis of rates of return by abolishing the 12 per cent ceiling on profits that CA 146 imposed. Instead, the law provided that the NTC would establish rates that sustain the economic viability of telecommunications entities and fair returns on investments.[66] RA 7925 also required telecommunications companies to publicly list and sell at least 30 per cent of their common stock within five years of operations.

RA 7925 affirmed the power of Congress as the sole franchise-giving body. As a reaction to the SAS, the law reduced the timeframe for rollout compliance and provided for the cancellation of a company's authority to operate if it failed to comply with the obligation within three years.[67]

Yet, RA 7925 was quiet on two important matters. First, while it mentioned the need to set "a fair and reasonable interconnection of facilities of authorized public network operators,"[68] the law did not specify the NTC's role in fostering interconnection. This deliberate omission was significant, given that there were already bills filed in both Houses of Congress calling for a clearer NTC role on interconnection. Thus, instead of bolstering the provisions of EO 59, which mandated interconnection, RA 7925 removed the regulator in the interconnection process.

A second issue that the authors of RA 7925 deliberately avoided was the strengthening of the NTC as a regulatory body. A bill had already been filed as early as 1987 calling for the reorganisation and professionalisation of the NTC, to make it more independent and autonomous from political pressures. One proposal was to institute a fixed term of office for the Commissioner and the two deputies, instead of them serving at the pleasure of the President. However, the authors of RA 7925 purposely wanted to set parameters and boundaries that would limit the power of the NTC. Otherwise, as Senator John Osmeña stated, "the NTC would go wild and they could do whatever they want."[69]

Clearly, RA 7925, while it etched into law the principles of competition and liberalisation, showed how Congress could be obstructionist. Far from encouraging reform, the authors of RA 7925 sought to reverse the process that the executive had laboured to create. The new telecommunications law was quiet about things that it could have been categorical about because its authors had private considerations in mind. Despite the existence of bills on interconnection and strengthening the NTC, which could have been incorporated into the law, the lobbying of PLDT coincided with a senator's self-interest and won the day. The ensuing insertions in the law are evidence of the anomaly of the bicameral conference committee structure, where after a law has gone through three readings and a lengthy process of public hearings to accommodate various points of view, the final law can be entirely different, with watered down provisions or insertions that negate the legislation's original intent.[70] This is clearly demonstrated by RA 7925, where the authors succeeded in setting back reforms that the executive had initiated.

SUMMARY

Table 8-2 summarizes the main actors and their positions during the reform of the telecommunication sectors in the Philippines.

The liberalisation of the telecommunications industry in the Philippines was initiated by a coalition for reform, which strategised the stages that led to market opening. PLDT did not welcome the attack on its cosy position and counteracted with its own publicity campaign. However, when it became clear that President Ramos was committed to breaking its monopolistic control, the Cojuangcos accepted a negotiated settlement. Once this compromise was reached, the specifics of liberalisation were left to the NTC and the DOTC. Legislators who felt that the executive had appropriated legislative authority tried to negate what the executive had done. Their actions benefited PLDT, which even though it agreed to allow new entrants into the market, nonetheless lobbied to influence legislators to shape laws in its favour.

Liberalisation in the Philippines was made possible by the President's commitment and the work of an ad-hoc group of reformers. These reformers, as will be shown in the next section, were not the direct beneficiaries of market entry, in contrast to the Malaysian case. In the Philippines, the beneficiaries of market entry were mostly big businessmen. Once policy credibility was established and the state demonstrated its determination to open the sector, these businessmen enthusiastically entered the telecommunications market. This had the effect of diffusing market entry rents, in contrast to the handpicked new players in Malaysia.

Unlike in Malaysia, many actors were involved in the reform efforts in the Philippines. In the Malaysian case, where power was concentrated in the hands of the Prime Minister, the facilitation of policy change and the introduction of reforms were relatively easier. In the Philippines, where power is diffused in three different branches of government and where a powerful economic elite influences segments of the state bureaucracy, policy reform is more complicated, needing more pressure from outside government as well as backing of key

Table 8-2: Philippine Actors and Their Positions During Telecommunications Reform

Actors	Role and Position in the Issue
1. President Fidel Ramos	Committed to introducing liberalisation and improving economy by "breaking down monopolies and cartels controlled by the oligarchy"; PLDT as test case
2. Jose Almonte, Antonio Carpio, Peoples' 2000 and the "Coalition for Reform"	Almonte strategised phase of reform, Carpio ironed-out legal issues and coalition composed of reform-minded individuals who agreed with the reform track and supported the Ramos agenda for change
3. MORE Phones	Composed of individuals and groups pushing for liberalisation of the telecommunications industry and became the public face of support for liberalisation
3. DOTC	Initiated policy of liberalisation as early as 1987 but had no power to enforce it; drafted a 20-year telecommunications development plan which was superseded by the SAS
4. NTC	Under Aquino administration, initiated liberalisation by licensing existing companies to offer international gateway and cellular services but its decisions and mandate to interconnect opposed by PLDT; under Ramos, oversaw liberalisation through the SAS
5. Supreme Court (Judiciary)	Ruled inconsistently on the issue of liberalisation; one of its decision favouring PLDT was exposed to have been written by a PLDT lawyer
5. Congress (lower and upper house)	Legislated telecommunications liberalisation as a knee-jerk reaction to executive-led liberalisation, in some ways reversing the actions of the President
6. PLDT and the Cojuangcos	Opposed liberalisation using its political connections at various fronts; eventually agreed to liberalisation after the Ramos government used issue of company ownership as a bargaining chip; nonetheless, made interconnection difficult for new entrants
7. Big businessmen and prospective entrants to the industry	Lukewarm support for liberalisation and were wary of the actions of the coalition for reform. Some even supported the Cojuangcos but enthusiastically entered the industry once it was opened
8. WB, ADB, and other aid donors	Source of loan for PLDT and despite PLDT's inefficiencies, did not call for the opening of the industry
9. Lee Kuan Yew	Comment on poor state of telecommunications reportedly irked President Ramos, adding to his resolve to break down PLDT's monopoly; comment a form of competitive liberalisation pressure

officials. The more diffused character of state power in the Philippines created bottlenecks and sometimes led to one branch of government negating the actions of the other, as demonstrated by the insertions into the telecommunications law at the Bicameral Conference Committee level. These types of deadlocks were absent in Malaysia, where the Barisan Nasional controlled the Parliament.

The next chapter considers the impact of liberalisation on the structure of the telecommunications market, examining the new entrants, the direct beneficiaries of liberalisation, and how the policy changes discussed here affected the economy and society in general.

NOTES

1. *Manila Times*, 3 August 2006.
2. Fidel V. Ramos, *Developing as a Democracy: Reform and Recovery in the Philippines 1992–1998* (Hong Kong: Macmillan Publishers, 1998), p. xii.
3. Fidel Ramos, "To Win the Future", inaugural address as President of the Philippines, Manila, 30 June 1992, in Ramos (1998), p. 4. Various studies since the 1960s have documented how economic power is concentrated in the hands of a few families in the Philippines. See, for instance, Dante Simbulan, *A Study of the Socio-Economic Elite in Philippine Politics and Government, 1946–1963*, Ph.D. dissertation, Australian National University (1965); Temario Rivera, *Landlords and Capitalists: Class, Family and State in Philippine Manufacturing* (Quezon City: University of the Philippines Center for Integrative Studies and the University of the Philippines Press, 1994); John Doherty, *A Preliminary Study of Interlocking Directorates Among Financial, Commercial, Manufacturing and Service Enterprises in the Philippines* (Manila, 1979); Yoshihara Kunio, *The Rise of Ersatz Capitalism in Southeast Asia* (Singapore: Oxford University Press, 1988).
4. Ramos (1998), p. 31.
5. Interview with Antonio Carpio, Makati City, 3 May 2001.
6. See Comelec data on the 1992 Election Results, cited in Shiela Coronel, "Who Wants to be a President?" *imag* (October–December 2003), http://www.pcij.org/imag/PublicEye/presidency.html (accessed 20 January 2004. See also Julius Caesar Parreñas, "Transition and Continuity in the Philippines, 1992", in *Southeast Asian Affairs 1992* (Singapore: Institute for Southeast Asian Studies, 1992), pp. 271–74; *Time*, 15 May 1995, p. 14.

7. Michael Pinches, "The Philippines' New Rich: Capitalist Transformation amidst Economic Gloom", in *The New Rich in Asia: Mobile Phones, McDonalds and Middle-class Revolution*, edited by Richard Robison and David S.G. Goodman (London and New York: Routledge, 1996), p. 116.

8. *Far Eastern Economic Review*, 3 September 1992, p. 38. See also Parreñas (1992), p. 279.

9. Parreñas (1992), pp. 274–77.

10. Both the Almonte and Navarro quotes were taken from *Far Eastern Economic Review*, 3 September 1992, p. 38.

11. Jose Almonte developed a close professional and personal relationship with Ramos during the years of their military service, starting from the 1960s. Under Aquino, Almonte was head of the Economic Intelligence and Investigation Bureau.

12. Antonio Carpio was a senior partner in the former Carpio, Cruz, and Villaraza law office and one of the first supporters of Ramos in his presidential bid. Carpio headed Ramos' legal team during his candidacy. His partner, Avelino Cruz Jr. represented Ramos in the election protest challenge filed by losing presidential candidate Miriam Santiago in 1992. See *Manila Standard*, 15 July 1993, p. 9. Carpio and Cruz were major supporters of Gloria Macapagal-Arroyo in the move to depose Estrada from the presidency in 2001. Arroyo appointed Carpio to the Supreme Court and Cruz as her Presidential Legal Counsel, and eventually Secretary of National Defense.

13. *Far Eastern Economic Review*, 12 August 1993, p. 15.

14. The Philippines 2000 strategy came out of a series of meetings and consultations led by the NSC under Almonte. E-mail correspondence with Professor Serafin Talisayon, 9 May 2001.

15. Interview with Anthony Abad, Quezon City, 16 January 2001.

16. The MORE Phones coalition included the following groups: the Church-based Consumer Movement, the Institute for Popular Democracy, the Popular Education for People Empowerment, the Education for Life Foundation, the Women's Action Network for Development, the Movement for Popular Democracy, the Institute on Church and Social Issues, the Cooperative Foundation Philippines, Inc., Pandayan Para sa Sosyalistang Pilipinas, Women's Action, the Coalition of Youth Organizations, the Asian Institute of Management, the Freedom from Debt Coalition, the Lean Alejandro Foundation, and the National Economic Protectionism Association. See *Philippine Graphics*, 8 July 1994, p. 34.

17. As was discussed in Chapter 4, the Subscribers Investment Plan (SIP) required all PLDT subscribers to buy company shares before a subscriber can get a telephone connection. This case languished in court for over a decade. When I interviewed Dr Mendoza in May 2001, she had just received a court summons stating that the case would again be heard in court on June 2001. She said that "my purpose in filing the case was to shake them so that they would perform better in those times, it was not to get money. They were willing to give money if you would just shut your mouth. ..." Interview with Helen Mendoza, Quezon City, 11 May 2001.

18. Ibid.

19. See General Register (G.R.) no. 94374, PLDT vs. ETPI and NTC 27 August 1992.

20. Affidavit of Dr David Miles Yerkes, Expert in identifying authorship of an English Language Work, 14 September 1992, p. 1. Hereafter cited as Yerkes Affidavit.

21. Yerkes Affidavit, p. 3.

22. Ibid.

23. Almonte reportedly recommended Magtanggol Gunigundo for the position of PCGG Commissioner. See Shiela Coronel, "Monopoly", in *Pork and Other Perks: Corruption and Governance in the Philippines*, edited by Shiela Coronel (Manila: Philippine Center for Investigative Journalism, 1998), p. 137.

24. The Ramos administration had the power to do so because in 1986 the government sequestered 60 per cent of ETPI's shares due to their alleged ownership by Marcos cronies. See the discussion on ETPI in Chapter 9.

25. *Manila Bulletin*, 28 April 2000.

26. *Malaya*, 16 February 1993, p. 11.

27. Ibid.

28. *Philippine Daily Inquirer*, 2 February 1993.

29. *Manila Bulletin*, 26 October 1992, p. B-3.

30. *Philippine Daily Inquirer*, 17 February 1993.

31. Article 12, Section 11 of the 1987 Philippine Constitution.

32. PLDT's ally in the Senate, Senator John Osmeña, who was the Chairman of the Senate Committee on Franchises, sat on Digitel's franchise while issuing

60 other franchises. This delayed Digitel's operation for over five years. See *Business World*, 3–5 February 1993.

33. Ramos' EO 59 drew from two laws: EO 546, dated 23 July 1979, which created the NTC as the telecommunications regulatory body, and RA 6849, or the Municipal Telephone Act of 1989, which provided for interconnection among all domestic telecommunications carriers and utilities.

34. Interview with Antonio Carpio, Makati City, 3 May 2001.

35. *Far Eastern Economic Review*, 25 March 1993.

36. *Business World*, 2 March 1993.

37. *Far Eastern Economic Review*, 6 May 1993, p. 48. See also Coronel (1998), p. 138.

38. *Manila Times*, March 1993.

39. *Far Eastern Economic Review*, 6 May 1993, p. 48.

40. *Manila Times*, 19 July 1993.

41. *Far Eastern Economic Review*, 6 May 1993, p. 45. Carpio confirmed that the Ayalas opposed their actions and were very hostile. He pointed out the irony that once the industry was opened the Ayalas entered the sector and would later on own the company that would became PLDT's number one competitor, Globe Telecom. Thus, the Ayalas were one of the main beneficiaries of the opening of the industry. Interview with Antonio Carpio, Makati City, 3 May 2001.

42. De Villa was part of the military group that supported Ramos but he was at odds with Almonte.

43. Interview with Jose Almonte, Paranaque City, 4 April 2001. See also *Far Eastern Economic Review*, 17 June 1993, p. 4.

44. Interview with Tony Carpio, Makati City, 3 May 2001 and *Far Eastern Economic Review*, 6 May 1993, pp. 48–49. In 1993, SIP subscribers held 62 per cent of the total common preferred shares of the company.

45. Interview with a People's 2000 subgroup member, 16 January 2001, Quezon City.

46. *PLDT's Management's Discussion and Analysis of Financial Condition and Results of Operations for the Nine Months Ending 30 September 1998*, p. 6.

47. Almonte and Carpio explained that they were only involved in removing bottlenecks in the economy and not in the administrative implementation.

They moved on to help introduce competition to other economic sectors such as banking, insurance, shipping, and airline, with varying levels of success. Interview with Jose Almonte, Paranaque City, 4 April 2001; interview with Tony Carpio, Makati City, 3 May 2001.

48. Section 4 of EO 109 stated that "until universal access to basic telecommunications services is achieved, and such service is priced to reflect actual costs, local exchange service shall continue to be cross-subsidized by other telecommunications services within the same company".

49. Interview with Simeon Kintanar, Quezon City, 20 March 2001.

50. PLDT launched the Zero Backlog program, and cunningly prioritized the urban areas in which new operators were assigned to rollout fixed lines.

51. Of the 11 new market entrants, three did not participate in the SAS. Digitel and Bell Telecommunications were still securing their legislative franchises at the time, and Extelcom was having internal management troubles. Extelcom eventually agreed to roll out the 400,000 lines required in 1997, but it was not assigned a service area. Digitel was given authorization to put in 300,000 lines in Luzon, but Bell Telecommunications, which insisted on a national area, failed to get its service off the ground.

52. Interview with Josefina Lichauco, Makati City, 19 March 2001.

53. Privilege Speech of Rep. Jose Zubiri. Minutes of the House of Representative, 19 April 1994. Document found at the House of Representative Library, Quezon City.

54. PLDT's letter to Hon. Jose Zubiri on the "Issue regarding the constitutionality and legality of the DOTC/NTC allocation of services areas to 10 major telecommunications companies", 17 May 1994.

55. PAPTELCO's letter to Hon. Jose Zubiri on "Subdivision of the Philippines into 11 telecom service areas by the NTC as a National Policy", 9 June 1994.

56. Bell Telecommunications, "Position Paper Re: the NTC policy dividing the country into 11 service areas pursuant to EOs 59 and 109", no date. Bell Telecommunications eventually sued the NTC for the SAS, insisting on its plan to launch a national network. Four telecommunications companies petitioned the Supreme Court to oppose Bell's position. See *Business World*, 6 and 29 May 1997.

57. *Business World*, 3 June 1994.

58. Interpellations on Senate Bill no. 11, Development of Philippine Telecommunications, Records of the Senate, 17 January 1995.

59. Osmeña's Sponsorship Speech of Senate Bill 11, 20 April 1994.

60. Records of the Senate, 18 January 1995.

61. Records of the Senate, 17 January 1995. It was never established whether there was any credibility to Maceda's claims.

62. Ibid.

63. Bicameral Conference Committee on the Disagreeing Provisions of House Bill no. 14028 and Senate Bill no. 11, 20 February 1995. Document found at the Philippine Senate Library.

64. Interview with Dina Polo, Quezon City, 3 April 2001.

65. Coronel (1998), pp. 142–43.

66. Republic Act 7925, Section 17.

67. RA 7925, Sections 10 and 12.

68. RA 7925, Section 5c.

69. Records of the Senate, 18 January 1995.

70. I am grateful to Professor Emmanuel De Dios of the University of the Philippines School of Economics for this insight.

9

THE NEW PLAYERS AND THE SERVICE AREA SCHEME

From an industry dominated by an influential virtual monopoly, Philippine telecommunications after liberalisation was a flurry of activity, construction, and competition to provide for a market that has been starved of telephones for decades. This chapter discusses the outcomes of liberalisation by identifying the beneficiaries of market entry. The chapter then examines the impact of the 1997 financial crisis before considering the overall effects of liberalisation on the sector and in the economy. A final section summarises the chapter's arguments.

THE NEW PLAYERS

Table 9-1 shows the coverage of the SAS and the companies that were assigned to them. The SAS was the NTC's implementation of EO 109, President Ramos' edict to liberalise and expand telephone provision nationwide. EO 109 required duly enfranchised international gateway and cellular telephony companies to rollout 300,000 or 400,000 fixed lines, respectively. An additional condition was that one telephone line be installed in the rural areas for every ten lines in the urban areas.

Table 9-1: Service Area Scheme Assignments

Subdivision No.	Region	Coverage	Assigned Carrier
1	Region I	Abra, Ilocos Norte, Ilocos Sur, La Union,Pangasinan, Mt. Province, Benguet	Smart
	NCR D	Pasay City, Las Pinas, Paranaque, Pateros, Taguig, Muntinlupa	
2	Region II	Batanes, Cagayan Valley, Isabela, Quirino, Nueva Vizcaya, Ifugao, Kalinga-Apayao	ETPI/Teletech
	NCR A	Manila, Navotas, Caloocan City	
3	Region III	Tarlac, Pampanga, Zambales, Bataan, Bulacan, Nueva Ecija	Smart
4	Region IV-A	Aurora, Laguna, Quezon, Marinduque, Rizal, Romblon	PT&T/Capwire
5	Region IV-B	Cavite, Batangas, Occ. Mindoro, Or. Mindoro, Palawan	Globe Telecom
6	Region V	Albay, Camarines Norte, Camarines Sur, Catanduanes, Masbate, Sorsogon	ICC/Bayantel
	NCR B	Quezon City, Valenzuela, Malabon	
7	Region VI	Aklan, Antique, Capiz, Iloilo, Negros Occidental, Guimaras	Islacom
	Region VII-A	Negros Oriental, Siquijor	
8	Region VII-B	Bohol, Cebu	Islacom
	Region VIII	Eastern Samar, Leyte, Northern Samar, Southern Leyte, Samar, Biliran	

Table 9-1: Service Area Scheme Assignments (*continued*)

Subdivision No.	Region	Coverage	Assigned Carrier
9	Region IX-A	Zamboanga del Norte, Zamboanga del Sur	Piltel
	Region X	Agusan del Norte, Agusan del Sur, Bukidnon, Camiguin,	Philcom
		Misamis Occidental	Piltel
		Misamis Oriental, Surigao del Norte,	Philcom
	Region XI-B	Surigao del Sur, Davao Oriental	
10	Region XI-A	Davao del Norte, Davao del Sur, South Cotabato, Sarangani	Philcom Piltel
	Region IX-B	Basilan, Sulu, Tawi-Tawi	Philcom/Piltel Piltel
11	Region XII	Lanao del Norte, Lanao del Sur, Maguindanao, North Cotabato, Sultan Kudarat	Globe Telecom
	NCR C	Makati, San Juan, Mandaluyong, Marikina, Pasig	

Source: National Telecommunications Commission, 1998.

Nine companies joined the implementation of the SAS. Globe Telecommunications, Isla Communications, and Smart Communications proposed to offer international gateway and cellular services, and were required to rollout 700,000 lines each. Piltel, Bayantel, ETPI, Philcom, and PT&T planned to offer international gateway services only, and were required to rollout 300,000 lines each. Digitel finally received its legislative franchise in 1994 and was required to rollout 300,000 lines in Luzon. PLDT was not assigned a service area but was required to rollout at least 300,000 additional lines. Thus, the SAS was an ambitious plan that aimed to install at least 4.2 million telephone lines by 1999. The following section discusses the new corporate players, their main shareholders and foreign partners, and their areas of service. Table 9-2 summarises this information.

Table 9-2: Philippine Telecommunications Company Ownership, 1998

Telco	Filipino Owners	Franchise and Issue Date	Type of Services Offered	Foreign Partner	Foreign Equity
PLDT	Cojuangcos	Republic Act 7082, 3 August 1991	IGF, Cellular, Local exchange, interexchange services	None	Various funds
Globe Telecom	Ayalas (36.1%)	RA 7229, 19 March 1992	IGF, CMTS, local exchange services	Singapore Telecoms	35.60%
Smart	Vea and Fernando	RA 7294, 27 March 1992	IGF, CMTS, and local exchange services	First Pacific NTT	40.00% 12.00%
Piltel	PLDT (57%)	RA 7293, 27 March 1992	IGF, CMTS, and local exchange services	None	
Islacom	Delgados	RA 7372, 10 April 1992	IGF, CMTS, and local exchange services	Shinawatra Deutsche Telecoms	30.00% 10.00%
ICC/Bayantel	Lopezes	RA 7633, 20 July 1992	IGF and local exchange services	Nynex (US), Telekom Asia (Thai), Chase Manhattan AIF	15.00% 10.00% 0.93% 6.00%

Table 9-2: Philippine Telecommunications Company Ownership, 1998 (*continued*)

Telco	Filipino Owners	Franchise and Issue Date	Type of Services Offered	Foreign Partner	Foreign Equity
Digitel	Gokongweis	RA 7678, 17 February 1994	IGF, local exchange, and CMTS	Telia AB (Sweden) Jasmine Int'l of Thailand	11.80% 3.00%
Philcom	Willy Ocier	RA 4617, 19 June 1965	IGF and local exchange services	Comsat	21.00%
PT&T	Santiagos	RA 6970, 15 November 1990	IGF and local exchange services	Korea Telekom	20.00%
ETPI	Benedicto, Africa, Nieto	RA 5002, 17 June 1967	IGF and local exchange services	C&W	40.00%
Extelcom	Bayantel (46.6%) Mayon Holdings (13.4%)	RA 2090, 22 June 1958	CMTS	Millicom Int'l Cellular	40.00%

Source: The table was constructed from Republic Acts and various company annual reports and statements.

Globe Telecom

Globe Telecommunications was a joint venture between the Ayala Corporation and Singapore Telecom International Pte. Ltd. (SingTel), a wholly owned subsidiary of Singapore Telecommunications, Singapore's leading full-service telecommunications provider. Singtel's entry into the Philippine telecommunications market was significant not only because it was the first foreign company to be granted entry into the liberalised market, but also in the light of its former Prime Minister's comment on the state of Philippine telecommunications. In a speech to a group of businessmen during his visit to the Philippines in December 1992, Lee Kuan Yew joked, "99 per cent of Filipinos are waiting for a telephone and the other 1 per cent for a dial tone."[1] The comment reportedly irked President Ramos and added to his resolve to break the PLDT monopoly. The symbolic meaning of the joint venture was not lost on President Ramos, who was the proud guest at the signing of the Memorandum of Understanding between the Ayala Corporation and Singtel International on 13 February 1993.[2]

Under the SAS, Globe Telecom, which applied to operate international gateway and cellular services, was required to rollout 700,000 landlines. It was assigned to install and operate telephone lines in the Region IV provinces of Cavite, Batangas, Occidental Mindoro, Oriental Mindoro, and Palawan, and in the Region XII provinces of Lanao del Norte, Lanao del Sur, Maguindanao, North Cotabato, and Sultan Kudarat. These provinces were paired with cities in the National Capital Region such as Makati, San Juan, Mandaluyong, Marikina, and Pasig.

The Ayala family controlled 58 per cent of the Ayala Corporation. They are one of the oldest and richest business families in the country, consistently listed in Forbes magazine's list of world billionaires.[3] The other substantial shareholders of the Ayala Corporation were Mitsubishi Corporation of Japan (with 19.1 per cent of the shares), Shoemart Incorporated (with 4.4 per cent), and the public (with 18.5 per cent).[4]

By December 2000, Globe installed 790,291 fixed lines, surpassing its 700,000-line commitment under the SAS. However, the company was better known for its mobile business. In September 1994, Globe launched its digital cellular telephone services, being the first to use the Global System for Mobile Communications (GSM) technology. It introduced and popularised the short messaging service (SMS) by bundling it free of charge with its basic service.[5] In 2002, the company had 6.6 million mobile subscribers, 92 per cent of whom were prepaid subscribers, while 8 per cent were post-paid subscribers. An additional 130,000 were landline subscribers.[6]

As of 2005, Globe was second to Smart in terms of cellular market share with over 12.5 million subscribers. It emerged as one of the most profitable Philippine companies, with a net operating revenue of US$1.1 billion in 2005.

Smart Communications

Smart Communications was the first of two new companies that entered the liberalised telecommunications market. It was authorised to offer both international gateway and cellular telephone services and was required to lay 700,000 landlines in the Region I provinces of Abra, Ilocos Norte, Ilocos Sur, La Union, Pangasinan, Mt. Province, and Benguet, the Region III provinces of Tarlac, Pampanga, Zambales, Bataan, Bulacan, and Nueva Ecija, and in Pasay City, Las Pinas, Paranaque, Pateros, Taguig, and Muntinlupa in Metro Manila.

Smart was a company that evolved from the ideas of David Fernando, a telecommunications engineer, and Orlando Vea, a banker-entrepreneur. Fernando and Vea established the company in January 1991 and secured a 25-year authority to offer telecommunications services on 27 March 1992 via Republic Act 7294. Fernando and Vea advocated "for the release of the economy from family control" during government-industry consultations. They argued that the problem with PLDT was not its size or its monopoly position, but its inability to deliver quality services.[7]

Because it was a new company without any connection to established business families, Smart did not encounter opposition from PLDT in its franchise and license applications. The bigger problem was securing a strategic partner. Six foreign companies turned them down before First Pacific, through its Philippine affiliate Metro Pacific, agreed to invest in Smart. The Salim family of Indonesia, noted associates of former President Suharto, owns and controls the majority stake in First Pacific.[8] In 1995, Nippon Telephone and Telegraph Company (NTT), Japan's largest and the world's second largest telecommunications company, acquired a 15 per cent interest in Smart. NTT increased its equity interest in Smart from 15 to 37.2 per cent in 1999.

Smart launched its cellular service using Extended Total Access Communication System (ETACS) technology in February 1994. According to Fernando, its success was due to a number of factors. First, the company did not compromise on quality. Whereas PLDT management bought second hand equipment, which it passed as new, such a practice was unthinkable at Smart.[9] Second, Smart focused on providing service not merely in Metro Manila but also in major urban areas and other provinces.[10] Third, the Smart management model was based on selling airtime rather than expensive handsets. Thus, to attract subscribers, Smart aggressively advertised a package wherein it sold mobile phone units cheaply and recouped the loss from airtime usage. Fourth, the financial backing of First Pacific was crucial in Smart's rapid construction of its network, as well as in its subsidisation of handsets.[11]

With its subscribers reaching 300,000 by the end of 1997, the company had significant leverage in negotiating interconnection with PLDT. Nevertheless, PLDT, threatened with Smart's rise, once again delayed interconnection until First Pacific bought into the former monopolist and took control of its management from the Cojuangcos in November 1998. This merger will be discussed further below.

As of December 2000, Smart installed 866,954 telephone lines, exceeding its 700,000-line commitment. Yet, the bulk of Smart's business was in mobile telephony. At the end of 2005, Smart was the largest cellular operator in the Philippines, with over 15.4 million subscribers. It offered both analogue (ETACS) and digital (GSM) services nationwide. The company opened its international gateway facility in December 1994, and its local exchange services in Metro Manila and Regions 1 and 3 in July 1996. Subsequent to First Pacific's takeover of PLDT, on 24 March 2000, PLDT acquired Smart through a share-swap deal with the Metro Pacific Corporation.[12] Although this acquisition meant a windfall for Smart's original investors, the merger augured the return of a monopoly threat. This will be discussed further later in this chapter.

Bayantel

The Lopez family owned Bayan Telecommunications Incorporated. Bayantel received a 25-year franchise to operate a domestic communications system on 20 July 1992. The original company, International Communications Corporation (ICC) was incorporated on 18 April 1961, to engage in broadcasting, telecasting, and domestic and international communications.[13]

Alfred McCoy traced the "spectacular post-war climb of the Lopez brothers, Eugenio and Fernando, based on their masterful manipulation of the state's regulatory and financial powers."[14] He described the Lopezes as "the most successful rent seekers, prospering largely because they were skilled in extracting special privileges from the state apparatus … relying on state licenses that restricted access to the market."[15]

When President Marcos declared martial law in 1972, he targeted the Lopezes as one of the supposed oligarchs dominating the economy. Even though Fernando Lopez was Marcos' Vice-President, he shut down the ABS-CBN network and the *Manila Chronicle*, both owned by the Lopez family. In addition, Eugenio Lopez was forced to sell his shares in Meralco, the country's biggest electricity distributor, to a group led by Benjamin Romualdez and Roberto Benedicto, well-known Marcos cronies.[16] Only after Marcos was deposed in 1986 did the Lopezes regain control of their companies.[17]

After February 1986, the Lopezes worked towards restoring their vast family business, as well as entering new sectors that were being made available by the liberalisation of the economy. Benpres Holdings was incorporated in 1993 as the family's flagship company as they diversified into new business ventures.[18] In fact, the Lopezes was one of the biggest beneficiaries of economic liberalisation as it gained contracts to manage privatised services (water and toll-way), bolstered its control over its original businesses (media and power generation), and gained access to the telecommunications sector through Bayantel.

Before the launch of its services, Bayantel was reportedly in joint venture talks with Cable and Wireless and Telstra of Australia. Eventually, the company agreed to a partnership with Verizon (formerly Bell Atlantic), which owned 15 per cent of Bayantel through its wholly-owned Philippine subsidiary, Nynex Network Systems Company, and the Asian Infrastructure Fund (AIF), which held 10 per cent of the company.[19]

Bayantel proposed to operate an international gateway facility and was required to rollout 300,000 lines under the SAS. The company was assigned the Region V provinces of Albay, Camarines Norte, Camarines Sur, Catanduanes, Masbate, and Sorsogon, and Quezon City, Valenzuela, and Malabon in Metro Manila.

Unlike Smart, Islacom, and Globe, which were full service telecommunications companies, Bayantel did not offer cellular service. It applied to the NTC to operate a cellular service in December 1992, but the application was archived until frequencies become available. The company obtained a 47 per cent stake in Extelcom in 1996 by buying Marifil Holdings.[20]

Because Bayantel was solely reliant on fixed line revenues, it bore the brunt of interconnection problems with PLDT. Despite the Lopezes' political influence, it took almost six years before interconnection problems with the equally politically influential PLDT were ironed out. Due to its interconnection frustration, Bayantel led a consortium of seven companies to construct an alternative national backbone for domestic long distance calls. Telecom Infrastructure of the Philippines (TelicPhil) was owned by Bayantel (58 per cent), PT&T (10.5 per cent), Digitel (8.4 per cent), Smart (6.3 per cent), Globe (4.6 per cent), Extelcom (3.9 per cent), and ETPI (1.3 per cent).[21] The network cost US$70 million and became operational in 1999.[22]

In December 2000, the company surpassed its 300,000-line SAS commitment, having installed 466,493 telephone lines. Yet, Bayantel, unlike the other telecommunications companies, which enjoyed revenues from the high-growth cellular market, suffered heavily when the impact of the financial crisis was felt in the fixed-line market.[23]

Islacom

Isla Communications Corporation (Islacom) was the second new company to enter the liberalised telecommunications market. The Delgados, a well-known business family from Cebu involved in shipping and port operations, owned Islacom. It received its legislative franchise on 10 April 1992 under Republic Act 7372, which granted it the authority to operate and maintain telecommunications services nationwide.

Islacom applied to offer both international gateway and cellular telephony. It was required to install 700,000 lines in Regions VI, VII, and VIII, covering the provinces of Aklan, Antique, Capiz, Iloilo, Negros Occidental, Guimaras, Negros Oriental, Siquijor, Bohol, Cebu, Eastern Samar, Leyte, Northern Samar, Southern Leyte, Samar, and Biliran.

The Delgados controlled 60 per cent of Islacom through Visay-Tech, in joint venture with Deustche Telekom (30 per cent) and Shinawatra of Thailand (10 per cent). Islacom launched its GSM cellular service in 1993, and as of 2000 had 121,654 cellular subscribers and 175,000 fixed line subscribers in the Visayas. It was only able to rollout 488,531 of its 700,000-line commitment.[24]

In April 1999, in the face of mounting debts and slow subscriber uptake, the Delgados sold their share in Islacom to the Ayala Corporation, Globe's parent company. Deustche Telekom remained interested in Islacom, which the Ayalas planned to merge with Globe. Political manoeuvring slowed down the merger, which was completed in 2001. This will be discussed in more detail later.

Digitel

Digital Telecommunications Philippines Incorporated (Digitel) was established in 1987 as a 60:40 joint venture between ETPI and Cable and Wireless. Digitel planned to offer a domestic telephone service to complement ETPI's international operations. As was discussed in Chapter 4, ETPI became involved in legal wrangling with PLDT over its international gateway license. Moreover, PLDT opposed the award to Digitel of a contract to manage the government's Luzon telephone system. The contract was eventually awarded to Digitel in February 1993, as part of the efforts of the coalition for reform to liberalise the industry.

By 1993, however, Cable and Wireless grew tired of the legal wrangling with PLDT and decided to pull out of Digitel. In June that year, John Gokongwei of JG Summit bought Cable and Wireless' share in Digitel. Gokongwei reportedly provided P200 million to pay the DOTC for the P40 billion management contract. Gokongwei was also credited with helping the company secure its legislative franchise on 17 February 1994.[25]

Because it obtained its legislative franchise late, Digitel was not part of the original SAS. However, the company applied for a license to operate an international gateway facility and was required to install 300,000 lines in Luzon. By 2000 it had installed 611,166 lines, and had a subscriber base of 344,368, making it the second biggest fixed line operator in the country.

In August 2000, Digitel secured a license to operate cellular services, through its wholly-owned subsidiary, Digitel Mobile Philippines Incorporated (DMPI). Digitel Mobile received a 25-year congressional franchise on 3 January 2003 and launched its services in March 2003. [26]

John Gokongwei's highly diversified JG Summit Holdings holds 47.4 per cent stake in Digitel while Telia of Sweden and Jasmine International of Thailand were its strategic partners.

PT&T

The Santiago family established Philippine Telegraph and Telephone Company (PT&T) in 1962. The Santiagos other company, Retelco, then second biggest landline network in the Metro Manila area, was PLDT's earliest competitor in local telephony. The family sold Retelco to PLDT in 1981 due to pressure from Marcos' ostensible policy of integrating the national telephone system.

PT&T pioneered the establishment of public calling offices, initially offering telegraph and telex services, and later on, long distance and international telephone services. As of 2002, PT&T had 151 branches and 222 outlets. A publicly listed company, PT&T was a subsidiary of the Santiago family's Republic Telecommunications Holdings Incorporated (Retelcom), with 40 per cent stake controlled owned by the Korean Telecom Authority. The Santiagos also owned Capitol Wireless (Capwire), an international gateway facility operator, and Philippine Wireless Incorporated, which operated Pocketbell, the country's first radio paging company.[27]

In 1995, Capwire was licensed to operate an IGF, and in line with EO 109 its sister company PT&T took on the requirement to build and operate a 300,000 line local exchange carrier service in the Region IV provinces of Rizal, Laguna, Quezon, Aurora, Marinduque, and Romblon.[28]

At the end of 2000, PT&T was one of the four companies that were unable to meet their SAS commitments. Of its 300,000 required lines, PT&T was only able to rollout 190,456 in 396 out of 494 towns in its service area. Of the total lines installed, only 27 per cent had been subscribed.[29] Hence, in December 2001 PT&T announced it would shift its focus to providing Internet-based services and broadband technology in the Laguna-Rizal and Metro Manila area.[30]

ETPI

Eastern Telecommunications Philippines Incorporated (ETPI) was the oldest communications company in the Philippines. Established in 1878 and formerly known as Eastern Extension, ETPI was a subsidiary of Britain's Cable and Wireless. The company changed its name to Eastern Telecommunications Philippine Incorporated (ETPI) in 1974 when 60 per cent of the stock was sold to Filipino businessmen, while Cable and Wireless retained its 40 per cent stake.[31] The new Filipino stakeholders were Roberto Benedicto, Jose Africa, and Manuel

Nieto, each holding 20 per cent of the shares. All three businessmen were well-known Marcos cronies. Under martial law, ETPI's business expanded from data communications to an exclusive right to handle the ASEAN communications market.

When Marcos was deposed in 1986, the PCGG sequestered ETPI shares, but two months later it released the Cable and Wireless stake. The remaining shares stayed under PCGG control.[32]

In 1987, when the government announced its policy of introducing competition in the industry, ETPI, under the new PCGG and Cable and Wireless management, applied for a license to set-up an international gateway facility. PLDT blocked ETPI's application, suing the company in the Supreme Court. The matter was only finally settled in 1995, which meant that ETPI had lost years of operations.

Under the SAS, ETPI was required to install a minimum of 300,000 lines. The company's assigned service areas were the Region II provinces of Batanes, Cagayan Valley, Isabela, Quirino, Nueva Vizcaya, Ifugao, and Kalinga-Apayo, and the cities of Manila, Caloocan, and Navotas in Metro Manila. ETPI's rollout program was undertaken through a wholly owned subsidiary, Telecommunications Technologies Philippines, Inc. (Teletech), which started operations in June 1999.[33]

At year-end 2000, ETPI had only installed 69,085 landlines of its 300,000 commitment. Out of its installed capacity, only 18,000 lines had subscribers as of April 2002.[34]

Philcom

Philippine Global Communications Corporation (Philcom) was the successor company of the American-owned RCA Communications, which operated the first short-wave radio telegraph circuit between Manila and the United States in 1924.[35] In 1965, the company gained a 50-year legislative franchise to operate communications systems via Republic Act 4617. In 1977, the company changed its name to Philippine Global Communications when 60 per cent of the company was sold to Marcos' Ministers Juan Ponce Enrile and Geronimo Velasco.[36]

In October 1989, Philcom was the first company to gain a license from the NTC to enter the lucrative international communications business. Unlike the tough time that ETPI faced, PLDT did not oppose Philcom's entry and agreed to interconnect with it.[37] Reportedly, this was because some of Philcom's shareholders, notably the Yuchengcos, were also members of the PLDT Board.

In July 1994, Philcom was granted a new franchise under RA 7783 to operate domestic and international communications systems. Under the SAS, Philcom's service areas were in the Region X provinces of Agusan del Norte, Agusan del Sur, Bukidnon, Camiguin, Misamis Oriental, and Surigao del Norte, the Region XI provinces of Surigao del Sur, Davao Oriental, and Davao del Norte, and Basilan province in Region IX. It began commercial operations in July 1998.[38]

As of the end of 2000, Philcom had only installed 64,620 of its required 300,000 lines in 34 of the 164 towns in its area of service. The company blamed the onset of the 1997 economic crisis for its inability to complete its rollout obligation, citing the resultant credit crunch and slowdown in demand in Mindanao.[39]

In 2002, Philcom was majority-controlled by the Belle Corporation via its subsidiary, the APC Group Inc. Both APC and Belle were publicly listed companies controlled by Willy Ocier.[40]

Piltel

The Pilipino Telephone Corporation (Piltel) was incorporated in 1968, initially operating a local telephone exchange in General Santos City. PLDT bought the company in 1980.[41] As a PLDT subsidiary, Piltel was the first company granted provisional authority to offer a cellular mobile telephone service, which began operations in 1991. The company also offered paging services from August 1993. Ramon Cojuangco, Tonyboy's younger brother, headed Piltel's management.

Piltel was the third company to join the SAS. Before the scheme's launch it already operated local exchange services in the cities of Baguio, Olongapo, Subic, Boac, Puerto Princesa, and Masbate. Piltel received a 25-year franchise on 27 March 1992 under RA 7293.

Under the SAS, Piltel applied to construct its own international gateway facility and was required to install 300,000 lines in Regions IX, X, and XI, covering the provinces of Zamboanga del Norte, Zamboanga del Sur, Misamis Occidental, Davao del Sur, South Cotabato, Sarangani, Basilan Sulu, and Tawi-Tawi. By December 2000, Piltel had rolled out 463,541 lines, exceeding its commitment of 400,000 lines under the SAS.[42]

Extelcom

Express Telecommunications Company (Extelcom) was the second company licensed to operate a cellular telephone service. Extelcom launched its analogue cellular service in 1991 and became Piltel's number one competitor before the entry of Smart, Globe, and Islacom.

Extelcom's legislative franchise was based on Republic Act 2090 issued on 22 June 1958, granting authority to the Felix Alberto Company to establish radio stations for domestic and transoceanic telecommunications. The company changed its name to Extelcom in 1964. In 1987, the Albertos sold their stake in the company to Horacio Yalung, who in turn sold it to a group led by Ruby Tiong-Tan. Tiong-Tan took over management of Extelcom, raised the company's capital stock, and in joint venture with Millicom of Luxembourg, Motorola of the US, and Comvik of Sweden, proposed to build a cellular mobile telephone company.[43]

On 13 May 1987, Extelcom applied to the NTC to operate a cellular mobile telephone system and an Alpha Numeric Paging System in Metro Manila and Southern Luzon. Despite opposition from PLDT,[44] which had applied to offer similar services, the NTC granted Extelcom a provisional authority to operate. The NTC also ordered PLDT to interconnect with Extelcom. PLDT challenged the NTC decision at the Supreme Court. This legal move delayed the construction of Extelcom's network, as PLDT won a Temporary Restraining Order from a lower court order for Extelcom to "cease and desist from any acts towards the implementation of its provisional authority".[45] However, on 18 October 1990, the Supreme Court voted in favour of Extelcom, rejecting PLDT's opposition.

When PLDT lost the case in the Supreme Court, it stalled interconnection, making it impossible for Extelcom mobile subscribers to call PLDT subscribers. In early 1992, out of exasperation, Tiong-Tan sold her shares in Extelcom for US$10 million, but refused to reveal the identity of the buyer. In July 1992, it was reported that Marifil Holdings bought Tiong-Tan's 40 per cent share in Extelcom. Marifil's registration papers at the Securities and Exchange Commission identified the company's shareholders as Wigberto Clavecilla Jr. and lawyers from the De Borja, Medialdea, and Ata Law office. Clavecilla was a golfing partner of Ramon Cojuangco, President of PLDT's cellular subsidiary, Piltel. It was widely believed that Clavecilla and the lawyers were merely acting as nominees for the Cojuangcos because the shareholders of Marifil did not have enough assets to pay for Tiong-Tan's shares.[46] This belief was bolstered by the fact that after Marifil bought Tiong-Tan's shares, PLDT readily assented to interconnect with Extelcom.[47]

In 1996, Bayantel wholly acquired Marifil Holdings for P6 billion, ending up with 47 per cent of Extelcom. However, Extelcom's other shareholders, Millicom International Cellular of Luxembourg and Mayon Holdings, which owned 40 per cent and 12 per cent of Extelcom respectively, banded together to deny management control to Bayantel. The Gozon family owns Mayon Holdings, which is a part-owner of the Lopezes' media competitor GMA network

Extelcom purportedly did not join the SAS because it planned to assign its 400,000-line obligation to Bayantel. In 1997, NTC Commissioner Simeon Kintanar threatened sanctions against Extelcom unless the company fulfilled its rollout obligations.[48] However, by 1998 Extelcom had not been assigned an area of responsibility.

EFFECTS OF THE 1997 FINANCIAL CRISIS

Similar to Malaysia, Philippine telecommunications companies were badly affected by the 1997 financial crisis. They were left debt-laden, as their mostly dollar-denominated loans ballooned to twice their original peso value, when the peso depreciated to P50 to US$1, from

P26. The financial crisis led to two consolidations in the industry, one that was negotiated between the companies themselves and another where no less than President Joseph Estrada was involved.

The Globe-Islacom Merger

The Globe-Islacom merger was a straightforward corporate transaction between two companies and their foreign partners, although political manoeuvrings slowed it down. In April 1999, the Ayala Corporation, Globe's parent company, acquired Visay-Tech Inc., which owned the majority stake in Islacom. With the purchase, Globe and Islacom agreed to merge.

However, some politicians raised barriers to the merger. Islacom was one of four companies that were not able to fulfil their rollout commitments by 1999. Yet, Congress only investigated Islacom. In addition, the NTC did not renew Islacom's provisional authority to operate, which had lapsed on 29 March 1998. Islacom accused the NTC of delaying its license, due to pressure from certain government officials. In particular, John Osmeña (who authored the Senate version of the Telecommunications Act) ordered the NTC "to hold in abeyance any action on Islacom's application" for a license because of pending Senate investigations into the company's possible violation of the anti-dummy law.[49] Meanwhile, then NTC Commissioner Joseph Santiago declared that the NTC would only renew Islacom's license if the company proved that it had substantially complied with the SAS. Deutsche Telekom, Islacom's partner, sought audiences with the Secretaries of the Trade and Industry and Finance to seek assistance in clearing Islacom's license. This delayed the Globe-Islacom merger. The license was finally renewed in September 2000, paving the way for the deal.[50] A holding company, Asiacom, was created to own the shares of the combined company. The Ayala Corporation held a 60 per cent share, while SingTel and Deutsche Telekom each held 20 per cent stakes.

The Globe-Islacom merger was completed in June 2001 and resulted in the Philippine's 8th largest company, with a market capitalisation of US$1.8 billion.[51] The consolidation also resulted in improved services and wider coverage for the two companies' subscribers. Compared to this merger, the PLDT-Smart union was an intricate transaction that involved no less than then President Estrada's involvement as the alleged "de facto broker."[52]

The PLDT-Smart Merger

First Pacific announced the PLDT-Smart merger in November 1998. The merger came about not because PLDT was badly affected by the crisis. Rather, First Pacific, the parent company of Smart's major stockholder, was redirecting its investments to Southeast Asia. First Pacific was a Hong Kong-based holding company majority-owned by Suharto crony Liem Sieo Liong. Its Managing Director and CEO, Manuel Pangilinan, thought that it was a good idea to buy into large companies that were undervalued after the crisis. Thus, from January 1998 until December 2000, First Pacific sold its shares in various companies worldwide, leaving it with US$3.412 billion cash on hand.[53]

First Pacific's first target was the San Miguel Corporation, the Philippines' largest publicly listed company and producer of beverages, food, and packaging. However, Eduardo Cojuangco,[54] San Miguel's main stockholder, rebuffed First Pacific's attempt to take-over the company. Having failed in this bid, Pangilinan turned his attention to PLDT, which was giving Smart, First Pacific's cellular subsidiary in the Philippines, a difficult time with interconnection. First Pacific, via Metro Pacific, secretly bought shares in PLDT and announced in October 1998 that it had bought a strategic stake in the company. However, the company neither revealed the exact amount of shares nor from whom they were purchased.

First Pacific's bold move to take-over PLDT should be seen in the light of the backing it received from Joseph Estrada, who succeeded Fidel Ramos in the presidency on 30 June 1998. Estrada's term ended abruptly in January 2001, becoming the first impeached Philippine President and the second to be deposed by a "People Power" uprising. His presidency was marked by the excessive use of presidential discretion to benefit those allied with him. While such use of discretion is observable in earlier governments, Estrada brought it to a higher extent by intervening in corporate mergers and in the working of the stock market. In fact, it was because Estrada himself gave the green light that First Pacific was eventually able to purchase a controlling stake in PLDT.

Antonio Cojuangco and family did not welcome the hostile takeover bid that First Pacific was orchestrating. Being a publicly listed company, however, PLDT was open to takeovers by investors who could afford to buy controlling stake in the company given enough willing sellers. Cojuangco looked for a white knight to counter the move. He found an ally in Southwestern Bell Company (SBC), the US's second largest telephone company. In October 1998, SBC's officials visited the Philippines and, accompanied by then US Ambassador to the Philippines Thomas Hubbard, made a courtesy call to President Estrada.[55] The visit illustrates how the SBC knew that the blessing of the President was vital factor in deciding who would ultimately control PLDT.

Allegedly, Estrada brokers Mark Jimenez and then Executive Secretary Ronaldo Zamora convinced Antonio Cojuangco and family to sell their 44 per cent stake in PTIC to First Pacific for P17 billion (US$427 million).[56] President Estrada himself told the media before the deal's formal announcement that "he heard First Pacific bought Tonyboy out of PLDT."[57]

Various concerns were raised about the transaction. On the corporate side, it was portrayed as a simple deal leading to better value for the shareholders of both companies, including the subscribers of PLDT. Some industry analysts, however, were worried that the merger was the wrong kind of consolidation, effectively the merger of the number one landline provider and the number one cellular service provider. Others celebrated it as the coming home of a foreign-trained professional to dislodge one of the most influential and established families from control of its corporate jewel.

On 24 November 1998, First Pacific announced its acquisition of a 17.3 per cent stake in PLDT for P29.7 billion (US$749 million). This stake represented a 27.2 per cent voting interest in the company.[58] First Pacific revealed that it acquired control of PLDT in a two-step process. First, it bought 5.9 per cent stake in PLDT on the open market at a cost of US$197 million. Secondly, it bought an indirect 11.3 per cent stake by purchasing 52.7 per cent of PTIC, from Antonio Cojuangco's family, Nori Ongsiako and her family, Antonio Meer, and Alfonso Yuchengco, for US$552 million. In 1998, PTIC controlled 21.5 per cent of PLDT's voting stock. First Pacific paid about P1,420 per share, a 31 per cent premium on the then market rate, which was trading at P1,085.[59]

First Pacific's Managing Director, Manuel Pangilinan, became President and Chief Executive Officer of PLDT, replacing Antonio Cojuangco who became Chairman of the Board. First Pacific also appointed six new directors to the 11-member Board.[60] In addition, First Pacific announced its intention to inject Smart into PLDT, in exchange for new shares, which would further increase its stake in the company. First Pacific promised the result to be a stronger, more efficient PLDT that would bring direct benefit to its shareholders, consumers, and employees.[61]

This deal, however, was only possible with the blessing of Malacañang. More accurately, the First Pacific-PLDT deal was a case of executive intervention to facilitate a corporate merger in exchange for "cold cash." It was a step backward from the Ramos administration's intention of removing state involvement through liberalisation.

Why Estrada sided with First Pacific and how the transaction was consummated was later divulged by then Securities and Exchange Commission (SEC) Chairman Perfecto Yasay.[62] Less than a month into the Estrada presidency, Yasay was suspended from his post due to an alleged corruption case concerning his refusal to renew a contract of lease between the SEC and a private company. A couple of weeks into his suspension, Metro Pacific, First Pacific's Philippine subsidiary, started buying PLDT shares. Rumours of a First Pacific takeover of PLDT became rife. The main focus of the apprehension was a return to monopoly status. Yasay's objection to the deal, however, was the secrecy that surrounded the transaction in direct violation of the SEC's full disclosure policy. The shares acquisition were only announced after the fact.[63] How many shares, at what price, and from whom the shares were bought were not disclosed.

When his suspension ended in October 1998, Yasay investigated whether the Metro Pacific-PLDT transaction complied with the Securities Code. Before he could make any headway, another suspension order was handed down against him. Yasay was told that the second suspension was an extension of the first. It became evident to Yasay that his new suspension implied that strings were being pulled. Yasay contested the suspension in the Supreme Court, and was reinstated in January 1999. By then, however, the First Pacific-PLDT transaction was a done deal.

When he eventually resigned from his position after a more direct clash with President Estrada, Yasay announced what he knew about the First Pacific-PLDT transaction. He told the media in November 2000 that Estrada received US$20 million (P1 billion) as a kickback to facilitate the transaction. Yasay's revelation substantiated rumours that had already been circulating in the media, claiming that Estrada and his men received pay-offs that were integrated into the total purchase price and thus appeared as part of the transaction cost.[64] First Pacific denied Yasay's claim and categorically stated that neither the company nor Metro Pacific, its subsidiary, made payments to President Estrada.[65]

During the impeachment hearings against Estrada in January 2001, Yasay repeated his testimony under oath. The impeachment court summoned Manuel Pangilinan, First Pacific's Managing Director and PLDT CEO, to clarify the matter. However, on 17 January 2001 Estrada allies in the Senate forced a vote on whether to admit an envelope of evidence purportedly containing bank accounts of President Estrada, which triggered the 5-day mass action that led to the President's removal from office. If not for what is now called EDSA 2, Pangilinan would have been called to account for what actually took place during the PLDT sale.

Pangilinan's Explanation

Under the new administration of Gloria Macapagal-Arroyo, the Ombudsman and volunteer lawyers pursued the prosecution of Estrada. One of the first plunder and graft cases filed in the Sandiganbayan against Estrada included Pangilinan as a co-accused.[66] Allegedly,

Pangilinan issued a P20 million cheque to Estrada, and was one of 67 that were deposited into an account that Estrada held under the pseudonym Jose Velarde.

In response to the indictment, Pangilinan executed an affidavit on 2 May 2001 acknowledging that he did issue a P20-million pay-to-cash cheque for Estrada in November 1998, but claimed that the money had nothing to do with First Pacific's buy-in into PLDT. He explained that associates of President Estrada solicited financial support from businessmen to set up a new political party for the 2001 general election. A report on the Estrada plunder case in the Sandiganbayan noted the peculiar nature of the P20 million campaign donation, given 17 months before the May 2001 elections.[67] Carlos Arellano, then head of the Social Security System (SSS), revealed a more feasible reason. Arellano admitted that under Estrada's order, the SSS sold some of its stake in PLDT so that Pangilinan could buy it. These were probably the shares that Metro Pacific bought on the open market. This suggests that First Pacific's take-over of PLDT's involved Estrada's collusion.[68]

It is not clear which share in PTIC that the Cojuangcos sold. Ownership of these shares was still under investigation in the Sandiganbayan. Without waiting for a court decision whether his family owned the shares or not, Antonio Cojuangco, sold 44 per cent stake in PTIC and received US$460.9 million for it. One lawyer opined that "Metro Pacific is at risk of losing its PTIC shares in the event of a decision forfeiting the shares in favour of the government or ordering the conveyance of the shares to Yuchengco or to the Marcos family."[69] The case is currently in the Supreme Court awaiting resolution.

ASSESSMENT OF THE SERVICE AREA SCHEME

By 2000, the nine telecommunications companies had installed over four million landlines, increasing the available number of lines to 6.9 million from only 1.4 million in 1995. Installed telephone density increased from 2.01 in 1995 to 9.12 in 2000. However, only

Table 9-3: SAS Accomplishments, December 2000

Company	No. of Lines Commitment	No. of Lines Installed	No. of Lines Subscribed
PLDT*	300,000	2,623,797	1,701,607
Digitel*	300,000	611,166	344,368
Bayantel	300,000	466,493	219,082
Islacom	700,000	488,531	150,440
Globe	700,000	790,291	158,249
Smart	700,000	866,954	116,992
PT&T	300,000	190,456	50,678
Piltel	400,000	463,541	56,967
Philcom	300,000	64,620	38,539
ETPI	300,000	69,085	21,677
Total	4,300,000	6,634,934	2,858,599
Paptelcos		271,028	
Grand total		6,905,562	

* PLDT and Digitel were not part of the original 11 service area allocation. They were, however, required to put up a minimum of 300,000 in their existing service areas but opted to undergo bigger expansion programmes. PLDT on its own launched the Zero-Backlog programme in 1993, aiming to install 1.6 million lines in five years.

Source: "Assessment of the Implementation of the Service Area Scheme (SAS)" by DAI-Agile Consultants at the NTC, p. 4. Hereafter cited as Assessment of the SAS".

2.8 million lines were subscribed to, which constituted a subscribed teledensity of 3.44.[70]

Of the 11 telecommunications companies that existed when the scheme was designed, eight were assigned the original 11 service areas. PLDT and Digitel were not part of these area assignments, but were required to put up at least 300,000 lines each in their existing areas of service. Both companies decided to launch bigger expansion programs. In particular, under its Zero-Backlog program, PLDT installed 1.3 million new lines within five years[71] — more than twice the number of lines that it had rolled out in the previous eight decades! Extelcom did not initially join the program. The company was later required to rollout 400,000 lines, but it was never assigned an area of operation and did not actively pursue the requirement. A 12th company, Bell

Telecommunications, was granted a legislative franchise to operate in 1996, but its commercial operations never went off ground.

Only four of the eight companies accomplished their required fixed-line rollouts. These were Bayantel, Globe, Piltel, and Smart. Meanwhile, ETPI, Islacom, Philcom, and PT&T failed to meet their commitments. None of the telecommunications companies fulfilled the requirement of a one in every ten rural-urban deployment ratio. Like PLDT, the new entrants concentrated their rollout obligations in urban areas. Table 9-4 demonstrates the big regional disparity of telephone service availability in the country. About 72 per cent or 4.9 million lines were installed in 36 urban centres in the country, of which, 47 per cent or 3.2 million lines are in Metro Manila where only 14 per cent of the population resides.

Of the 1,609 towns and cities in the Philippines, only 52.4 per cent or 844 have fixed line coverage, while 40.6 per cent or 654 have cellular phone coverage, as Table 9-5 shows. The rest of the country relied on payphones or public calling offices (PCO) at the municipal level.[72] Thus, despite about four million fixed lines lying idle, 745 cities and towns were still without fixed line local exchange services. The pairing of lucrative and non-lucrative areas and the ten to one ratio of lines under the SAS did not provide enough compulsion to guide the private companies to build networks in unserved but low demand areas. Further complicating the story was how consumer demand declined as a result of the 1997 financial crisis.[73] As incomes shrank, people substituted fixed line telephones with prepaid cellular telephones, which offered the added features of mobility and services such as text messaging.

Telephone companies cited various reasons for their failure to fulfil their rollout commitments. Smart and Piltel cited peace and order problems in 37 towns in Mindanao. Other factors were delays of permit issuance at the local government level, environmental issues raised by residents of the area, and other local disputes. Only Philcom cited the financial crisis as a reason for its non-compliance, even though all companies experienced the resultant credit crunch.[74] In addition, the collection of termination rates from international calls, a traditional source of profit and cross-subsidisation for less profitable

Table 9-4: Concentration of Telephone Facilities in Urban Centres, 2001

Area	Number of Installed Lines	% of Total Installed Capacity
1. Metro Manila	3,248,046	47.03
2. Cebu, Mandaue and Lapu-Lapu City	322,951	4.68
3. Bacoor and Kawit Cavite	113,846	1.65
4. Davao City	85,757	1.24
5. Baguio City	75,406	1.09
6. Angeles City	71,116	1.03
7. Bacolod City	66,609	0.96
8. Malolos, Bulacan	60,218	0.87
9. Biñan, Laguna	58,224	0.84
10. Iloilo City	54,949	0.80
11. Antipolo City	51,398	0.74
12. General Santos, South Cotabato	49,348	0.71
13 Batangas City	47,132	0.68
14. Cabanatuan City	46,760	0.68
15. Cainta, Rizal	45,702	0.66
16. Imus, Cavite	40,693	0.59
17. Lipa City	39,148	0.57
18. Dagupan City	38,900	0.56
19. Iligan City	37,480	0.54
20. Naga City	37,100	0.54
21. Taytay, Rizal	36,608	0.53
22. Koronadal, South Cotabato	34,014	0.49
23. Tacloban City	30,794	0.45
24. Tarlac	30,612	0.44
25. Vigan	26,474	0.38
26. Meycauayan, Bulacan	22,340	0.32
27. Calamba	22,182	0.32
28. Tagbilaran City	21,234	0.31
29. San Fernando, La Union	20,776	0.30
30. Laoag City	18,020	0.26
31. Binangonan, Rizal	17,680	0.26
32. Zamboanga City	17,642	0.26
33. Baliuag, Bulacan	16,750	0.24
34. Legaspi City	16,088	0.23
35. Angono, Rizal	15,796	0.23
36. Mabalacat, Pampanga	11,000	0.16
Total	4,950,791	71.64

Source: "Assessment of the SAS", p. 11.

Table 9-5: Coverage of Telecommunications Services, December 2000 Total Cities and Towns in the Philippines (1,609)

Type of Service	No. of Cities and Town	As a % of Total
Local exchange service/fixed line service	844	52.4
Cellular service	654	40.6
Payphone/PCO service	1,417	88.1
Fixed lines/payphone/PCO	1,481	92.0
Fixed lines/cellular/payphone/PCO	1,495	92.9

Source: "Assessment of the SAS", Appendix A, p. 32.

operations such as fixed line operations, have been declining due to the US-imposed reduction on accounting rates for incoming calls from the United States and terminating in the Philippines.[75]

From the perspective of telecommunications executives, the SAS failed because of the mismatch between supply and demand. Today, over 50 per cent of the existing landlines were unsubscribed. Telecommunications companies criticised the NTC for not properly designing the program. In particular, they observed that no study was conducted to correlate the required number of fixed lines, the level of demand for telecommunications services, and the households' ability to pay. Yet, the companies themselves eagerly assented to the plan. Moreover, they concentrated on lucrative urban areas, which led to higher competitive pressures.

The biggest criticism of the SAS was how it preserved PLDT's dominant position. Although it might have averted the over-concentration of investment in urban areas and ensured investments in the provision of telephone services to unserved rural areas, some economists contend that the scheme ignored economies of scale.[76] New players were given geographically segregated areas, thus preventing them from realising economies of scale and scope, as well as enjoying positive network externalities. PLDT, in contrast, had a national network to which all of the new entrants needed to interconnect. All new entrants complained to the regulator that PLDT delayed interconnection, to the detriment of all subscribers. Smart and Bayantel, badly hit by interconnection problems with PLDT, were the most vocal in publicising their problems. Bayantel's owner, the Lopez family, used its ABS-CBN broadcasting network, to air their case against the PLDT. All of the new companies, which were owned by politically and economically influential families and their foreign business partners, also lobbied the government to solve the interconnection problem and reform the regulatory framework.[77]

Some also criticised the NTC for not acting decisively on complaints that PLDT refused to provide enough points of interconnection, which led to call traffic congestion and non-termination of calls. New companies also complained about PLDT's propensity to prolong disputes over revenues and charges. PLDT, however, claimed that by early 2001, the issue of interconnection had been resolved. For them, the only problem was commercial disagreements, especially when revenue-sharing disputes occur or when they cannot collect receivables due from another company. When this happens, PLDT either reduces the number of interconnection trunks or cuts them off completely. PLDT lamented that not enough attention was given to interconnection problems among new players, such as the interconnection problem between Globe and Bayantel.[78] Thus, PLDT tried to paint a broader picture, pointing out that interconnection problems did not merely involve PLDT.

Some analysts asserted that the SAS was a political scheme designed to accommodate competing interests rather than deal with the lack of telephones. The NTC, critics say, did not responsibly decide which players had the most efficient and viable network development plans.[79] Yet, this criticism ignored the fact that all aspiring entrants were granted entry, provided they demonstrate the capacity to fund their investments and exhibit a viable business plan. Allegations of favouritism and rewarding entry rents to influential rent-seekers was tenuous given the fact that none who sought entry, including those who were not linked to any established business families, were turned down. As mentioned earlier, without the SAS, new entrants would all have concentrated their deployment in urban areas.

When measured in terms of its overall goal of expanding and modernising the Philippine telecommunications infrastructure, the SAS was successful. It resulted in the speedy rollout of fixed lines and the provision of various types of communication services, increasing from 1.0 in 1991 to 9.12 in 1999 the available telephone lines per 100 persons.[80] Using another measure, as of 2000, the liberalisation of telecommunications in the Philippines attracted over P600 billion in foreign and local investments.[81] The use of mobile phones, which were only available to those with high incomes in 1991, reached about 34.8 million subscribers by the end of 2005.[82] In fact by 2005, 91.2 per cent of telephone subscribers were mobile phone users while only 8.8 per cent used fixed line phones. In effect, mobile phones became substitutes for fixed-line telephones as they became more affordable. This directly undermined the SAS, which was modelled on fixed-line telephone usage and teledensity count. In addition, unforeseen circumstances such as the financial crisis not only affected demand for telephone services but also their supply. These factors were beyond the control of both government and business.

ASSESSMENT OF PHILIPPINE TELECOMMUNICATIONS LIBERALISATION

The liberalisation of the telecommunications industry, implemented via the SAS, was successful in expanding the availability of fixed-line and mobile telephones at a very rapid rate. With the scheme, the government utilised liberalisation as a means to attain important social goals.

Tables in Appendix Three document the developments in the industry. Table A3-1 shows the spectacular growth of cellular telephony subscribers from about 10,000 in 1990 to 34.8 million in 2005. The commensurate decline in the cost of mobile telephony services can be seen in Table A3-2. Table A3-3 depicts the progressive decline in the number of people on the waiting list, demonstrating the availability of service due to liberalisation. Tables A3-4 and A3-5 map the growth of urban and rural fixed lines in the country, and the increasing availability of public payphones, although the growth rate seems slower than that of the population. Finally, Table A3-6 shows the declining cost of fixed line services for residential and business users from 1991 to 1999. In 2001, however, the cost of services increased, probably due to the high rate of unsubscribed lines rolled-out under the SAS.

The state-led break-up of PLDT's monopoly and the policy allowing entry of new players resulted in the expansion of the telecommunications industry in the 1990s. Except for one, the owners of telecommunications companies were well-known businessmen. This is understandable in the light of the fact that telecommunications is a very capital-intensive industry. Indeed, all new entrants took in foreign partners because of the need for capital and technical experience. Most of the owners were also politically influential. That, however, did not guarantee success in the business. Political influence may facilitate the release of a company's legislative franchise or permits for operation, but was not enough to guarantee profitability in a competitive environment. The Lopezes' Bayantel was the most adamant advocate of effective interconnection, utilising their television and radio network to publicise PLDT's anti-competitive actions. Yet, delays in interconnection and the onset of the financial crisis proved disastrous for the company. In 2003, Bayantel was in deep financial trouble, to the extent that the Lopezes were contemplating selling it. Such a situation, where a politically influential family was on the brink of losing one of its businesses, would have been unthinkable if the telecommunications environment was not competitive.

The liberalisation of the telecommunication industry in the Philippines has been successful in reforming a monopoly sector, leading to the availability of choice and lower prices for communications services. The success of the SAS, however, was qualified. Telecommunications companies have unanimously judged it a failure.[83] Yet, the program tried to ensure that new entrants would not ignore the rural areas. Despite the criticisms of the scheme, the Philippines avoided what happened to Malaysia, where most companies concentrated their networks in the Klang Valley and other urban centres. While most of the Philippine companies also focused on the urban centres of their assigned areas, the SAS requirement of one rural line to every ten urban lines helped tip the balance. A negative consequence of this, however, was the resulting incongruous areas of operation and a lack of economies of scale for the new market entrants, with only PLDT operating a national network until 1999. These were considered as developmental costs at a time when the state was undertaking reform and making difficult choices in the face of an urgent need to fast-track infrastructure development.

Like Malaysia, consolidation took place among the telecommunications companies in the Philippines. Although the Globe-Islacom merger was uncomplicated, that of PLDT and Smart pointed to a disturbing return of executive corruption.

SUMMARY

The liberalised telecommunications industry in both Malaysia and the Philippines swung from a position of monopoly to one with many corporate players. In Malaysia, eight new companies entered the telecommunications industry. In the Philippines, two new and eight existing companies were allowed full market entry. In Malaysia, only the politically well-connected were granted market entry. In the Philippines, all who were financially qualified and wanted market entry were allowed in. The new entrants in Malaysia had clear patronage linkages to the top three government officials. In the Philippines, well-known business families controlled all, except one, of the new entrants. Yet, political linkages did not seem to play a significant role in gaining market entry once the sector was opened.

A second difference was the presence of foreign joint-venture partners in the Philippines, whereas the Malaysian companies, with the exception of STW, did not initially have foreign partners. The Philippine companies, despite being owned by financially-capable families, readily sought foreign partners, citing the need for huge capital outlays and technological and management expertise in running a telecommunications business. Third, despite being licensed to offer fixed, mobile, and international gateway services, the Malaysian telecommunications companies concentrated on the lucrative mobile segment. This was due to the lack of a general plan to guide the companies' deployment. In contrast, with the SAS, new entrants in the Philippine telecommunications market were required to install fixed lines nationwide. Thus, market liberalisation in the case of the Philippines contributed to accomplishing the social goal of expanding telephone access. Liberalisation in Malaysia was implemented for a different social goal: the creation of a bumiputera business class.

The Malaysian state politically determined the distribution of market entry licenses for redistributive and/or patronage reasons. In contrast, the Philippines used liberalisation to remove monopoly rents. Nevertheless, liberalisation in both countries had expansionary effects on the economy as the market opening exercise led to huge investments in the sector and the provision of formerly unavailable services. The most important social gain from competition in both countries was the expansion of consumer choice, improved quality of service, and lower prices.

The Malaysian telecommunications market underwent more consolidation than that of the Philippines. Malaysia now has five fixed-line providers (Telekom Malaysia, Celcom, Digi, Maxis, and Time) and three cellular providers (Telekom-Mobikom-Celcom, Maxis-Time, and Digi). The Philippines as Table 9-6 shows, in contrast, still has eight players, with two major companies. The PLDT-Smart-Piltel combine is the clear dominant player in both the fixed line and cellular markets, followed closely by the merged Globe-Islacom. Digitel and Bayantel are catching up as fixed line providers, with Digitel recently launching its mobile service. Finally, Philcom, ETPI, and PT&T are trying to develop niche markets in data communications to ensure their survival.

The financial crisis negatively affected the industry in both countries, leaving companies in positions of indebtedness. However, the Malaysian and Philippine states responded differently. In Malaysia, the state under Mahathir got directly involved in two ways. First, it bailed out the companies that were owned by bumiputera. Second, the state had the final say on the foreign investors that would be allowed entry, which led to the scuttling of several deals.

The Philippine state responded differently, which generally reflected a more hands-off policy. Two consolidations took place after 1997. One was a straightforward market transaction between the Ayalas

Table 9-6: Philippine Telecommunications Players, as of 2002

Telco		Filipino Owners	Foreign Partner	Foreign Equity
PLDT	PLDT	Cojuangcos, Manuel Pangilinan	First Pacific NTT	24.70% 15.00%
	Smart	100% PLDT		
	Piltel	45% PLDT	AIG (US)	
Globe Telecom (merged with Islacom, 29 June 2001)		Ayalas (46.8%)	Singapore Telecoms Deutsche Telecoms	28.01% 24.80%
Bayantel		Lopezes (66.5%)	Verizon AIF	19.36% 10.30% 10.30%
Digitel		Gokongweis (51.85%)	Telia (Sweden)	10.00%
Philcom		APC	Comsat	16.80%
PT&T		Santiagos	Korean Telekom	20.00%
ETPI		Phillippine government (50.2%), Smart (9.8%), Manuel Nieto (20% via Aerocom Inc.), and Jose Africa (20% via Polygon Investment and Mngmt Company)	None	
Extelcom		Bayantel (40%), Mayon Holdings (13.4%)	Millicom	40.00%

Source: The table was constructed from various company annual reports and press statements.

and SingTel, which owned Globe, and the Delgados and Deutsche Telekom, which owned Islacom. However, when Estrada took over the Presidency in mid-1998, he allegedly become involved in the take-over of PLDT for the sake of "cold cash." This relationship was transaction-based, as opposed to traditional patronage links. On the surface, both the Malaysian state under Mahathir and the Philippine state under Estrada employed favouritism and cronyism. Yet, the intervention by the Philippine state under Estrada was more damaging. In fact, the use of state resources and authority for personal enrichment is the worst possible type of intervention. For Mahathir, the bailing out of Malay capitalists was part of an overall developmental state project. The state responses also had different costs. Malaysia lost public money in saving crony-owned or ruling party-linked companies, but Estrada's intervention cost First Pacific its own capital, effectively raising the purchase price of PLDT. This, among other things, is why Estrada is on trial for graft and plunder.

NOTES

1. Lee Kuan Yew added, "I don't understand why there should only be one telephone company here when you have a low telephone density". See *Philippine Daily Inquirer*, 23 November 1992.
2. *The Globe Spectrum*, October 2000, p. 46.
3. *Forbes Magazine*, 20 August 2001.
4. http://www.ayala-group.com (accessed 15 February 2003).
5. In 2001, Globe's network processed an estimated 40 million messages per day, or approximately ten messages per subscriber per day. SMS, popularly known as texting, played a key mobilizing role in the rallies that eventually toppled Estrada in January 2001.
6. Globe SEC Form 17-Q, "Quarterly Report Pursuant to Section 17 of the Securities Regulation Code and SRC Rule 17 (2) (b) Thereunder", 30 September 2002, p. 8.
7. Interview with David Fernando, Quezon City, 16 February 2001.
8. The First Pacific Group, established in 1981, is a Hong Kong–based investment and management company with operations and investments in Europe, Asia, the

United States, and Australia. Unlike the situation in Indonesia, where there is heavy involvement by the Salim family, professionals led by its Filipino Managing Director, Manuel V. Pangilinan, manage the diversified conglomerate.

9. Interview with David Fernando, Quezon City, 16 February 2001. See also Manapat, *Wrong Number*, pp. 28–32.

10. The company shot to national prominence when a Cebu Pacific airplane crashed 30 miles northeast of Cagayan de Oro in February 1998, and the only communication links that were usable in the search for victims in the mountainous terrain were Smart mobile phones. See *Asiaweek*, 8 October 1999.

11. Interview with David Fernando, Quezon City, 16 February 2001.

12. PLDT Notice of Special Meeting of Stockholders and Proxy Statement on the Acquisition of Smart, Strategic Investment by NTT Communications and Certain Other Corporate Actions, 22 October 1999, p. 53.

13. Bayantel Company profile, http://www.bayantel.com.ph (accessed 20 February 2003).

14. Alfred McCoy, "Rent-seeking Families and the State", in *An Anarchy of Families: State and Family in the Philippines*, edited, by Alfred McCoy (Wisconsin: University of Wisconsin Center for Southeast Asian Studies, 1993) p. 435.

15. Fernando Lopez was mayor of Iloilo, a three-term senator, and a three-term Vice-President. His political influence ensured that their companies gained privileged access to business opportunities. Meanwhile, Eugenio Lopez led the family's business and orchestrated the family's move to Manila from Iloilo in the 1960s. See McCoy (1993).

16. Oscar Lopez, "An Open Letter to Juan Ponce Enrile, 2 October 2002", http://www.benpres-holdings.com/press-10202.html (accessed 20 February 2003).

17. Ibid. The Aquino government returned control of ABS-CBN, Meralco, and PCIBank to the Lopez family.

18. Benpres Profile, http://www.benpres-holdings.com/h-heritage2.html (accessed 20 February 2003).

19. http://www.bayantel.com.ph/index.jsp (accessed 20 February 2003).

20. As was mentioned in Chapter 3, Antonio Cojuangco reportedly owned Marifil Holdings. See Manapat, *Wrong Number*, p. 37.

21. http://www.phix.net.ph/phix/network.html (accessed 2 December 1999).

22. http://www.abs-cbnnews.com/abs/inews-sep2000.nsf/business/20000913044; *Philippine Daily Inquirer*, 14 September 2000.

23. "NTC Assessment of EO 109 Implementation, Conclusions and Future Sector Development Planning", p. 6. Hereafter cited as "Assessment of EO 109".

24. Ibid.

25. As early as March 1988, Digitel had applied for a national franchise. The House version, House Bill 27107, was approved on 5 April 1990. However, the Senate version was stuck at the Committee on Public Services chaired by Senator John Osmeña, who reportedly sat on the bill. See *Globe Spectrum*, October 2000, p. 89; *Philippine Daily Inquirer*, 28 October 1992; Manapat, *Wrong Number*, p. 40.

26. *ABS-CBN.com* news, 3 February 2003.

27. http://www.ptt.com.ph/ptt1a.htm#rdn (accessed 20 February 2003).

28. "Assessment of EO" *109*.

29. *Philippine Daily Inquirer*, 6 November 2000.

30. *Business World*, 3 December 2001.

31. http://www.etpi.com.ph/aboutus/index.asp (20 February 2003).

32. G.R. No. 124478, "Victor Africa versus The Honorable Sandiganbayan (Third Division), Roman Mabanta and Eduardo De Los Angeles", 11 March 1998; G.R. No. 83831, "Victor Africa versus PCGG et al.", 9 January 1992.

33. "Assessment of EO 109".

34. *Business World*, 18 April 2002.

35. http://www.philcom.com/corp_profile.html (accessed 20 February 2003).

36. Gerald Sussman, "Telecommunications Transfers: Transnational Corporations, the Philippines and Structures of Domination", Third World Studies Center, University of the Philippines Dependency Series no. 35 (June 1981), p. 52.

37. G.R. No. 94374, "PLDT versus NTC and ETPI", Dissenting Opinion, 27 August 1992, p. 20.

38. http://www.philcom.com/corp_profile.html (accessed 20 February 2003).

39. *Philippine Daily Inquirer*, 11 September 2001.

40. Willy Ocier was reportedly one of Estrada's close associates during his term. He turned state witness against Estrada in the corruption trial against the latter. See Yvonne T. Chua, "The Company He Keeps", *Investigative Reporting Magazine Special Report* (October–December 2000) at http://www.pcij.org/imag/latest/cronies.html (accessed 15 February 2003).

41. *Daily Globe*, 29 October 1992.

42. "Assessment of EO 109".

43. *Newsday*, 27 October 1990, p. 9; Manapat, *Wrong Number*, p. 36.

44. See the Supreme Court's Decision on G.R. No. 88404, PLDT versus the NTC and ETCI, *Newsday*, 26 October 1990, p. 9.

45. Ibid.

46. Manapat, *Wrong Number*, p. 37.

47. Manapat, *Wrong Number*, p. 38. On 8 November 2000, a *Philippine Daily Inquirer* columnist categorically stated that the De Borja, Medialdea, and Ata law office represented Tonyboy Cojuangco Jr. when he secretly acquired Extelcom from Ruby Tiong-Tan. See *Philippine Daily Inquirer*, 8 November 2000.

48. *Business World*, 7 and 16 April 1997.

49. The anti-dummy law prohibits investors from owning more than 40 per cent of any utility company. See *Philippine Daily Inquirer*, 26 June 2000.

50. *Business World*, 18 October 2000.

51. *Philippine Daily Inquirer*, 8 February 2000.

52. *Philippine Daily Inquirer*, 23 August 2001.

53. 'First Pacific Company Limited Strategic Review and Outlook 2001", p. 16, http://www.firstpacco.com/invrshp/frame.htm (accessed 20 February 2003).

54. Eduardo Cojuangco, uncle of PLDT President Tonyboy Cojuangco, was one of former President Marcos' top cronies. The PCGG contends that Danding's shares in San Miguel were ill-gotten. A case is pending in the Sandiganbayan.

55. *Asiaweek*, 20 November 1998.

56. It should be recalled that the Cojuangcos controlled 44 per cent of PTIC, the holding company that in turn controlled about 21.5 per cent of the total voting shares in PLDT in 1998.

57. *Asiaweek*, 20 November 1998.

58. For additional information, see Appendix 2 for a timeline of the PLDT ownership question, especially, point no. 16 onwards.

59. "First Pacific Acquires PLDT Stake for US 749 Million", First Pacific Press Release, 24 November 1998.

60. PLDT's Management's Discussion and Analysis of Financial Condition and Results of Operations for the Year Ended 31 December, 1998, p. 5.

61. "First Pacific Acquires PLDT Stake for US 749 Million", First Pacific Press Release, 24 November 1998; PLDT's press release "Explanatory Memorandum on the Acquisition of Smart and Strategic Investment by NTT Communications", 22 October 1999.

62. President Ramos appointed Perfecto Yasay Chairman of the Securities and Exchange Commission in October 1995 to serve a nine-year term until 2004. The SEC is mandated as the regulator of publicly listed companies.

63. Interview with Perfecto Yasay, Pasig City, 8 March 2001. Yasay revealed that when he was under suspension, then acting SEC Chairman, Fe Gloria, did nothing to investigate the issue. He explained that this was because Ms Gloria was a friend and law school classmate of Estrada's Executive Secretary, Ronaldo Zamora.

64. Yasay claimed that based on raw data that he received but was unable to verify because of presidential intervention, the total kickback was P3 billion, with Estrada allegedly pocketing P2 billion and Mark Jimenez and others sharing P1 billion. Because of his role in facilitating the transaction, Estrada described Mark Jimenez as a "corporate genius" and appointed him as special adviser on Economic Affairs to Latin America (interview with Perfecto Yasay, Pasig City, 8 March 2001. *Philippine Daily Inquirer*, 9 November 2000 and 14 November 2000).

65. First Pacific media report, 11 November 2000.

66. *Philippine Star*, 11 April 2001.

67. Malou Mangahas, "The Estrada Plunder Case Year 1: Politics and Other Nightmares Hound Prosecution", 16–18 January 2002, http://www.pcij.org/stories/2002/eraptrial3.html.

68. Ibid.

69. *Philippine Daily Inquirer*, 23 August 2001.

70. Teledensity is the standard set by the International Telecommunications Union as measure for the number of telephone lines per 100 people. The NTC argues that the use of teledensity as a measure of telephone access is misleading in developing countries such as the Philippines because it is based on the entire population count. Instead, the NTC proposes the use of households or families as a more accurate count of subscriber access to telephone services. As of 2000, 13 per cent (2 million) of the 15 million households in the Philippines had subscribed to a telephone service. In terms of regional composition, 40 per cent of these households were in the national capital region (NCR) while less than 10 per cent outside of the NCR were telephone subscribers. See "Assessment of EO 109", p. 6.

74. Ibid., p. 5.

72. 1,495 or 93 per cent of the country's towns and cities have fixed line, payphone, PCO, or cellular telephone services. However, 80 per cent of these were installed through the Ramos government's Telepono Para sa Barangay (Telephones for the Village) programme, and not the provision of universal access to communications services by the new telephone companies.

73. As of 2000, a monthly household income of P10,000 or less for the average Filipino family meant that less that 25 per cent of households could afford the basic telephone service which cost between P300 and P500 per month. See "Assessment of EO 109", p. 6.

74. Ibid., p. 4.

75. US companies cited increasing global competition and the availability of cheaper options such as Voice over the Internet, callback, and international simple resale as the reasons behind this reduction in rates. See "Assessment of SAS", p. 1.

76. Ma. Joy Abrenica and Gilberto Llanto, "Services", in *Philippine Economy: Developments, Policies and Challenges*, edited by Arsenio Balisacan and Hal Hill (New York: Oxford University Press, 2003), pp. 254–82.

77. Interviews with Rodolfo Salazar, former CEO of Bayantel, Ortigas City, 9 February 2001; and with David Fernando, founder of Smart Communications, Libis, Quezon City, 16 February 2001.

78. Interview with Antonio Samson, PLDT Vice-President, Makati City, 6 March 2001.

79. See Abrenica and Llanto (2003). In addition, it should be noted that before the NTC issues a licence, the company has to obtain a franchise from Congress.

80. See *National Telecommunications Commission Annual Report 2002*, p. 29, http://www.ntc.gov.ph/consumer-frame.html (accessed 10 December 2003). In 2002, teledensity was down to 8.70 per 100 persons, because of a faster population growth.

81. "Assessment of SAS", p. 6.

82. Clearly, with changes in technology and types of telephone services available, the use of teledensity based on fixed line telephones is no longer the best measure of the state of telecommunications in developing countries such as the Philippines, where the number of mobile phone users grew to more than fivefold the number of fixed line subscribers by 2002.

83. Various interviews with representatives of all telecommunications companies and AGILE Report to the NTC.

10

REGULATORY REFORMS IN THE PHILIPPINES

As the experience of most countries attest, re-regulation was necessary immediately after the introduction of competition to ensure that the market works. This is all the more true in the provision of public utilities like telecommunications.

This chapter focuses on regulatory reforms in the Philippines. The first two parts review the pre-liberalisation regulatory structure. A third looks at the regulator's failed attempts at liberalisation under the Aquino government. A fourth focuses on regulatory problems, specifically the weakness of the regulatory body, its lack of independence and resources, and its capture by the regulated. Next, the role of the regulator during liberalisation under the Ramos administration is considered as well as the regulatory impact of RA 7925. A sixth part looks into the problem of introducing regulatory reform and the role of a group of consultants in pushing for them. A seventh section summarises the chapter's discussion and arguments. Finally, the experience of the Philippine telecommunications reforms is summarised.

THE PUBLIC SERVICES COMMISSION

The Philippine Legislature passed Commonwealth Act 146 (CA 146) on 7 November 1936 creating the Public Services Commission (PSC) and empowering it to regulate the operation of public utilities such as telephone systems for the promotion of public welfare. CA 146 stated that a legislative franchise from Congress and a Certificate of Public Convenience and Necessity (CPCN) from the PSC were required for the operation of a public utility. The law also provided that permission to operate public utilities would only be granted to Philippine or US citizens, or to companies organised under Philippine laws that were at least 60 per cent owned by Filipino or US citizens.[1]

Aside from controlling entry into the industry via the issuance of a CPCN, the PSC was vested with the authority to: (1) fix and determine "just and reasonable" rates; (2) require any public service company to construct, maintain, and operate any reasonable extension of its existing facilities; (3) require any public service to keep its books and to furnish the PSC with annual reports of finances and operations; and (4) penalise any company that violated or failed to comply with its CPCN.[2] Because the law did not quantify what was "just and reasonable rate," the Supreme Court established a 12 per cent rate of return on investments and assets as a fair level of profit.[3]

The PSC served as the regulator of the telecommunications industry from its inception until 1972, when Marcos launched an overall reorganisation of the Philippine bureaucracy. Under the Integrated Reorganisation Plan, the PSC was abolished. The responsibility of regulating telecommunications was transferred to a new body, the Board of Communications. In addition, a Telecommunications Control Bureau was established in 1974 to deal with technical issues.

THE CREATION OF THE NATIONAL TELECOMMUNICATIONS COMMISSION

On 23 July 1979, Marcos abolished the Board of Communications and the Telecommunications Control Bureau. Their functions were integrated under a new body, the National Telecommunications Commission (NTC). Executive Order (EO) 546 established the NTC, provided for its structure and composition, and defined it as a quasi-judicial body whose decisions were appealable only at the Supreme Court. The NTC was placed under the supervision and control of the Ministry of Transportation and Communications (MOTC).

Although the broader powers of the NTC are traceable to CA 146, EO 536 redefined the NTC's mandate to cover the supervisory, regulatory, and control functions of telecommunications, broadcasting, and radio.

In 1987, Aquino changed the former bureaucratic ministries into departments. Through EO 125-A, the MOTC was renamed Department of Transportation and Communications (DOTC), and was mandated as the primary policy-making body while the NTC was designated as the industry regulator.[4] No mention was made of the role or functions of the NTC, which were presumably carried over from the provisions of CA 146 and Marcos' EO 546.

DOTC AND NTC ATTEMPTS AT LIBERALISATION DURING THE AQUINO ADMINISTRATION

As was discussed in Chapter Four, both the DOTC and the NTC sought to liberalise the telecommunications industry under the Aquino government. In particular, the DOTC issued Department Circular (DC) 87-188 on 22 May 1987, the first indicator of the move towards "regulated competition" in the industry. The circular, which consisted of fourteen policy statements, was a result of a dialogue between the DOTC and the private sector.

A second DOTC circular pertinent to liberalisation was DC 90-248, issued on 14 June 1990. DC 90-248 prescribed a policy of fair and non-discriminatory interconnection of all public telecommunications facilities within the shortest practicable time. The circular mandated the NTC to define the points of interconnection between service operators, and to set an interconnection rate based on the "recovery of costs and a fair return on investment in the facilities employed."[5]

Third, DC 90-252, issued on 10 August 1990, stated the government's plans to privatise its fledging operation of telephone and telegraph services in rural areas. The circular also ventured the idea of encouraging and allowing the private sector to tap Official Development Assistance funding for its communications expansion programs.[6]

Finally, in October 1990, the DOTC launched the National Telecommunications Development Plan (NTDP) 1991–2010. The NTDP superseded the 1982 document of the same title that placed the burden of telecommunications provision on the shoulders of government. The plan set out the overall policy for telecommunications and established a comprehensive set of service-oriented targets. It articulated a private sector-driven development policy, identified cross-subsidies within the sector to fuel development and achievement of universal access goals, and the interconnection of all private networks.[7]

The NTDP analysed the state of the sector in terms of the quality and availability of service, market structure, and the role of government in improving the sector. It noted that there was large unmet telephone demand and that only 320 of the country's 1600 municipalities (20 per cent) had access to telephone services. In addition, in areas where services were available, service quality did not usually meet reasonable standards. There were even situations where existing exchanges were not interconnected to the public switched network.[8] It also noted that while the NTC was authorised to monitor and enforce quality of service, it does not collect performance statistics from operators. Thus, it has failed to developed performance standards.[9]

The second section of the plan contained strategies for addressing the sector's problems. The NTDP set out physical targets, as shown in Table 10-1, for the industry's development. These targets, established in 1990, were considerably lower than what the industry achieved after less than a decade of liberalisation. For instance, the targeted teledensity was merely 2.4 by 2000 and 3.5 by 2010. This target was surpassed in 1998 when fixed-line teledensity reached 9.18. Meanwhile, mobile telephone users reached 12.1 million in 2001. In addition, the total amount of investment during the sector's liberalisation surpassed almost five times the expected investment of US$9 billion for the 20-year period.[10]

The third section of the NTDP enumerated reforms necessary to achieve developmental goals in the sector. The most important of these was the NTC's reorganisation to enable it to deal with

Table 10-1: Summary of Physical Targets, National Telecommunications Development Plan, 1991–2010

Criterion	1989 Status	Target Status	Target Year	Additional Manpower Required	Investment Cost (US$ Millions)
1. Main Station Density per 100 people	1.0	1.9	1995		
		2.4	2000		
		3.5	2010	39,854	8,375
2. Telephone Quality of Service — service application response within 4 weeks	69%	98%	2005		
a. monthly trouble rate	17%	5%	2005		
b. trouble response within 2 days	89%	98%	2005		
c. other standards	Not available	To be set			
3. Percentage of Local Exchanges with Long Distance Interconnection	85%	100%	1991	153	27
4. Percentage of Municipalities with Public Calling Office (PCO) Services	22%	100%	1997	4,531	117
5. Mobile Telephone Service Availability				86	20
a. Metro Manila	Available, 1989	Available	1992		
b. Cebu	Not available	Available	1992		
c. Davao, Cagayan De Oro, Zamboanga	Not available	Available	1995		

6. Percentage of Municipalities and Cities with Access to Switched Data Networks	0.6%	15% 35% 50%	1995 2000 2010		
7. Percentage of Municipalities with Access to Non-switched Digital Data Circuits	0.6%	25% 45% 55%	1995 2000 2010	5,405	211
8. Percentage of Municipalities with Telegraph Service via Cable or Radio	89%	100%	1995	Minimal	5
9. Integrated Services Digital Network (ISDN) Trial Exchanges					
a. Manila	None	Operational	1993		
b. Cebu	None	Operational	1995		
Total				50,029	8,799

Source: NTDP, 1990–2010

specialised, technical issues. Bills on NTC reorganisation had been filed in the House of Representatives since 1988 but none have ever been legislated.

On its part, the NTC also made decisions towards liberalisation. Under the leadership of Jose Luis Alcuaz in 1988, the NTC moved to introduce competition by issuing one cellular and two international gateway licenses to existing companies. It is not clear whether the NTC Commissioner acted in conjunction with the DOTC's efforts to liberalise. What is evident, however, is that the issuance of new licenses that would compete with PLDT signalled the start of liberalisation. As was discussed in Chapter Four, the lobbying of Aquino's relatives acting on behalf of PLDT led to Alcuaz's removal from office. This showed how politically vulnerable and dependent on the President the position of NTC Commissioner was. Yet, the Commissioner was also powerful enough to make a difference, which, in Alcuaz's case, initiated the process that led to the entry of competition into the cellular and international gateway market segments.

Fortunately, Josefina Lichauco, who replaced Alcuaz as Commissioner, affirmed the legality of the latter's decisions about the licenses. Lichauco was concurrently DOTC Undersecretary for Telecommunications during her appointment as NTC Commissioner from 1989 to 1991. She was actively involved in drafting the NTDP 1991–2010 and was thus in the best position to synchronise the efforts of the two bodies.[11] In addition, Lichauco's experience at the DOTC convinced her that the only way to improve the state of telecommunications was to demonopolise the industry. During Lichauco's term as NTC Commissioner, radio-paging, video satellite transmission, and cable TV were opened to market competition.[12]

On 13 July 1990, the NTC issued Memorandum Circular (MC) 7-13-90, which called for the interconnection of all local exchanges to a national network. Interconnecting parties had to share the cost of interconnection, with the NTC considering recommendations from players and prescribing a traffic agreement format. The memorandum provided that traffic settlement agreements had to be based on the recovery of toll-related costs, a fair return on investment for both companies, and a subsidy to local exchange providers.[13]

Thus, both the DOTC and the NTC under the Aquino goverment worked towards the goal of opening the industry, with the persuasion that only by doing so would telecommunications services improve. Yet, introducing reform in an industry monopolised by a company owned by a politically powerful family was not an easy task. As discussed earlier, PLDT legally contested the decisions to license new competitors and the directive to interconnect with them. These legal challenges effectively crippled its competitors until President Ramos intervened in 1993.

NTC REGULATORY PROBLEMS

Marcos' EO 546 granted the NTC quasi-judicial powers that could only be challenged in the Supreme Court. Yet, its regulatory capacity was weak because it lacked funding, equipment, and sufficient staff members. Compared to PLDT, a profitable company that could afford to hire the best engineers and accountants, the NTC was badly under-staffed.

In a 1990 study, Jacinto Gavino described the NTC as an ineffective regulator for a number of reasons.[14] First, the President appointed the Commissioner and the Deputy Commissioners, and no policy governed their terms of office. Thus, regulatory decisions were captive to, or heavily influenced by, political pressures. This was clearly evident in the case of Jose Luis Alcuaz.[15]

Second, the NTC had inadequate technical and commercial expertise due to the agency's low budgetary allocations. The NTC collected license fees from telephone and broadcasting companies, but these collections were transferred to the central treasury.

A third problem was regulatory capture. The NTC was reliant on PLDT for information.[16] The best example of this was the calculation of the allowable 12 per cent rate of return on investments. Because it does not have the expertise to audit the financial statements of the company and independently verify them, the NTC decides on the rate based on financial information that PLDT submits. The NTC was thus caught in a bind where the regulated company controlled the information flowing to the regulator, hindering its capacity to make independent decisions.

Fourth, the legislation on telecommunications regulation was too general, outdated, and insufficient. The laws did not give the NTC concrete enforcement powers to ensure that regulated companies obeyed its policies. Gavino argued, "the government must decide on the level of priority it will give telecommunications, given the high demand, and translate these to new policies to give the NTC more independence."[17]

As early as 1988, bills on telecommunications policies and industry structure (Senate Bill 1353 and House Bill 32327[18]) and on the reorganisation of the NTC (House Bill 20989[19]) were filed in Congress. However, they were not considered urgent and were not enacted. Gavino correctly concluded that the success or failure of regulatory reform did not rest with the regulatory body, but in the broader political processes.

THE TELECOMMUNICATIONS ACT OF 1995 AND ITS IMPACT ON REGULATION

In contrast to the tenuous positions of NTC Commissioners under earlier administrations, Simeon Kintanar, who headed the NTC from 1993 to 1998, enjoyed greater independence.[20] The creation of the SAS to implement EO 109, which aimed to improve local exchange services to promote universal access, was proof that the NTC under Kintanar was given leeway to manage the liberalisation of the industry. The SAS was an indicator that the regulator can play a key role in determining the direction of the industry's development given proper support from the executive. Yet, independence and support from the executive was not enough to settle regulatory problems that were institutional in nature and required legislation for their resolution. A distinction between institutional regulatory capacity and independence of the commissioner due to executive support has to be made. The latter is not assured of institutionalisation after the term of the executive, unless it is embedded in the law.

Republic Act (RA) 7925 appointed the NTC as the principal administrator of the law and the DOTC as responsible for developing a long-term national development plan for the industry. The law distinguished between development plan-making and policy setting, which were assigned to the DOTC, and the regulatory function, which the NTC would conduct. This distinction helped to clarify areas of responsibility. A closer look at the list of functions expected of the NTC, however, reveals a serious problem in the post-liberalisation scenario, where the NTC was saddled with new responsibilities without a commensurate provision for an increase in its financial allocation or institutional reorganisation. Thus, instead of resolving them, RA 7925 left many regulatory issues unresolved.

RA 7925 strengthened instead of simplified the two-step process of entry into the industry. First, firms had to secure a legislative franchise from Congress and then a Certificate of Public Convenience and Necessity (CPCN) from the NTC. Before the issuance of a CPCN, the NTC had to issue a Provisional Authority (PA) to allow a company to operate until such time that it demonstrated its technical and financial capacity and that there was sufficient demand for its services.[21] This cumbersome process can be contrasted to the relatively easier but non-transparent manner of licensing in Malaysia, where a new entrant only had to secure a license from the Minister.

Also, RA 7925 removed the 12 per cent ceiling on return on investments. Section 17 directed the NTC to establish rates and tariffs that were "fair and reasonable, for the economic viability of telecommunications firms, and a fair return on their investments considering the prevailing cost of capital in the domestic and international markets". A floor or ceiling on tariffs will be set to prevent ruinous competition, monopoly, cartel, or a combination of these that restrains free competition. However, the law was silent on what "fair and reasonable rate of return" meant.

A related issue was access charge or the rate that companies pay to each other for interconnection. Section 18 provided for revenue-sharing as basis of the access charge which had to be negotiated between interconnecting parties. If the parties failed to agree within a reasonable period, the NTC was empowered to resolve the dispute. The law further provided that the NTC would ensure equity, reciprocity, and fairness among the parties concerned when approving the rates for interconnection. Three factors were considered in rate setting: the costs of the facilities for interconnection, cross-subsidies for local service provision, and a rate of return on network investment at parity with those earned by other segments of the telecommunications industry.[22] Finally, Section 19 of RA 7925 mandated the establishment of a Uniform Charter of Accounts for all companies as the basis of establishing rates and tariffs.

Given the weakness of the regulatory body and the existence of bills for its reorganisation and strengthening, it is curious that these were never mentioned in a law that enlarged the NTC's functions. The reason for this can only be gleaned from the comments of the law's main author, Senator John Osmeña. Osmeña openly admitted that he inserted several provisions into the law to limit the power and discretion of the NTC. He reasoned that because appointees of his brother controlled the NTC and the DOTC, he needed to lessen the NTC's areas of control and discretion to reduce his brother's moneymaking opportunities. Having interconnection agreements negotiated bilaterally rather than mandated by the NTC was PLDT's suggestion, because according to Osmeña, PLDT was afraid that it would not be given a fair treatment if the NTC dictated the terms of interconnection.[23]

Another important issue that RA 7925 failed to address was technological convergence. Unlike Malaysia, which was the first country to legislate a law that recognised the convergence of telecommunications, computing, and broadcasting technologies, RA 7925 stated that "no single franchise shall authorize an entity to engage in both telecommunications and broadcasting, either through the airwaves or by cable."[24] This provision was designed to prevent one company from monopolizing telecommunications and broadcasting. Nevertheless, it barred firms from maximising the use of available technology.

Finally RA 7925 embedded into law the principle of cross-subsiding unprofitable services (i.e., services in rural areas) with revenues from the more profitable segments (international long distance, long distance, or cellular business). Some economists were doubtful of the viability of such a regulatory regime given the decline in international accounting rates. They argued that reforming the system of cross subsidy was crucial to resolving the issue of achieving universal access as well as establishing a cost-based access and interconnection regime.[25]

Thus, while liberalisation opened the telecommunications industry to competition, the new telecommunications law left many issues unresolved, making regulation more difficult, and with the net effect of benefiting the dominant player, PLDT.

AGILE AND ITS ROLE IN REGULATORY REFORM

The liberalised telecommunications industry needed a strong, independent regulator and a clear regulatory framework to ensure competition and protect consumer welfare. Yet, the NTC had weak regulatory capabilities due to its lack of political independence and human, financial, and technical resources. The independence from the executive that the NTC enjoyed during the Ramos administration was not replicated under Estrada. The Telecommunications Act of 1995 could have addressed these issues. However, due to the correspondence of the interests of PLDT and the author of the law, the law failed to endow the regulator with a clear role in resolving regulatory issues.

Despite the lack of legislative interest in strengthening the NTC, the lack of policy continuity between executives (from Ramos to Estrada in 1998, and then Arroyo in 2001), and the continued market dominance of PLDT, this section examines how regulatory reforms are being put in place. The focus is on how the NTC, through the support of a USAID-funded (United States Aid for International Development) project, had been instituting regulatory rules to solve thorny technical regulatory issues such as interconnection and pricing.

Since the 1980s, both the DOTC and the NTC have received technical and financial support from various aid donors such as the World Bank, USAID, Canadian International Development Aid (CIDA), Japan International Cooperation Agency (JICA), and Asian Development Bank (ADB). Most aid was directed at bureaucratic capacity building, human resource development, technical support, and equipment procurement.[26] There was also some official development assistance directed towards telecommunications infrastructure expansion and the provision of telephony services to unserved areas.[27]

The Estrada administration did not seem to be interested in furthering regulatory reform at the NTC.[28] Thus, a USAID-funded project in July 1999, which included the provision of technical support to the NTC filled a void when the Ramos administration's term of office ended.

The telecommunications sector received several types of USAID support. In 1999, a Carana Corporation report provided a history of USAID's involvement in infrastructure in developing countries. In the 1960s, USAID was heavily involved in directly financing capital projects for infrastructure and industry, based on the idea that this would trigger economic growth. In the 1970s, USAID shifted focus toward small-scale projects in nutrition, health, and education, emphasising provision for basic human needs. This was a result of recognition that investments in infrastructure projects were not alleviating poverty because of weak institutions, inadequate policy environments, and low human resource capacities in developing countries. In the 1980s, USAID emphasised institution building, policy dialogue, private sector development, and technology transfers, and left the role of financing infrastructure projects to the World Bank and regional banks. In the 1990s, USAID introduced the Business and Development Partnership initiative with the aim of "engaging the American private sector in efforts to develop and sustain free-market principles and broad-base economic growth" by providing support for sound capital investments of "direct relevance to US trade competitiveness." More recent thinking within USAID perceived infrastructure development not as an end in itself but as a means of attaining priority goals identified as economic growth, sustainable democracy, population and health, the environment, and humanitarian support.[29]

In the 1990s, USAID involvement in infrastructure in various developing countries included the provision of technical assistance on reforming and restructuring the infrastructure sector such as telecommunications, power, railroads, and water and wastewater management. These projects were linked to financing mechanisms to procure US technology, promote US private sector investment, and provide commodity assistance.[30] In effect, USAID provides technical support to developing countries with the aim of promoting US private sector interests and to promulgate "free market principles."

In the Philippines, USAID undertook four major projects in telecommunications after 1991. The first was the writing of the *Barriers to Entry Study* in April 1992. This report investigated entry restrictions in various economic sectors, analysed their effects in the economy, and proposed policies to minimise or remove these barriers. The USAID commissioned SyCip Gorres and Velayo (SGV), a Filipino consulting company, to prepare the study. The study's release was very timely as the "coalition for reform" used it to identify sectors needing liberalisation, the first of which was telecommunications.[31]

A second project was the provision of a mixed credit facility from September 1990 to August 1992, with the goal of promoting the sale of US manufactured goods and services to the Philippines. A total of US$32.9 million was lent to telecommunications, of which US$22.863 million went to the DOTC and US$12.04 million to Capwire. Capwire bought Motorola and GTE satellite and trunk radio systems with this concessional loan.[32] A third project from May 1995 to June 1997 was the provision of US$2.5 million technical assistance to the NTC, to improve its regulatory efficiency and reliability. This project provided technical and commercial guidelines on monitoring the achievements of the basic telephone program, tariff rebalancing, emerging technologies, and interconnections. It also provided NTC staff members with training and study tours and funded the acquisition of computers and radio monitoring equipment. The fourth project consisted of a technical assistance package that amounted to US$249,900 for the DOTC from 1997 to 1998, which resulted in reports on universal access, broadband services, global mobile personal communications by satellite, and tariff rebalancing.[33]

In 1998, bilateral negotiations between the Ramos government and USAID officials resulted to the creation of an umbrella project known as Accelerating Growth, Investment, and Liberalisation with

Equity (AGILE).[34] AGILE was a set of project activities that aimed to accelerate economic policy reforms, generate growth, create jobs, and reduce poverty. According to its press release, AGILE provided "demand driven technical assistance to Philippine government agencies and non-governmental organization in forwarding the national agenda for economy policy reform in several areas like (1) stabilizing and deepening the financial sector, (2) strengthening fiscal policy, (3) facilitating international trade, (4) making investment more competitive, and (5) improving economic governance."[35] A consortium of three American and one Filipino consulting firms managed and implemented the projects. These were: Development Alternatives, Inc. (DAI), the Harvard Institute for International Development, PricewaterhouseCoopers, and Cesar Virata and Associates. The main contractor, DAI, a Maryland-based agency, won the public bidding to administer the project.[36] The project was expected to last for 7 years, from June 1998 to June 2005. USAID provided 75 per cent of the project's funding and the Philippine government was expected to put up the remaining 25 per cent. Already, the US Congress allotted $41.2 million for the project.[37]

The programs that AGILE undertook were negotiated between the Philippine government, led by the National Economic and Development Administration (NEDA) and the Department of Finance, and USAID. A Steering Committee consisting of representatives from both the public and private sectors reviewed and evaluated the work plans. Once a project was approved by the respective government agency, AGILE usually created a "satellite office" in that government unit, enabling project consultants to work directly with the client. AGILE then provided technical analysis and policy advocacy to help identify, legally enable, and implement economic policy reforms. Advisors and consultants from leading universities and firms in the Philippines, as well as international consultants, were tapped to come up with technical advice and policy papers.[38]

Under the goal of liberalising and expanding trade and investment, the NTC received AGILE assistance with the general aim of opening up trade and investment opportunities, increasing competition, and reducing the infrastructure-related costs of doing business in the Philippines.[39] AGILE's technical support for the NTC began in July 1999 with the NTC's approval of its work plan entitled "Telecommunications Deregulation Program for the 21st Century: Strategic Plan to Facilitate the Transition to a Fully Competitive Marketplace." The paper outlined a ten-point program to restructure the Philippine telecommunications industry's regulatory framework and facilitate the transition towards a competitive market.[40]

Of ten regulatory issues, AGILE concentrated on five: interconnection, wholesale pricing, retail pricing, accounting regime, and increasing public awareness. According to Edgardo Cabarios, Head of the NTC's Common Carriers Authorization Department, the NTC welcomed technical assistance from aid agencies because it did not have budgetary allocation to hire consultants. So far, NTC's budgetary increases only occur in line with inflation. Cabarios clarified that it is the NTC, which sets the agenda of what technical assistance it received. In the case of AGILE, the work plan was agreed upon between the two bodies.[41]

Jaime Faustino, AGILE's task manager in charge of telecommunications reform, argued that AGILE's support was specifically designed within the Philippine political economy context. The basic assumption was that the bureaucracy had weak capabilities and that legal loopholes existed giving rise to corruption and inefficiency. Since the NTC's reorganisation has been on the agenda for more than 15 years without real action, AGILE's strategy was to create clear and transparent rules and processes in the face of regulatory weakness. Faustino asserted that AGILE was designed to work within the given constraints, aware that establishing a credible regulatory body does not happen overnight. This approach resulted in codifying clear rules and transparent procedures to minimise bureaucratic discretion and corruption. Thus, the focal point of AGILE's work in telecommunications was in creating policies to resolve competition bottlenecks, establish a framework for negotiation between players, and provide a clear timeframe for negotiation.

Faustino clarified that the biggest lesson learned over the years on technical assistance was that it was easy to get the technical answers right. However, the key problem was political. Good technical knowledge should therefore be matched with good political strategy.

AGILE's technical support for the NTC resulted in clear benefits for the regulatory body and the public. As of the end of 2002, the project had helped the NTC to come up with several consultative documents and regulatory determinations.[42] In the Philippines where there was no continuity in the executive agenda for reform, a USAID-funded project played a key role in putting in place much-needed regulatory reforms.

Some sectors of Philippine society, however, consider such technical assistance as a form of neo-colonial intervention in the affairs of a sovereign state. AGILE has been criticised for embedding US interests in Philippine policies.[43] However, Filipino technocrats and academics, as well as staff members of the NTC, welcomed AGILE's technical support to push for crucial and long overdue reforms that have weak or no domestic constituency.

Faustino admitted that AGILE directly intervened in the workings of government because the technical input that the consultants provided directly influenced the regulatory framework adopted. Yet, he emphasised that most of the staff are Filipinos. In addition, AGILE's work converged with the interest of the Philippine government and the broad US interest of promoting competitive market-based economies. At the end of the day, it was the prerogative of the Philippine government to say no to the aid or agendas that were offered. In this particular case, the US economic interest in keeping the Philippine telecommunications market competitive converged with the public interest, resulting to efforts towards establishing a clear and transparent regulatory framework in telecommunications.

The AGILE group composed of local and international consultants provided technical knowledge and benchmarks of good international practice. They played a crucial role in pushing for the deepening of reforms that would otherwise not have domestic support. Before AGILE's intervention, the second phase of reform was in limbo because the coalition for reform had moved on to work on other economic sectors and left the responsibility of implementing liberalisation to the administrative agencies.[44] Thus, the essential regulatory reforms were left to the existing institutions and new companies to pursue. Given the lack of a clear constituency for the deepening of reform and an obstructionist Congress used to horse-trading and receiving kickbacks, AGILE filled a void by providing the NTC with consultative documents, research, and analysis that became the basis of regulatory memorandum circulars. Thus, a new regulatory framework is in the offing at the NTC, despite the unwillingness of Congress to strengthen the regulatory body.

SUMMARY

Table 10-2 summarises the major actors involved during the regulatory reform process in the Philippines. Meanwhile, Tables A3-7 and A3-8 in Appendix Three summarise the existing regulatory regime, the powers and functions of the NTC, and the accountability, autonomy, and competency of the regulatory process in the Philippines.

This chapter looked at regulatory reforms in Philippine telecommunications. In contrast to pre-reform Malaysia, where a government department provided the service and also regulated it, the Philippines had a separate body for the regulation of public utilities while utilities provision was in the hands of the private sector. Yet, regulatory capacity was weak, and the NTC was captured by the regulated. Also, the NTC commissioners served at the pleasure of the President, the appointing power, and thus, lacked independence in their decision-making.

Under the Ramos administration, the NTC enjoyed the President's backing, and led the implementation of liberalisation in the industry. Yet, after Ramos' term, the continuation of such executive backing was not guaranteed. RA 7925 could have solved this problem. Instead, the law gave the NTC a vague mandate in resolving regulatory issues, as a result of PLDT's lobbying. Thus, in a situation where state power was dispersed, Congress, which was prone to the influence of rent-seeking lobbies, could undo or water down reform initiated by the Executive. In addition, when the Executive only has a fixed six-year term, the continuation of regulatory reform is in doubt. Instead, a USAID-funded project managed by Filipino and foreign consultants pushed for regulatory reforms. Unlike in Malaysia, where the Prime Minister was on top

Table 10-2
Philippine Actors and Their Positions During Regulatory Reforms

Actors	Role and Position in the Issue
1. Ramos' Finance Secretary Roberto De Ocampo	Signed agreement with USAID for liberalisation support package.
2. USAID/AGILE project	Provided funding that established AGILE, which among other things is involved in helping NTC draft a new regulatory framework.
3. Academics and technocrats	Worked with AGILE in pushing for reforms that may not have easily passed through Congress or not been carried forward from one administration to another.
3. NTC	Ill-equipped and ill-funded; given many responsibilities by RA 7925 without commensurate increase in budgetary allocation; accepted AGILE support and tried to codify clear regulations given that they were not part of the legislature's priorities.
4. Congress	Bill on NTC reorganisation and regulatory issues were filed, some over a decade ago but were never legislated; regulatory reforms not Congress priority; some legislators openly expressed support for PLDT.
5. PLDT	Agreed to liberalisation but used its dominant position to delay interconnection; used its political connections in Congress to insert provisions in laws favourable to it.
6. New entrants	Mostly owned by established and influential business families, they nonetheless had difficulty competing with PLDT; were supportive of clarifying competition rules but only when it was within their interest.

of the creation of the CMC, the continuation of the second stage of reforms was uncertain in the Philippines.

The Philippines and Malaysia, like countries that liberalised their telecommunications industries, were faced with regulatory issues such as resolving interconnection, setting wholesale and retail prices, establishing universal service funds, and responding to the convergence of technologies. The two countries responded to regulatory reform demands in different ways, thus illustrating the type of state in each country, and in particular how and where power was concentrated.

While the Philippines had a regulatory body before liberalisation, it was captured by a politically influential monopoly. The creation of a strong and credible institution to regulate the newly liberalised industry proved to be difficult, especially because doing so needed congressional intervention and not merely an executive order. Thus, it took the NTC, in collaboration with a USAID-funded group, to finally work on regulatory reforms. Both Malaysia and the Philippines relied on the technical support of international consultants to draw up new regulatory policies, but unlike the Philippines, the Malaysian government hired consultants and did not rely on aid from international or multilateral agencies.

A SUMMARY OF PHILIPPINES' TELECOMMUNICATIONS REFORMS

Table 10-3 summarises the highlights of telecommunications reforms in the Philippines.

Chapter 8 detailed how the telecommunications sector was liberalised following the failure of attempts to do so under the Aquino administration. The discussion focused on Ramos' key role and support for reform, enabling the adoption of liberalisation as policy in this sector. In addition to the President, the work of the coalition for reform that pushed for liberalisation in the sector was important. The chapter also looked at the responses of PLDT, which opposed attempts to liberalise the industry. Given this, the discussion highlighted the crucial role of Almonte in convincing Ramos that liberalisation of the telecommunications industry was necessary.

Table 10-3: Highlights of Telecommunications Reform in the Philippines

	Philippines
Year of reform	1993
Type of reform	Liberalisation
Regulatory body	PSC (1936) NTC (1979 to present)
Pre-reform market structure	Private monopoly (PLDT) but existence of data carriers and small provincial companies
Post-reform market structure (2002)	Competitive, but existence of dominant carrier (PLDT); 8 players (PLDT, Globe, Digitel, Bayantel, EPTI, Philcom, PT&T, and Extelcom)
Major telecommunications laws	EO 59, EO 109, and RA 7925 of 1995
Interconnection	Mandatory but problematic
Universal service obligation	Service area scheme, based on cross-subsidy
Convergence	Under study

After the removal of opposition to liberalisation, the NTC oversaw the implementation of reform by creating the Service Area Scheme, which became the over-all plan for liberalisation. Finally, the chapter discussed Republic Act 7925, and how, because of the equal power of Congress, and PLDT's influence on the law's major authors, some of the gains of liberalisation were tempered. This showed how in a situation where vested interests were influential and state power was shared among three co-equal branches, reform was possible but its sustainability was tenuous.

The work of a coalition for reform and the President's exercise of political will made possible the occurrence of reform in the Philippines. This defied expectations that the state was incapable of introducing policy changes that go against vested interests. Yet, the sustainability of reform was tenuous because state power was dispersed, and thus, there were other areas for rent-seekers to influence the process. While almost all of the new entrants were also economically powerful, the battle over regulatory rules favoured the former monopolist, which delayed interconnection. By not specifying in RA 7925 what the NTC's role should have been, Congress gave PLDT areas to manoeuvre.

The nature of groups that support reform directly influenced whether or not rents were created after market liberalisation. If the support for liberalisation came from businessmen who wanted market entry, and the state held licensing power to allow that entry without any clear criterion, such a process created rents for those who lobbied for market opening. This was the case in Malaysia. In contrast, the group that supported market reform in the Philippines was composed of advisers to the President who believed that market liberalisation was necessary for the economy to grow. The coalition planned and strategised how reform would take place but were not the beneficiaries of market entry. Liberalisation allowed the entry of anyone who wanted market access as long as they demonstrated financial capacity. Market reform in the Philippines eliminated PLDT's monopoly rent, which was one of the goals of the reformers. In contrast, liberalisation in Malaysia created market entry rent for the politically well-connected businessmen who lobbied for liberalisation.

Chapter 9 introduced the new entrants to the sector in the Philippines. All, except two, were existing companies owned by politically influential and established business families. All new entrants to the sector linked up with foreign partners before starting operations. The section also presented an assessment of the SAS, enumerating the criticisms about, and the gains of, the scheme. All of the new entrants faced debt problems and a decline in demand after the 1997 financial crisis. Two mergers took place, one negotiated privately, while the other involved the intervention of no less than then President Estrada. Estrada's intervention meant a windfall of kickbacks for him and his men.

Finally, this chapter analysed regulatory reform in Philippine telecommunications. The regulatory body was weak, did not have enough human and financial resources, and was captured by the regulated. The NTC was also politically compliant, with the regulators serving at the pleasure of the President. Presently, regulatory reform is being put in place by an aid-funded project.

In sum, this study on Philippine telecommunications reform has presented three major arguments. First, despite the weakness of the Philippine state, the adoption of market-oriented reform that went against vested interests was made possible through the work of a coalition for reform that had the support of President Ramos. The occurrence of reform in the face of intense opposition from influential vested interests shows that reform can take place even when the state is weak and captured. Such situations challenge the current dismissal of the Philippine state as incapable of adopting policy reform. Second, those who supported the liberalisation of the industry were not businessmen or rent-seekers who wanted market entry. The Philippine case demonstrates that it is possible to have a group of reformists who are neither rent-seekers nor direct beneficiaries of market entry. The coalition for reform benefited only in so far as the prices for telecommunications services declined, phones became available, and teledensity increased. Finally, telecommunications market liberalisation in the Philippines removed monopoly rents. Market licenses were secured by anyone who wanted entry to the sector and had the financial qualifications to do so. Market reform removed the rents enjoyed by an influential family, improved the sector's performance, provided better infrastructure for the country, and boosted the economy.

NOTES

1. Section 15 and Section 16-a of Commonwealth Act 146. Until 1946, the Philippines was under American colonial rule, thus the favourable treatment of American citizens and their economic interests.
2. Sections 16-c, 16-h, Section 17-f, h, and Section 21 of Commonwealth Act 146. Section 21 provided for a fine of no more than P200 per day for the duration of the violation.
3. Ahmed Galal and Bharat Nauriyal, "Regulating Telecommunications in Developing Countries", *World Bank Policy Research Working Paper 1520* (October 1995): p. 12.

4. EO 125-A, Section 4 and 14-a.
5. Department of Transportation and Communications, Department Circular 90-248: Policy on Interconnection and Revenue Sharing by Public Communications Carriers (14 June 1990).
6. Department of Transportation and Communications, Department Circular 90-252: Policies to Promote the Rapid Expansion of Telecommunications in all Regions (10 August 1990).
7. Ibid., pp. 6–7.
8. Department of Transportation and Telecommunications, *National Telecommunications Development Plan, 1991–2010, Executive Summary*, October 1990, p. 1.
9. Ibid., p. 3.
10. http://www.ntc.gov.ph/consumer-frame.html (accessed on 17 June 2003).
11. Given Lichauco's involvement with the drafting of the NTDP, it is not surprising that she was very critical of the SAS that the NTC under her successor Simeon Kintanar drew up to implement EO 109.
12. Interview with Josefina Lichauco, 19 March 2001, Makati City; See also Coronel (1998), pp. 132–33.
13. National Telecommunications Commission Memorandum Circular 7-13-90, "Rules and Regulations Governing the Interconnection of Local Telephone Exchanges and Public Calling Offices with the Nationwide Telecommunications Networks, the Sharing of Revenues derived therefrom and for other purposes", Sections 2-1, 2-5, 2-6, 3-1, and 3-4.
14. Gavino, Jacinto, *A Critical Study of the Regulation of the Telephone Utility: Some Options for Policy Development*, Ph.D. dissertation, College of Public Administration, University of the Philippines (1992).
15. *Daily Globe*, 15 November 1989.
16. Gavino, 1992, p. 108.
17. Ibid., p. iii.
18. Senate Bill 1353, and its counterpart, House Bill 32327, entitled "An act to promote the development of Philippine telecom and the delivery of Public Telecom services," were filed on 19 October 1989 and emphasized the idea that telecommunications development was a private sector function, thus calling

for deregulation, a clarification of the NTC's responsibilities and independence, and the unbundling of the rates charged by telephone companies to make them more transparent.

19. House Bill 20989, "An Act to Reorganize and Strengthen the NTC", advocated minimum regulation to encourage private sector participation in telecommunications development.

20. Interview with various telecommunications executives and with Simeon Kintanar, Quezon City, 20 March 2001.

21. Firms interested in offering radio paging and value-added services were exempted from acquiring a congressional franchise and were only required to register with the NTC, provided that they did not construct their own telecommunications infrastructure.

22. Republic Act 7925, Section 18.

23. Coronel (1998), p. 143.

24. Republic Act 7925, Section 4-j.

25. Abrenica and Llanto (2003), p. 268. The international accounting rate is the fee paid by an originating international carrier to the receiving carrier for completing a call. PLDT traditionally relied on this segment of telephone service for its dollar earnings, as there are more incoming than outgoing international calls. This rate, however, has been declining due to competitive pressures and new technologies. See *Philippine Daily Inquirer*, 20 February 2003.

26. For instance, the DOTC drafted the National Telecommunications Development Plan 1991–2010 with Canadian consultants, Teleconsult and Canadian International Development Aid's financial support. On the part of the NTC, funding was used to help carry out a Frequency Management study in the mid-1980s and draw up a Manual of Practices and Procedures in 1990. Interview with Josefina Lichauco, Makati City, 19 March 2001.

27. Various aid grants such as from Japan and Italy funded the phases of the Regional Telephone Development Project (RTDP).

28. Instead, Jaime Dichaves, an Estrada crony, who was in the business of supplying telecommunications equipment, reportedly influenced the appointment of NTC commissioners in the anticipation of influencing their decisions. Dichaves is said to have been responsible for the early dismissal from office of Ponciano Cruz and his two deputy commissioners, who led the NTC from June to December

1998, because they had not acted according to his bidding. Dichaves' modus operandi was allegedly to make sure that he knew ahead of time who would obtain a new permit, authorization, or licence from the NTC so that he could make them buy overpriced equipment from his company.

29. Carana Corporation, "A Framework for Examining USAID Involvement in Infrastructure Final Report Submitted to United States Agency for International Development", 1999, http://www.carana.com/knowledge/infrastruct.pdf (accessed 15 January 2003).
30. Ibid., p. 26.
31. According to Mario Lamberte, one of the consultants in the SGV project, researchers from Secretary Almonte's office approached him for a copy of the *Barriers to Entry Study*. Interview with Dr Mario Lamberte, Makati City, 21 December 2001.
32. Ricardo Monge and Francisco Galema, *The Philippines Reform on Telecommunications: The Impact of USAID's Support* (Carana Corporation, 1999), p. 21.
33. Ibid., pp. 23–29.
34. Then Finance Secretary Roberto de Ocampo and USAID Director Kenneth G. Schofieled signed the "Joint Project Implementation Letter No. 55 (JPIL No. 55)" to the "Philippine Assistance Program Support (PAPS) Project No. 492-0452" on 21 November 1997, which created the AGILE project. See *Manila Times*, 23 February 2003.
35. Interview with Mr Jaime Faustino, Investment Policy Advisor, AGILE, Pasig City, 23 April 2001.
36. http://www.dai.com (accessed on 6 June 2003).
37. *Manila Times*, 23 February 2003.
38. AGILE brochure, no date.
39. Aside from the NTC, other government agencies such as the Departments of Finance, Budget and Management, Trade and Industry, Agriculture, Transportation and Communications, Tourism, the National Economic and Development Authority, the Bureau of Internal Revenue, the Bureau of Customs, the Bangko Sentral ng Pilipinas (Central Bank of the Philippines), the Securities and Exchange Commission, and the Philippine Stock Exchange received technical support from AGILE.

40. The ten-point programme is summarized in Appendix 5.

41. Interview with Edgardo Cabarios, Quezon City, 31 January 2001.

42. Appendix 5 provides a summary of the outcomes of the support that AGILE had given to the NTC as of 2002.

43. *Philippine Daily Inquirer*, 19–23 March 2003; *Manila Times*, 24 February 2003, 28 February–2 March 2003.

44. Interview with Jose Almonte, Paranaque City, 4 April 2001; interview with Antonio Carpio, Makati City, 3 May 2001.

11

CONCLUSIONS

The puzzle of why the Philippines was a laggard in terms of economic development, relative to its neighbours in Southeast Asia provided the backdrop to this study. Its poor economic performance had been attributed to a weak state captured by a strong oligarchic class. In contrast, Malaysia's capable and strong state was considered a major factor in its more successful development. Although the post-colonial political and social systems in both countries were substantially different, patronage and rent-seeking stood out as clear similarities. Yet, they seemed to have had different effects. While the persistence of patronage and rent-seeking had apparently not prevented growth in Malaysia, they were an important explanation for the economic underdevelopment of the Philippines.

With this as background, this book's goal was to address debates on: the possibility and sustainability of policy reform in states with varying capacities; the impact of different coalitions or groups supporting reform; and the consequences of rent-seeking and rent-outcomes during market liberalisation and their role in promoting or hindering economic development. To this end, the study answered three major questions: First, how did the strength or weakness of the Malaysian and Philippine states and the presence of rent-seeking and patronage networks affect the possibility and sustainability of policy reform? Second, who made up the crucial constituencies for market-oriented reform and to what extent did they affect the outcomes of reform? Third, what were the outcomes of market liberalisation? Did market liberalisation invariably remove rents? If not, were the effects of policy change in the face of rent-seeking always deleterious? What sorts of rents were created, who captured them, and how?

To address these questions, this book focused on the reform of the telecommunications sector — one of the first to be liberalised in both countries — as a source of empirical material. In analysing the case studies, the study used a political economy approach to investigate how reform took place and what outcomes it had in each country. The aim of using a political economy approach to studying policy reform was not so much to analyse the intended economic effects of policies, which is of prime interest to economists. Rather, policy changes were viewed as outcomes of interactions among politicians, bureaucrats, and social actors operating within a given set of institutional constraints.

This chapter summarises the study's central arguments and assesses the empirical evidence provided by the reform of the telecommunications sector in Malaysia and the Philippines. While generalisations gleaned from case studies only have limited applicability, they nonetheless provide a glimpse into the political economy dynamics of both countries. Such insights can lead to a greater understanding of why one state had been more capable of pursuing developmental goals than the other.

ON THE MALAYSIAN AND PHILIPPINE STATES AND POLICY REFORM

Chapter 3 discussed Malaysia's colonial heritage, the rise of a strong state with authoritarian legal instruments, and a racially divided society. Political power was in the hands of Malay elites while Chinese business occupied a strong but subordinate position in a foreign-dominated economy. The 1969 racial riots marked a turning point, where Malay domination of the state was enhanced. After 1981, with Mahathir Mohamad as Prime Minister, the state became more pronouncedly Malay and authoritarian, with state power increasingly centralised in the hands of the Prime Minister. Chapter 3 also described the development of a weaker Philippine state, alongside the emergence of an oligarchy that penetrated and captured it. In contrast to Malaysia, the same elite group controlled both economic and political power in the Philippines. This group's control of political power, which was institutionally dispersed in three different branches of the state, enabled it to use the state apparatus for private wealth accumulation and expansion.

The post-colonial states in both Malaysia and the Philippines played an active role in their economies. However, the type of intervention and their outcomes were different. The Malaysian state can be considered strong and developmental, as it used its power to intervene in the economy for developmental and redistributive reasons, purportedly for the creation of a new bumiputera economic class. In the Philippines, the oligarchic elite used their privileged access to the weak state for the purposes of private wealth accumulation. Given this backdrop, the adoption of similar market-oriented economic policies was analysed to gauge the importance of differing levels of state capacity and the composition of the business class. Looking at the first puzzle, the type and capacity of the states in each country shaped how reforms were adopted and whether or not they were sustainable in the long run.

While the Malaysian state may not fit the developmental state mould of Japan, Korea, and Taiwan, it nonetheless exhibited characteristics that indicated its developmentalism and strength. After 1969, the Malaysian state articulated clear developmental goals in the NEP and demonstrated determination and commitment in pursuing them. In 1981, Mahathir inherited a strong state with a working bureaucracy and authoritarian legal instruments, which he employed to bolster his control over his party and the state. He used existing laws to limit the power of the political opposition and other branches of government. While consolidating his political position, Mahathir invigorated the pursuit of the NEP goals and articulated a corollary set of strategies to achieve them. He introduced privatisation, Malaysia Inc., Look East, and the heavy industrialisation program to industrialise the country's economy and to create a bumiputera business class. Of these policies, privatisation has been one of the most controversial and far-reaching, especially with regards to creating a class of bumiputera businessmen. The way privatisation was adopted and implemented demonstrated the centralisation of decision-making in the Prime Minister's hands. This pointed to a strong state, dominated by a powerful executive, with the capacity to set the policy agenda and actively pursue it. It helped that UMNO was in firm control of the BN government, providing Mahathir a ready-made coalition of supporters. Furthermore, it gave him a long time frame in office, ensuring the continuity and pursuit of pet policies and developmental goals.

The Malaysian state demonstrated strength and autonomy in its adoption of the general privatisation framework. Yet, while it can decide general policies autonomously, rent-seeking drove its implementation in the privatisation of JTM and the liberalisation of the telecommunications industry. Rent-seeking businessmen did not formulate the general policy. Yet they used their political influence to ensure that they became the beneficiaries as they saw opportunities in the state's declaration of privatisation. They proposed the privatisation of JTM even before the overall policy was fleshed out. In the same manner, liberalisation was also carried out, partly in response to the rent-seeking lobby. The telecommunications sector was privatised and liberalised without clear policy guidelines as a result of lobbying by politically-influential businessmen.

The existence of this lobby and its influence on policy implementation did not negate the fact that the Malaysian state was strong and autonomous in its adoption of the general policy framework. The state was in control and allowed rent-seeking to take place. Mahathir's ultimate control over the state was shown by how, after the 1997 financial crisis and the fall-out with Daim, he personally got involved in the renationalisation and restructuring of big privatised companies and conglomerates owned by the formerly politically well-connected. The goal of creating successful bumiputera businessmen through the award of licenses and privatisation contracts largely failed in the case of the telecommunications sector. Yet, Mahathir had sufficient power and political will to reverse policies even against the wishes of the rent-seeking lobby. He could do so because of the inherent strength of the Malaysian state and his own unchallenged position in it.

Thus, the Malaysian state was strong and developmental and, at the same time, cultivated the support of a rent-seeking, predominantly Bumiputera, business class. The case study illustrated that the state adopted liberalisation as a means of fomenting a bumiputera business class through purposive rent-creation. While the state set the general policy for privatisation, rent-seekers trouped to maximise the new opportunities and shaped the implementation process of telecommunications privatisation and liberalisation. This took place in the face of opposition from the bureaucracy, notably the JTM and the employee's union, and from consumer groups, academics, and the political opposition.

In contrast to the consensus in academic circles that the Malaysian state was strong, there was a consensus among analysts and within a section of Philippine society that the Philippine state was weak and captured by an oligarchic class. The Philippine oligarchy used the state to expand its businesses and at the same time, protect itself from both domestic and external competition. Marcos promised to reform this situation when he declared martial law but ended up using the state for the enrichment of his family and his cronies. After the downfall of Marcos, resilient oligarchs and cronies continued to be influential. The Cojuangcos of PLDT are a celebrated example of this capacity to remain influential from one administration to the next. Thus, the reform of the telecommunications industry was problematic — given the existence of politically influential owners that had so far successfully resisted competition.

The ability of a weak, captured state to introduce policy reforms that challenge the vested interest of influential oligarchs was surprising. This was possible under the Ramos administration, through the president's political will and decisiveness. Ramos started his presidency promising to introduce competition and break down the oligarchy's control of the economy. The President found support from a coalition of reformers that agreed with his diagnoses of the country's needs. Unlike Malaysia, where the Prime Minister had a ready coalition of supporters and weak opposition to policy change, Ramos needed to create an atmosphere of desirability of reform. As was detailed in Chapter 8, PLDT did not take the efforts to liberalise the industry lightly. It flexed its economic and political muscle to defend its position. The owners of the company attempted to foil the market-opening exercise by attacking the reformers themselves. The efforts of the coalition for reform and, ultimately, the President's decision finally led to the opening-up of the telecommunications industry. This situation showed that, despite a weak state and the strength of an influential family, liberalisation took place in the Philippines.

The chief executives of both the Malaysian and Philippine states played key roles in introducing market reforms. Yet, Mahathir and Ramos had different circumstances and power bases. In Malaysia, the state was strong, with its inheritance of a working bureaucracy and authoritarian laws. Mahathir successfully used existing state powers and moved to centralise it even more, in the context of the Barisan Nasional's (BN) unchallenged control of the state. In Malaysia's parliamentary system, he could rely on the BN with its overwhelming majority in parliament and his control of UMNO as the BN's dominant party. Thus, given the existing institutional structure, a modernising Prime Minister with developmental goals like Mahathir had a ready coalition of supporters within BN and the bureaucracy. In addition, he recruited his own set of advisers to generate ideas for him.

In contrast, Ramos was faced with a more limited set of choices and had to create a clear constituency for reform. The latter was crucial in the move to break-up a politically influential monopoly. Thus, the coalition for reform played an important role in making apparent a public that was clamouring for change. Meanwhile, President Ramos and the coalition for reform knew that their actions could only go so far given the uncertainty of gaining the support of Congress and the prospect of challenges in the Courts. As the case had demonstrated, other branches of government were open to rent-seeking influences and watered down or reversed some of the executive's actions. Nonetheless, without the executive-led reform, industry liberalisation would not have proceeded, despite the unfolding technological changes in the industry. Other things being equal, there was nothing to stop PLDT from blocking new entrants as it had done successfully during the Aquino administration. The unique combination of the coalition for reform's actions and the President's commitment to open the market made the liberalisation of the telecommunications industry possible.

Despite the weakness of the Philippine state, the case study challenged the usual expectation that a captive and weak state was incapable of going against vested interests. However, the question of continuity of reform and its replication in other sectors and at other times was constrained by the presidential structure of the Philippines. Unlike Malaysia where the parliamentary majority routinely supported the Prime Minister, Congress can reverse the actions of the President. Moreover, the President, unlike the Prime Minister in Malaysia, had a fixed six-year term of office. This made continuity of policies difficult, as the next President was not bound to pursue his predecessor's policies. Yet, the occurrence of reform pointed to the possibility of change even where the state was weak and captured.

To summarise, the evidence supported the first argument of this study. The conventional depiction of the Malaysian and Philippine states as respectively strong and weak were challenged by the experience of reform in the telecommunications sector. More specifically, while the Malaysian state was considered strong and capable, and while there was a general policy framework for privatisation, the liberalisation of the telecommunications industry showed the influence of politically well-connected rent-seekers. The influence of these networks raised doubts as to whether the Malaysian state was really as strong and insulated as it was usually portrayed. The study demonstrated that the Malaysian state was simultaneously strong and developmental and, at the same time, penetrated by rent-seeking. In the case of the Philippines, despite the weakness of the state and the absence of a crisis or international pressure compelling liberalisation, reform occurred in an industry controlled by a member of the oligarchy. This was due to the President's commitment and the work of a coalition for reform that out-manoeuvred vested interests. This finding cast doubts about the usual portrayal of the Philippine state as weak and incapable, and pointed to the possibilities of reform despite the expectations to the contrary.

ON SUPPORT FOR REFORM

The second issue that this study looked at was the role and type of support for reform. In Malaysia, rent-seeking, politically well-connected businessmen lobbied for liberalisation of the telecommunications industry. They technically cannot be called a coalition, as they competed with each other for market entry licenses. Nonetheless, these rent-seekers were the source of a major push for the liberalisation of the industry, aware of the potential benefits that they could enjoy with the policy change. There was no transparency in the license-awarding process, and all but one of the licensees were politically well-connected businessmen. The non-transparent process of liberalisation precluded the possibility of non-politically well-connected businessmen entering the industry. Furthermore, such a possibility was remote because part of the policy's purpose was to create a bumiputera business class while rewarding government supporters. As the cases showed, one of the licensees did not have patronage links to top politicians. Despite the company being owned by businessmen with telecommunications expertise, it was the only company required to apply through the EPU, show that it provided a technological solution to the problem of low telephone coverage, and demonstrate their proposal's viability. None of the politically well-connected businessmen went through this process. Moreover, the company, STW, was only granted a pilot license subject to a performance audit. When this was achieved, STW was the first to be crowded-out of the industry.

At the start, the government created a duopoly (TM and Celcom), but this lasted only five years. When more businessmen lobbied to enter, the government awarded licenses to all who sought entry without giving much consideration to what the market could bear. This arose because each of the politically influential businessmen wanted a slice of the market. Thus, even though the Minister of Energy, Telecommunications, and Posts knew that too many licenses were issued and called for rationalisation, it was difficult, if not impossible, to impose limitations due to the players' political influence. Whether there were other potential entrants to the industry who lacked political influence was difficult to establish because the liberalisation process was not transparent. The award of licenses itself was only announced after the fact. Thus, potential entrants would not have known of opportunities, except the politically well-connected that lobbied for it. Secondly, the fact that many licenses were issued instead of maintaining a duopoly, showed that there was competition among influential rent-seeking businessmen who were linked to the top three politicians. The proliferation of licenses reflected the influence of these businessmen and took place despite the advice of the regulatory body and the Ministry.

In the Philippines, the group that supported liberalisation reform was composed of academics and professionals who were advisers to then President Ramos. National Security Adviser Jose Almonte tapped reform-minded Filipinos to concretise and put flesh into Ramos' promise of freeing the economy from oligarchic control. This group can be categorised as a coalition for reform, which worked to pursue policy changes that they believed were necessary for the country. Most of them remained anonymous after the reform. In comparison to the Malaysian case, the members of the coalition were neither the direct beneficiaries of market entry nor were they stakeholders in the new companies that were granted entry. The businessmen who became beneficiaries of market entry initially viewed the actions of the coalition with suspicion, worried that they might be the next targets. The main goal of liberalisation was to destroy monopoly rents enjoyed by an influential company and to introduce competition in an underdeveloped industry.

Thus, the cases supported this study's second argument. The nature of groups that supported reform shaped the policy's implementation and outcomes, specifically in determining whether rents were created in the liberalisation process. In Malaysia, the promoters of liberalisation sought market access while the supporters of liberalisation were not the market entry beneficiaries in the Philippines.

ON RENT OUTCOMES

With regards to the third issue, the study provided evidence supporting the argument that market reforms do not always remove rents. In situations where the main support for liberalisation comes from rent-seekers, market-entry rents are likely to be created. As the Malaysian case demonstrated, market liberalisation created rents obtained by businessmen who lobbied for them. Meanwhile, because the main push for telecommunications liberalisation in the Philippines came from actors who had no stake in the industry and whose main goal was to dismantle a monopoly, liberalisation removed monopoly rents.

The Malaysian state purposively created market-entry licenses as "political transfer rents" and awarded them to politically influential businessmen connected to the top three politicians. These licenses in turn were actively sought by the politically well-connected through existing patronage networks. Following Khan's categories of rents, a license to enter the market in this case was considered a political transfer rent created by the state to redistribute income, with the aim of creating a new economic class.[1] The state controlled the distribution of these licenses and the criteria for distribution were not made public. However the telecommunications industry was a network industry that was highly competitive, as well as capital- and technology-intensive. Thus, the license, even if it was a type of rent, did not automatically translate into profit in the classic sense as defined by Buchanan: an income above the normal in a competitive market. Instead the potentially lucrative license can be considered a hybrid type of rent that had to be reaped in a competitive market.[2] Because many influential businessmen were allowed access to the potentially lucrative business, the industry became overcrowded and highly competitive. Some licenses were managed well, but those without expertise and foreign partners, mostly bumiputera and UMNO-linked, had to be bailed out after the 1997 financial crisis. The state played an active role in what was a costly exercise. The fact that the Ministry initiated a call for mergers as early as 1995 shows that the licensees were quite influential and could deflect calls for mergers.

Meanwhile, market entry licenses in the Philippines were not considered rents, because the entry and exit process was transparent and open to anyone who satisfied the requirements. Liberalisation destroyed monopoly rent that was enjoyed by an influential family through the introduction of competition. The market became very competitive although the former monopolist still held dominant position. The creation of service areas for new entrants would technically have been monopoly areas for them, but the need to interconnect to a national network controlled by the former monopolist kept this in check. Thus, all players were competing in all market segments (fixed, mobile, IGF, and value added), and though the former monopolist was still dominant (bolstered by its merger with Smart, the number one mobile company), the existence of other companies kept the dominant player in check and led to lower costs and more services.

The cases under examination supported Khan's arguments that not all rents were monopoly rents, and that rent outcomes or effects in terms of growth and efficiency were not as clear-cut as the neo-classical model held.[3] Both the Malaysian and Philippine telecommunications reform led to positive outcomes in the sector in particular, and to the economy in general. The Philippine case had a more straightforward economic effect, with the removal of monopoly rent as soon as liberalisation introduced competition to the industry. In the case of Malaysia, however, the state had to bail out those who failed, especially the bumiputera-owned companies. Such an exercise was costly. Yet, notwithstanding rent-seeking and the bailouts, one cannot deny that the entry of new players to the industry had generated investments, increased teledensity, reduced prices, and improved services. In the Malaysian case, the rent outcomes after liberalisation were mixed. This is because the Malaysian state had two goals: economic development and redistribution in favour of the Malays. The latter goal overrode the former. Thus, the rent-outcomes in this case were not easily assessed in terms of gains and losses to the economy per se. While economic improvements were present, there were also heavy costs for the state, which was not merely interested in economic efficiency considerations but also redistributive goals.

In assessing the outcomes of reform, both cases similarly evolved from a monopoly position (albeit one public and the other private) into a competitive industry. Both countries also experienced interconnection problems, with the dominant player reluctant to interconnect with new entrants.

Nevertheless, there were more differences than similarities in their experiences. Firstly, in the Philippines, the process of entry and exit was clear and transparent. In Malaysia, it was not a transparent process, and entry was only accessible to the politically well-connected. Second, in the Philippines, all new entrants, except one, came from established business families. In this regard, liberalisation saw the re-emergence of the old business elite in a newly opened market sector. Meanwhile, the state's overriding goal in Malaysia was to create a new business class. Third, given the capital- and technology-intensive nature of the business, all new entrants in the Philippines took foreign partners before starting operations. In Malaysia, however, only two companies sought foreign partners at the outset. Finally, in the Philippines, the Service Area Scheme (SAS) guided the liberalisation of the industry. Despite the criticisms and limitations of the SAS, it provided a clear deployment plan that helped to minimise overlapping and ensured that rural areas were not left out. In Malaysia, by contrast, there was no similar plan. Despite the fact that new companies had licenses to operate fixed and mobile services, all new entrants (except STW and Time Telekom) concentrated on mobile telephones, leading to duplication of networks in urban areas, mainly the Klang Valley. Also, new companies only had token investment in fixed lines, which Telekom continued to control.

To summarise, this book had shown that over the period under examination, the Malaysian state was developmental and strong but, at the same time, penetrated by a rent-seeking lobby. The Philippine state meanwhile was weak, and strong oligarchic families preyed on it. Yet, significant reform took place in both countries. The goal of the study was to explain the similarities and differences of the reform process and outcomes in different political environments.

The Malaysian state adopted liberalisation to purposively create rents that were distributed to a well-connected few — something that neoliberals who prescribe liberalisation would not expect to be a major objective of market opening exercises. That said, some scholars have highlighted this possibility as liberalisation meant changing the property rights regime, thus creating venues for rent-seeking.[4] The Philippine case, meanwhile, provided evidence consistent with neoliberal expectations, where the introduction of competition in a formerly monopoly situation removed monopoly rent, and improved industry performance and service quality. Khan argued that the growth effects of rent-outcomes vary from case to case, depending on a country's "political settlement" or the balance of power among state and social actors.[5] In the case of the Malaysian telecommunications sector reform, the creation of many new licenses led to both positive and negative growth effects. The positive impacts were the increase in teledensity, presence of choice, and decline in prices. However, the non-quantifiable and so far concealed negative costs of liberalisation were state bailouts and hidden subsidies. Thus, liberalisation of telecommunications in Malaysia had mixed economic effects. One of the reasons is that the state used liberalisation not merely for conventional reasons but more importantly, for redistributive purposes, which at that time contradicted the economic efficiency goals.

This study of liberalisation in telecommunications was placed in the context of the big question of why Malaysia' economic performance had been impressive while the Philippines had fallen behind. It provided empirical answers by looking at the cases and outcomes of similar policies in the telecommunications sectors of the two countries. The answers provided were partial but indicative of the richness and messiness of reality. The Malaysian state overall still enjoys better economic growth rates than the Philippines, despite costly bail outs of favoured businessmen as seen in the telecommunications industry. Meanwhile, the Philippines, has experienced a higher economic growth rate since the 1990s, but one that is still slow compared to its Southeast Asian neighbours. The state continued to be weak, the bureaucracy needed strengthening, and politically influential families continued to attempt to sway state decisions in their favour. Yet, the post-reform telecommunications industry, in particular, was a bright spot in the economy and perhaps one of the most competitive and fastest-growing sectors.

NOTES

1. Mustaq Khan, "Rents, Efficiency and Growth", in *Rents, Rent-Seeking and Economic Development: Theory and Evidence in Asia,* edited by Mustaq Khan and K.S. Jomo (Singapore: Cambridge University Press, 2000), p. 25.
2. Huizhong Zhou, "Rent-seeking and Market Competition", *Public Choice* 82, Nos. 3–4 (1995): 225–41.
3. Khan (1996), pp. 21–25. Ha-Joon Chang also makes the same argument when he argues that "rent-seeking may also be indirectly productive". See Ha-Joon Chang, *The Political Economy of Industrial Policy* (New York: St Martin's Press, 1994), pp. 27–31.
4. Chang (1944), p. 44. See also John Vickers and George Yarrow, "Economic Perspectives on Privatisation", *Journal of Economic Perspectives* 5, no. 2 (Spring 1991): 130.
5. Mustaq Khan, "A Typology of Corrupt Transactions in Developing Countries",s *IDS Bulletin* 27, no. 2 (1996): 18. Chang argues along the same line when he wrote that "whether competition, be it economic or political, is wasteful or not … can only be determined with reference to the concrete political economy of the society concerned, and not on the basis of simplistic model of political economy envisioned by the proponents of rent-seeking …" See Chang (1994), p. 45.

Appendices

APPENDIX 1

Selected Economic Indicators for Malaysia and the Philippines

A1-1: GDP, GDP Growth, and GDP per Capita

	GDP (Constant 1995 US$)		GDP Growth (Annual %)		GDP per Capita (Constant 1995 US$)	
	Malaysia	Philippines	Malaysia	Philippines	Malaysia	Philippines
1960	7,936,468,000	19,606,760,000	—	—	975	725
1961	8,539,481,000	20,707,990,000	7.60	5.62	1,015	742
1962	9,087,803,000	21,696,410,000	6.42	4.77	1,046	753
1963	9,754,740,000	23,228,280,000	7.34	7.06	1,088	782
1964	10,277,490,000	24,028,960,000	5.36	3.45	1,113	785
1965	11,067,300,000	25,294,280,000	7.68	5.27	1,165	801
1966	11,932,400,000	26,413,800,000	7.82	4.43	1,221	812
1967	12,392,650,000	27,820,020,000	3.86	5.32	1,234	830
1968	13,381,340,000	29,196,120,000	7.98	4.95	1,298	846
1969	14,035,490,000	30,554,940,000	4.89	4.65	1,327	860
1970	14,875,730,000	31,705,780,000	5.99	3.77	1,371	867
1971	15,731,600,000	33,426,970,000	5.75	5.43	1,414	889
1972	17,207,520,000	35,247,750,000	9.38	5.45	1,508	911
1973	19,223,240,000	38,391,500,000	11.71	8.92	1,644	965
1974	20,820,620,000	39,757,810,000	8.31	3.56	1,739	972
1975	20,987,540,000	41,970,710,000	0.80	5.57	1,712	999
1976	23,414,500,000	45,666,230,000	11.56	8.81	1,866	1,058

1977	25,230,640,000	48,224,590,000	7.76	5.60	1,966	1,087
1978	26,908,990,000	50,718,730,000	6.65	5.17	2,049	1,113
1979	29,424,740,000	53,579,490,000	9.35	5.64	2,189	1,145
1980	31,615,190,000	56,338,150,000	7.44	5.15	2,297	1,173
1981	33,809,900,000	58,266,760,000	6.94	3.42	2,397	1,184
1982	35,818,520,000	60,375,630,000	5.94	3.62	2,476	1,197
1983	38,057,270,000	61,507,440,000	6.25	1.87	2,563	1,190
1984	41,011,240,000	57,002,830,000	7.76	–7.32	2,689	1,077
1985	40,550,990,000	52,837,860,000	–1.12	–7.31	2,587	974
1986	41,018,350,000	54,643,210,000	1.15	3.42	2,541	984
1987	43,228,680,000	56,999,220,000	5.39	4.31	2,599	1,002
1988	47,524,630,000	60,848,120,000	9.94	6.75	2,772	1,045
1989	51,829,640,000	64,623,940,000	9.06	6.21	2,933	1,084
1990	56,499,310,000	66,586,550,000	9.01	3.04	3,104	1,091
1991	61,892,430,000	66,201,450,000	9.55	–0.58	3,317	1,060
1992	67,391,640,000	66,424,950,000	8.89	0.34	3,523	1,040
1993	74,060,010,000	67,830,710,000	9.89	2.12	3,777	1,038
1994	80,882,450,000	70,806,860,000	9.21	4.39	4,023	1,060
1995	88,832,450,000	74,119,700,000	9.83	4.68	4,310	1,085
1996	97,718,100,000	78,452,640,000	10.00	5.85	4,625	1,122
1997	104,873,700,000	82,520,690,000	7.32	5.19	4,840	1,154
1998	97,155,650,000	82,044,780,000	–7.36	–0.58	4,380	1,121
1999	103,118,700,000	84,831,350,000	6.14	3.40	4,541	1,133
2000	111,616,900,000	88,231,580,000	8.24	4.01	4,797	1,151
2001	112,057,100,000	91,231,670,000	0.39	3.40	4,708	1,165

— = No data.

Source: *World Development Indicators*, World Bank at http://www.worldbank.org/data/wdi2002/, accessed on 10 November 2003.

A1-2: Total Population, Telephone Mainlines, and Official Exchange Rate

	Total Population		Telephone Mainlines (per 1,000 People)		Official Exchange Rate (LCU per US$, Period Average)	
	Malaysia	Philippines	Malaysia	Philippines	Malaysia	Philippines
1960	8,140,000	27,055,000	5.8		3.06	2.01
1961	8,417,040	27,920,610	—		3.06	2.02
1962	8,691,070	28,804,420	—		3.06	3.73
1963	8,962,800	29,706,740	—		3.06	3.91
1964	9,232,890	30,627,830	—		3.06	3.91
1965	9,502,000	31,568,000	8.3	2.5	3.06	3.91
1966	9,770,730	32,530,120	—		3.06	3.90
1967	10,039,650	33,508,240	—		3.06	3.90
1968	10,309,330	34,503,800	—		3.06	3.90
1969	10,580,290	35,518,240	—		3.06	3.90
1970	10,853,000	36,553,000	9.5	4.7	3.06	5.90
1971	11,128,390	37,607,120	—		3.05	6.43
1972	11,407,770	38,679,640	—		2.82	6.67
1973	11,690,210	39,772,000	—		2.44	6.76
1974	11,973,810	40,885,640	—		2.41	6.79
1975	12,258,000	42,022,000	13.7	6.6	2.39	7.25
1976	12,544,900	43,180,340	15.4	6.9	2.54	7.44
1977	12,836,590	44,360,750	17.7	8.0	2.46	7.40
1978	13,135,120	45,563,290	20.6	8.0	2.32	7.37

1979	13,442,960	46,788,030	24.1	8.2	2.19	7.38
1980	13,763,000	48,035,000	28.7	8.6	2.18	7.51
1981	14,105,080	49,217,420	34.6	8.9	2.30	7.90
1982	14,465,810	50,431,130	40.4	9.4	2.34	8.54
1983	14,847,000	51,673,230	47.1	9.4	2.32	11.11
1984	15,250,250	52,940,820	55.8	9.4	2.34	16.70
1985	15,677,000	54,231,000	61.4	9.3	2.48	18.61
1986	16,143,010	55,545,700	65.2	9.5	2.58	20.39
1987	16,633,580	56,886,850	68.4	9.5	2.52	20.57
1988	17,144,390	58,251,550	73.6	9.7	2.62	21.09
1989	17,669,600	59,636,900	80.0	9.9	2.71	21.74
1990	18,201,900	61,040,000	89.2	10.0	2.70	24.31
1991	18,656,950	62,434,970	99.1	10.4	2.75	27.48
1992	19,127,100	63,861,810	111.4	10.3	2.55	25.51
1993	19,609,110	65,321,270	125.4	13.1	2.57	27.12
1994	20,103,260	66,814,080	145.6	16.5	2.62	26.42
1995	20,609,860	68,341,000	165.7	20.5	2.50	25.71
1996	21,129,230	69,909,590	178.1	25.5	2.52	26.22
1997	21,667,000	71,521,110	194.8	28.6	2.81	29.47
1998	22,180,000	73,176,860	201.5	34.1	3.92	40.89
1999	22,710,000	74,878,190	202.9	38.8	3.80	39.09
2000	23,270,000	76,626,500	199.2	40.0	3.80	44.19
2001	23,802,360	78,317,030	195.8	42.4	3.80	50.99

— = No data.
Source: World Development Indicators, World Bank at http://www.worldbank.org/data/wdi2002/, accessed on 10 Novermber 2003.

A1-3: Malaysian and Philippine GDP, 1960–2001

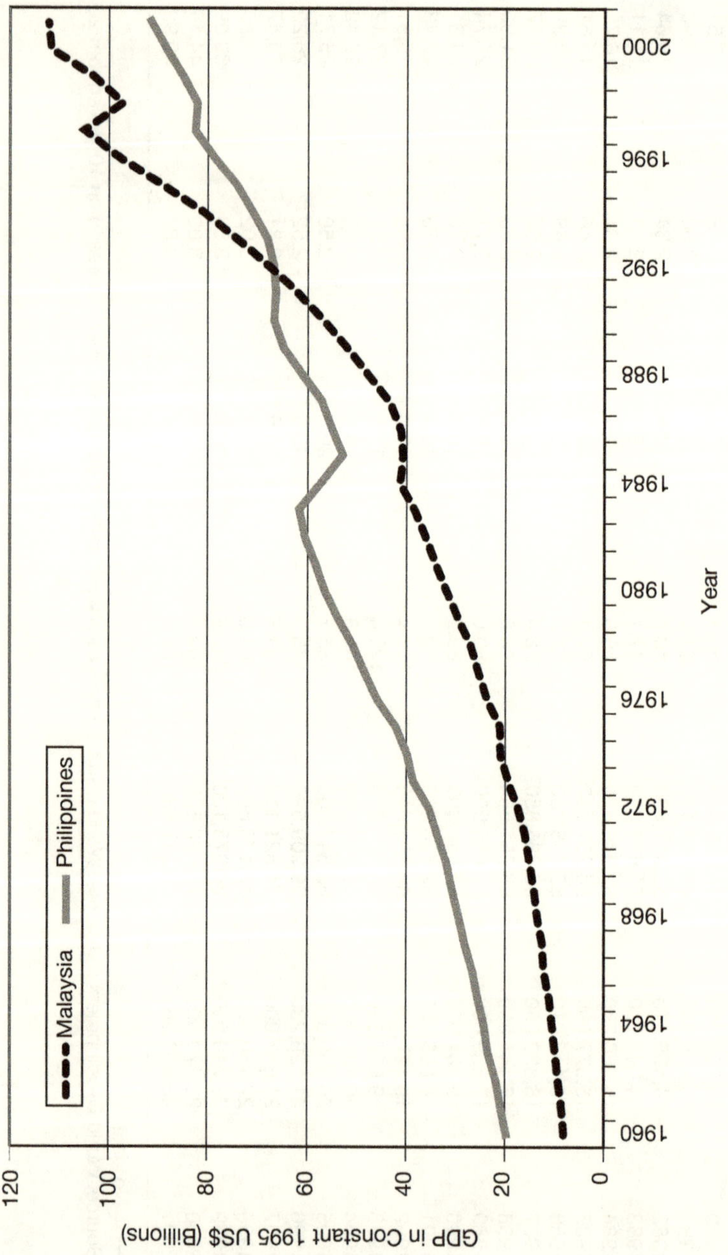

A1-4: Malaysian and Philippine GDP Growth Rate, 1960–2001

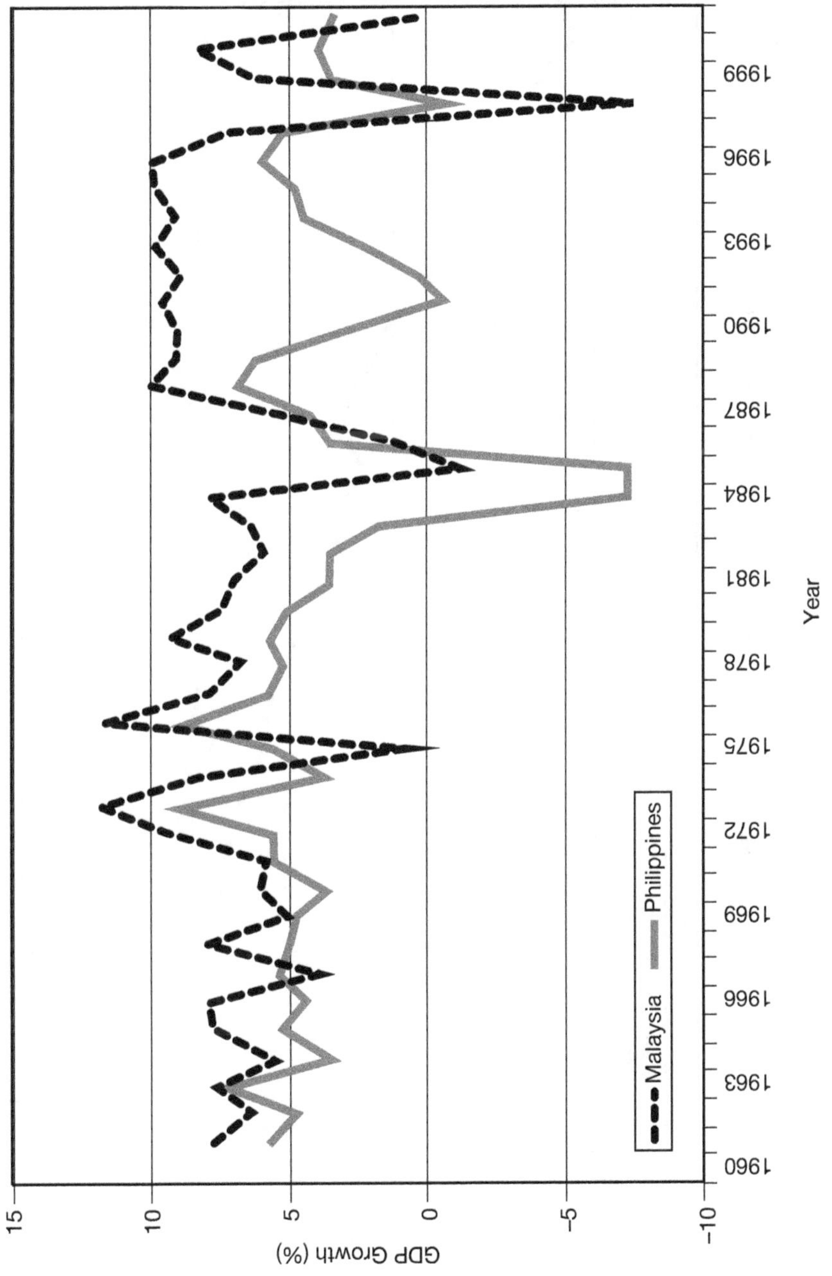

A1-5: Malaysian and Philippine GDP per Capita, 1960–2001

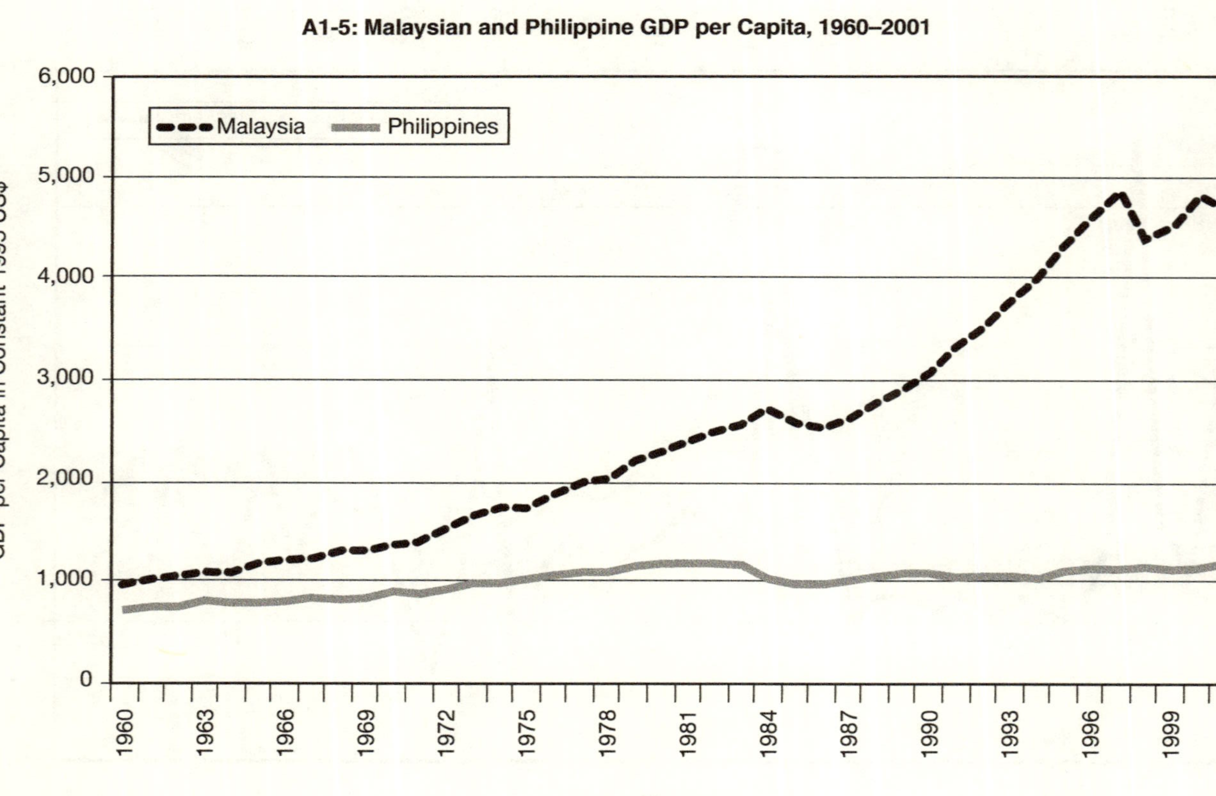

A1-6: Malaysian and Philippine Population, 1960–2001

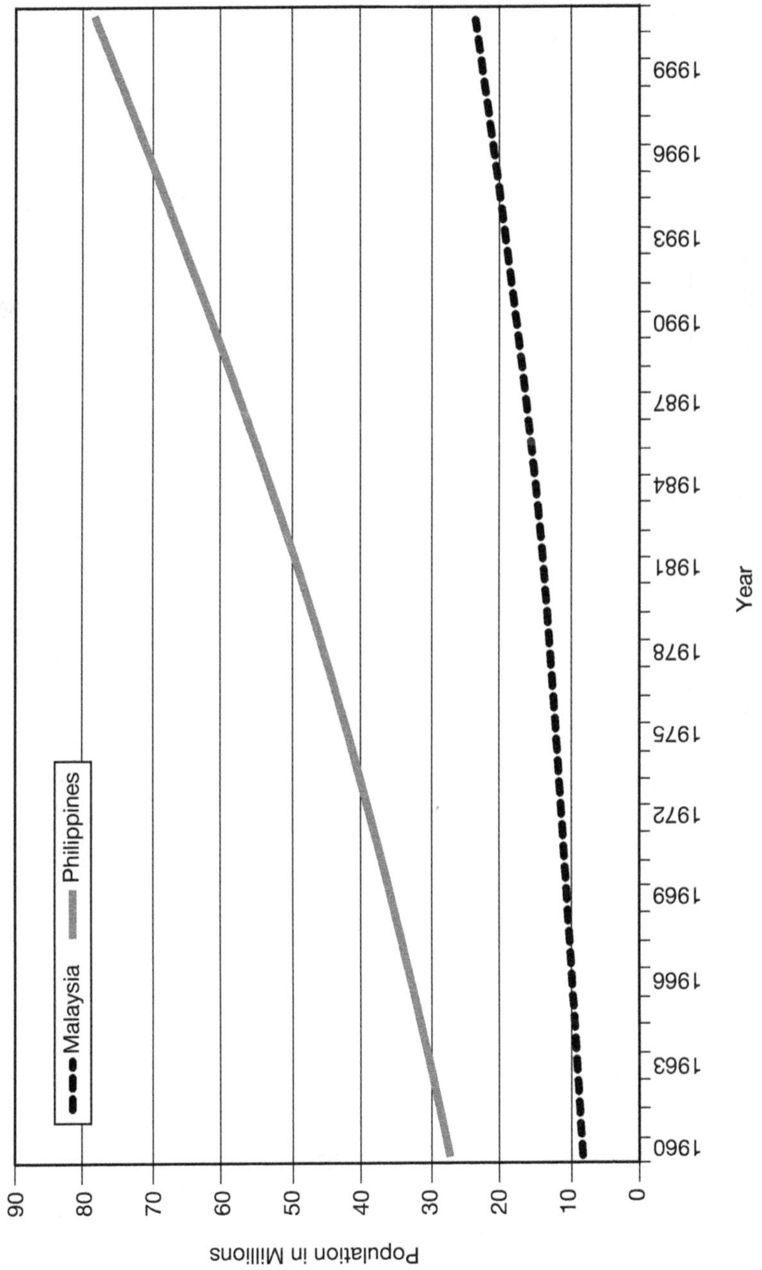

A1-7: Malaysian and Philippine Telephone Mainlines per 1,000 People, 1960–2001

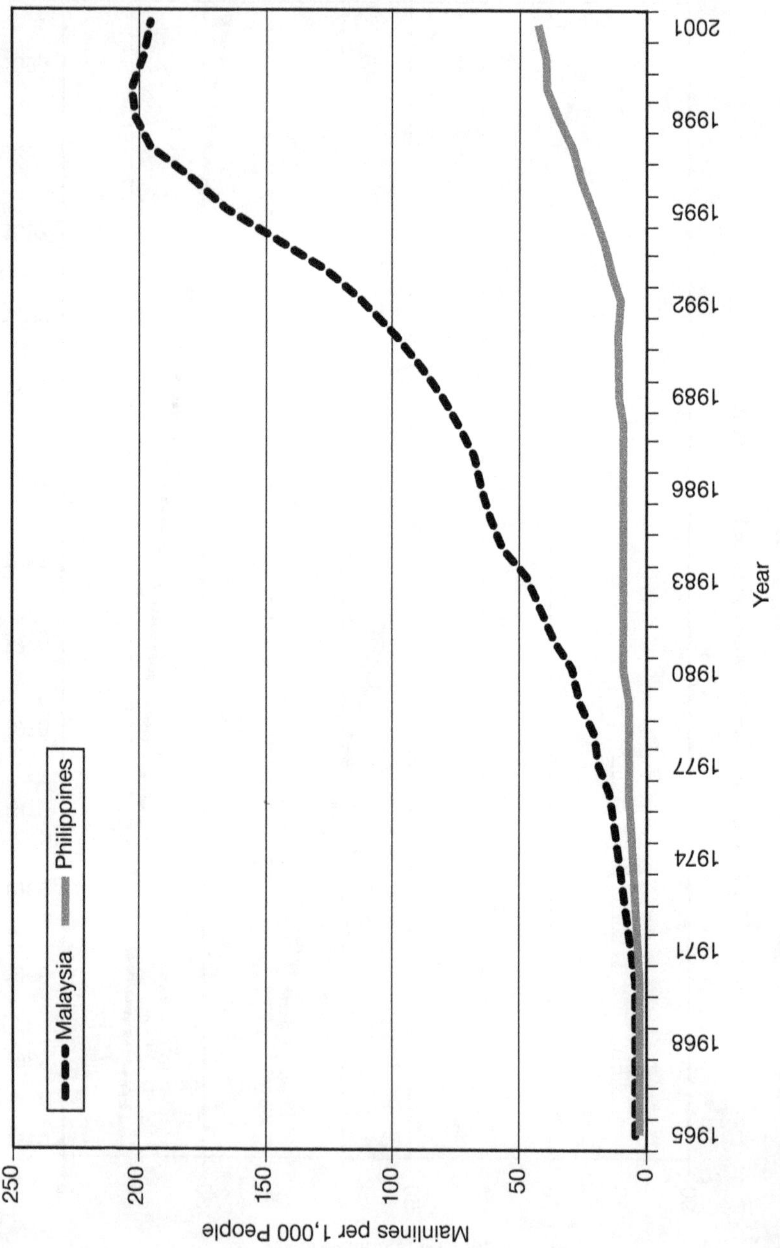

A1-8: Malaysian and Philippine Official Exchange Rate to the US$, 1960–2001

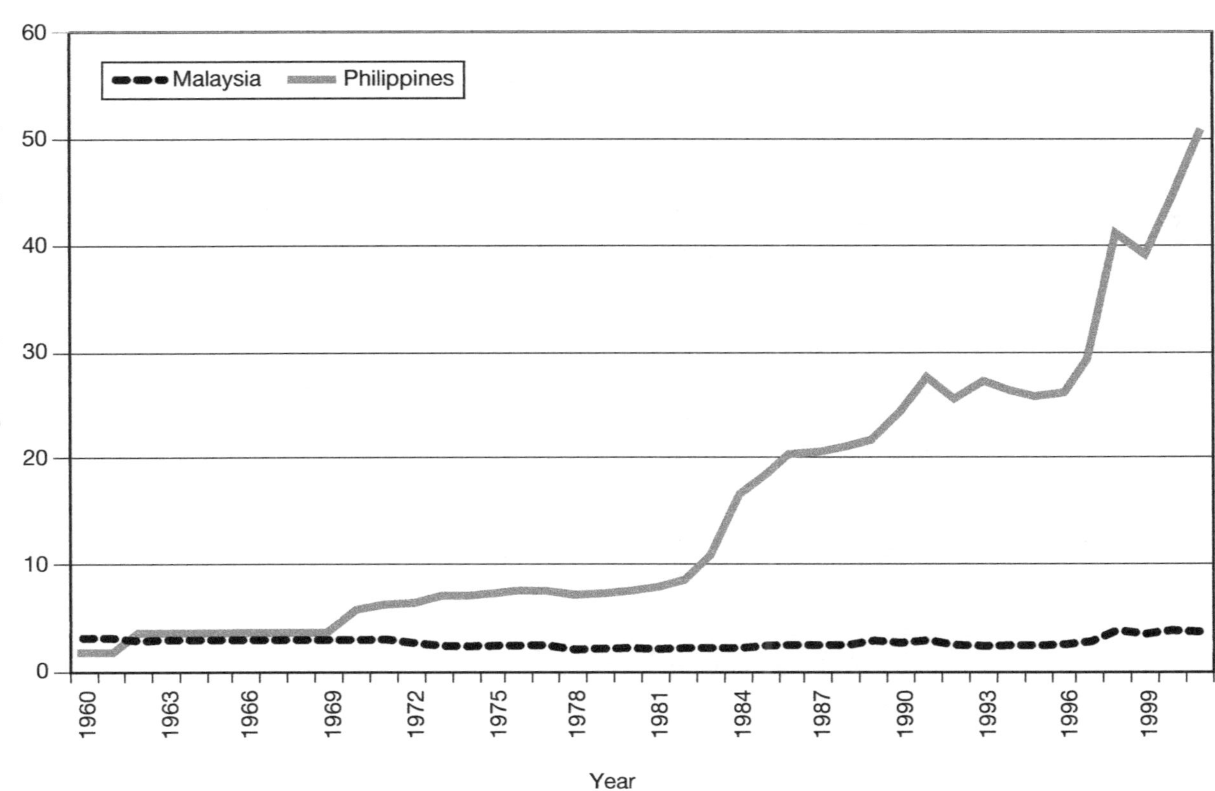

APPENDIX 2

PLDT Ownership Issue Timeline

1. In 1967, the Philippine Telecommunications Investment Company (PTIC) bought GTE's 28 per cent controlling share in PLDT.
2. At the start, Ramon Cojuangco, his wife, and brother-in-law Luis Tirso Rivilla held 57 per cent of PTIC. GTE held 22.5 per cent of the company as partial payment for its PLDT stock; Yuchengco owned 7 per cent; while Leonides Virata, Gregorio Romulo, and Antonio Meer were each given 3 per cent stake in the company for helping to organize the deal. PTIC formally took control of PLDT on 1 January 1968.
3. In February 1976, without any explanation GTE abandoned its 22.5 per cent stake in PTIC, the shell company that controls PLDT. None of the PTIC stockholders seemed to be interested in the free shares, except Ramon Cojuangco, who ended up owning an additional 22.5 per cent stake in PTIC.
4. On 5 May 1978, Ramon Cojuangco and Luis Tirso Rivilla transferred 46 per cent of the PTIC shares to a newly created company, Prime Holdings Incorporated (PHI). Jose Campos and Rolando Gapud, who were well-known cronies of President Marcos, established Prime Holdings Incorporated in 1977.

PTIC Ownership, May 1978[1]

Stockholders	Percentage
Prime Holdings Incorporated	46.0
Ramon Cojuangco and family	44.0
Alfonso Yuchengco	7.0
Antonio Meer	3.0
Total	100.0

5. On 21 March 1986, Jose Campos executed a sworn statement in Vancouver, Canada, where he fled after the fall of Marcos,

that he and his financial adviser, Rolando Gapud, and business associates who incorporated PHI were acting on behalf of President Marcos. Campos included in his sworn statement a list of companies that he held in trust for President Marcos, among them PHI. On 8 April 1986, Campos executed a second affidavit turning over to the government shares of stocks and the titles of properties and assets that he claimed were only in his custody on behalf of the former president.

6. In March 1986, PLDT became one of the first companies that PCGG took over on the basis of the Campos affidavit. A PCGG task force was appointed to investigate the ownership structure of PLDT and the extent of Marcos' involvement.
7. On 2 May 1986, PCGG lifted the sequestration of PLDT, with the exception of the PHI shares, even before the task force finished its investigation.
8. On 9 May 1986, PCGG Commissioner Mary Concepcion Bautista issued a sequestration order for PHI and PHI's shares in PTIC. The PCGG orders were composed of three documents: (a) an order of sequestration against all properties, assets, records and documents of PHI, (b) an order sequestering 111,415 shares of stock of PTIC registered on the PTIC books in the name of PHI, and (c) a letter addressed to Siguion Reyna, Montecillo, and Ongsiako law office advising the law firm that the PCGG, in its session on 2 May 1986, resolved among other things, "to order the sequestration of all the shareholdings of PHI which owns approximately 46 per cent of PTIC, which in turn owns approximately 26 per cent of PLDT". The late PCGG Comissioner Mary Concepcion Bautista signed the first two orders, while both Commissioner Bautista and then PCGG Commissioner Raul Daza signed the third document.
9. On 27 May 1986, the PCGG task force led by Luis Sison recommended, among other things, to sequester not only PHI shares but also those of the Cojuangco family because they were clearly Marcos' associates and dummies in PLDT. The Aquino government ignored these recommendations.

10. On 16 July 1987, the PCGG filed before the Sandiganbayan, the Philippines' graft court, a complaint for "Rescission, Reconveyance, Restitution, Accounting and Damages against Ferdinand and Imelda Marcos, Prime Holdings Inc., and others …" docketed as Civil Case No. 0002. The case sought to recover from the Marcoses and associates their alleged ill-gotten wealth consisting of funds and property, which was manifestly out of proportion to their salaries and other lawful income, having been allegedly acquired during Marcos's term as President of the Philippines. Among the properties mentioned in the complaint were shares in PTIC and PLDT.
11. On 11 August 1988, Alfonso Yuchengco, another businessman who since the 1960s had been a minority investor in PLDT and was also interested in taking over control of the company, filed a motion for intervention and complaint-in-intervention with the Sandiganbayan against the PCGG case versus Ferdinand and Imelda Marcos and PHI. Yuchengco claimed that not all of the PHI and Cojuangco shares were rightfully theirs. He alleged that he purchased a 6 per cent share in PTIC from Gregorio Romulo and Leonides Virata in 1968 but he was compelled to give up his shares to Cojuangco under coercion and duress from Marcos. Secondly, Yuchengco claimed that the 22.5 per cent GTE share in PTIC rightfully belonged to him by virtue of a "put and call" agreement that he signed with GTE in 1967. However, President Marcos had prevented him from exercising that right. Yuchengco was therefore contesting ownership of 28.5 per cent in PTIC, part of which was held by the Cojuangcos and part held by PHI, which were the object of government confiscations.[2]
12. On 23 April 1990, the PCGG amended its complaint against the Marcoses and added Imelda Cojuangco, the estate of Ramon Cojuangco, and Prime Holdings, Inc. as defendants. The PCGG charged that these new defendants held shares of stock in PLDT belonging to Ferdinand Marcos and his family.
13. In April 1993, the Ramos Administration, using the voting right of the sequestered PHI shares in PTIC, appointed five members to the PLDT board, and in tandem with Alfonso Yuchengco controlled 6 of the 11 seats on the PLDT board. However, it was agreed upon that Antonio Cojuangco, the son of Ramon Cojuangco

who took over management of PLDT when Ramon died in 1984, would remain as President and Chairman of the Board. Earlier in January 1993, the PCGG also revived the case against the Cojuangcos, claiming that not only the PHI shares but also the Cojuangco family shares in PTIC were ill-gotten wealth. This issue was perhaps used as a bargaining chip to make the Cojuangcos agree to the entry of new players in the industry in exchange for their continued control of the company. Throughout the Ramos administration years from 1992 to 1998, the case remained pending at the Sandiganbayan. This compromise agreement with the government was recorded in the PLDT minutes as follows: "in April 1993, Antonio Cojuangco, President of PLDT and acting as representative of the heirs of the late Ramon Cojuangco, entered into an agreement with the PCGG to avoid the postponement of the Company's annual meeting. The agreement, among other things, authorized then President Fidel V. Ramos to nominate five of the eleven directors to the company's board that year."[3]

PLDT Ownership Structure, 1993[4]

	No. of Shares (Millions)	Percentage
Total No.of shares	204.3	100.0
Voting shares	53.5	26.2
Non-voting shares	150.8	73.8

Breakdown of Voting Stock (26.2% of Total No. of Shares), 1993

Stockholders	Percentage
PTIC	24.2
Social Security System	10.6
PLDT staff	6.0
Others	11.1
Filipino Owned	51.9
Capital International Emerging Funds	6.0
First Philippines Fund	2.0
Others	40.1
Foreign-owned	48.1

PTIC Share Ownership, 1993

Shareholders	Percentage
Prime Holdings	46
Cojuangco Family	44
Alfonso Yuchengco	7
Antonio Meer	3
Total	100

14. On 15 April 1993, Antonio Cojuangco challenged the government sequestration of the PHI shares and claimed that his family had in its possession two documents executed by Jose Campos and Rolando Gapud on 18 February 1981, turning over control of PHI to Ramon Cojuangco and Oscar Africa. In effect, Antonio Cojuangco claimed that his family also owned the PHI shares in PTIC. This meant that the Cojuangcos were claiming ownership of 90 per cent of PTIC.[5]

    a. On 4 May 1993, the Cojuangcos filed a motion with the Sandiganbayan, seeking to automatically lift the order of sequestration against PHI, arguing that: (a) PCGG did not observe its own rules and regulations in the issuance of the order of sequestration against PHI shares in PTIC, because the order was signed by only one PCGG commissioner, and that (b) the PCGG failed to file the appropriate judicial action within the prescribed period, which should have been no later than 2 August 1987 (with the case against the Cojuangcos only filed on 23 April 1990).
    b. The Cojuangcos filed a second motion on 25 August 1993, requesting the dismissal of the Yuchengco lawsuit because of the alleged failure to pay the correct docket fee for the lodging of a case, wherein because Yuchengco was claiming 28.5 per cent of PTIC with a stated value of P1.6 billion in 1993, the docket fee to be paid should have been P5.4 million and not P400.00.

15. On 20 December 1993, the Sandiganbayan granted the Cojuangco motion and lifted the sequestration order against PHI and its

shares in PTIC because the order of sequestration was defective due to the fact that only one commissioner signed the order. The PCGG appealed the decision but was rejected with finality by the Sandiganbayan. Thus, the PCGG elevated the motion to the Supreme Court.

16. On 31 July 1998, the Supreme Court (G.R. 119292) ruled that the PCGG sequestration order was illegal and unconstitutional. It was signed by only one commissioner and contravened PCGG's own rules and regulations for operation, and the PCGG had failed to commence proper judicial action against the Cojuangcos within the prescribed period of time.[6] While the PCGG was granted the power and authority "to sequester or place or cause to be placed under its control or possession any building or office wherein any ill-gotten wealth or properties may be found, and any records pertaining thereto, in order to prevent their destruction, concealment or disappearance, which would frustrate the investigation or otherwise prevent the Commission from accomplishing its task", the PCGG was not given a general or blanket writ of sequestration, but had to go about it within the given limit of the law. This was because "sequestration is an extraordinary, harsh and even severe remedy that should be confined to its lawful parameters and exercised with due regard in the words of its enabling law, to the requirements of fairness, due process and justice".[7] The PCGG appealed the court's decision.
17. The 31 July 1998 Supreme Court decision lifted the sequestration of the PHI shares in PTIC. This decision is significant because in November that year it allowed the Cojuangcos to sell 52.7 per cent of the PTIC shareholding in PLDT to First Pacific.

    a. The Supreme Court, however, pointed out that the lifting of the writ of sequestration was not necessarily fatal to the main case of pursuing the ill-gotten wealth, where the deposition of Jose Yao Campos, incorporator of PHI, who declared that former President Marcos was the true owner of PHI, was strong evidence. "The lifting of the write of sequestration will not necessarily be fatal to the main case. It is in the latter case that the Campos testimony may be properly offered and its value and credit-worthiness appreciated. Even with the lifting of the sequestration orders against PHI and the PTIC shares, these properties may still be recovered by the government upon showing substantial proof, submitted in the proper law suit, that they indeed constitute unlawfully amassed wealth of the Marcoses or their conduits. The lifting of the subject orders does not *ipso facto* mean that the sequestered properties are not ill-gotten; neither does it pre-empt a finding to that effect. The effect of the lifting of sequestration against PHI and the subject PTIC shares will merely be the termination of the role of government as conservator thereof. In other words, the PCGG may no longer exercise administrative or housekeeping powers and its nominees may no longer vote the heretofore sequestered shares to enable them to sit on the corporate board of the subject firm."[8]

b. The Court further opined that "sequestration is not the be-all and end-all of the efforts of the government to recover unlawfully amassed wealth. The PCGG may still proceed to prove in the main suit who the real owners of these assets are … with the use of proper remedies and upon substantial proof, properties in litigation may, when necessary, be placed in *custodia legis* for the complete determination of the controversy or for the effective use of enforcement of the judgment. Yet, for violating the Constitution and its own rules, the PCGG may no longer exercise dominion and custody over PLDT and the shares it owns in PTIC."[9]

c. In the decision's epilogue, the Court scolded the PCGG, declaring that instead of "voting on the sequestered shares and sitting on the boards of private corporation for the purpose of safeguarding or preserving the sequestered assets until they are finally adjudicated, the PCGG must probe the ill-gotten nature of the sequestered assets and of causing their reversion to the people. Yet, 12 years have passed since most of the sequestration orders were issued but the substantiation of the claim that they are in fact ill-gotten remain pending. In fact, what is still being discussed is the validity of the sequestration orders ... The PCGG is yet to show determination to prosecute to final resolution any of the cases it has filed in Philippine courts over a decade ago ... It is about time that the PCGG created with the primary and paramount task of recovering ill-gotten wealth, act with deliberate dispatch on its primordial work substantiating its claims and thereby perform its duty of rendering justice to all."[10]

18. On 7 September 1998, the Supreme Court nullified with finality the government sequestration of PHI and PHI shares in PTIC because the sequestration order was signed by only one of the commissioners, Mary Concepcion Bautista. The court thus upheld the lifting of sequestration on the PTIC shares held by PHI. However, the issue of who owned the various shares of PTIC was still unresolved.
19. On 24 November 1998, First Pacific announced its acquisition of 17.3 per cent of PLDT's total common stock, which constituted a 27.2 per cent voting interest in the company. First Pacific revealed in a press statement that it acquired control of PLDT in a two-step process. First, it bought a 5.9 per cent stake in PLDT on the open market at a cost of US$197 million. Second, it bought an indirect 11.3 per cent stake in PLDT by buying 52.7 per cent of PTIC from Antonio Cojuangco's family, Nori Ongsiako and her family, Antonio Meer, and Alfonso Yuchengco, for US$552 million. In 1998, PTIC controlled 21.5 per cent of PLDT's voting stock. In effect, First Pacific paid about P1,420 per share, a 31 per cent premium to the market rate of PLDT stocks at time which was trading at P1,085.[11]
20. Reportedly, Estrada brokers led by Mark Jimenez and then Executive Secretary Ronaldo Zamora allegedly convinced Antonio Cojuangco and family to sell their 44 per cent stake in PTIC to First Pacific for P17 billion (US$427 million). No less than President Estrada announced to the media before the deal's formal announcement that "he heard First Pacific bought Tonyboy out of PLDT".[12]

21. First Pacific stated that it bought 52.7 per cent of PTIC, 44 per cent of which came from the Cojuangco family. It was not clear which shares in PTIC the Cojuangcos sold — the 44 per cent that they owned or part of the 46 per cent in PTIC that PHI controlled. It is to be recalled that the PCGG claimed that both the PHI and the Cojuangco shares in PTIC were ill-gotten, and thus its ownership was still legally unresolved. Therefore, the Cojuangcos received US$460.9 million for selling 44 per cent of PTIC, shares that are still subject to contestation in the Sandiganbayan and are being claimed by the government, Yuchengco, and Imelda Marcos. Yuchengco's lawyer opined that "Metro Pacific is at risk of losing its PTIC shares in the event of a decision forfeiting the shares in favour of the government or ordering the conveyance of the shares to Y Realty Corp. or to the Marcos family".[13]
22. In December 1998, Imelda Marcos claimed in a newspaper interview that "they practically owned everything in the Philippines"[14] during his husband's term as president, and that the big cronies who controlled the Philippines' major corporation were "trustees" of her husband. Imelda's outburst came a week after the announcement of the First Pacific-PLDT deal. In 1999, Imelda filed with the Sandiganbayan a counter-suit to the PCGG, Yuchengco, and Cojuangco cases, claiming that she and her husband owned a bloc of PLDT shares through PTIC.[15] The Sandiganbayan rejected with finality Marcos' claim in May 2000, dismissing Marcos' cross-claim. Imelda appealed the decision to the Supreme Court. In November 2002, it was reported that Imelda received a P2 billion quit-claim settlement from the Cojuangcos with regards to PLDT.[16] Both Imelda and PLDT denied the purported pay-off.[17]
23. In May 2002, the Sandiganbayan's Fourth Division rejected the government's claim that the Cojuangcos and PHI shares in PTIC were held for Ferdinand Marcos and his family. The anti-graft court ruled that the government had failed to present substantial evidence to prove that the Cojuangco shares in PLDT were illegally obtained or that Marcos owned the PHI shares. In addition, the court berated the PCGG, which filed the case in

July 1987 (Civil Case 0002), for using photocopied documents as proof. The court ruled that there was no indication that Marcos and the Cojuangcos shared a "relationship of trust" in connection with the PLDT shares. The ruling stated that "there is no competent evidence to tie defendant Ferdinand Marcos with PTIC, or to establish that presidential concessions, benefits or other incentives that could have improved the financial and operational situation of PTIC, PLDT, and PHI, were accorded said companies by defendant Ferdinand Marcos. Accordingly, there is no competent evidence to prove the Republic's allegation that the PLDT shares herein were ill-gotten." The Sandiganbayan also criticized the government for relying on former First Lady Imelda Marcos' claim that the said PLDT shares were owned by her family to prove that the Cojuangcos and PHI were merely dummies for the Marcoses. These claims, according to the court, were merely self-serving. The Sandiganbayan also dismissed the intervention petition filed by former ambassador Alfredo Yuchengco because he had failed to show proof that he was coerced into transferring PTIC shares of stock to the Marcoses.[18] The PCGG, then headed by Haydee Yorac, declared that it would appeal the Sandiganbayan ruling in the Supreme Court.[19]

24. In June 1999, PLDT announced its plan to merge with Smart and take in Smart's partner, Japan's NTT, as a strategic partner. With the merger, First Pacific's shareholding in PLDT increased to 31.5 per cent. As of 2002, PLDT's stockholding structure was as follows.

PLDT Ownership Structure, 2002[20]

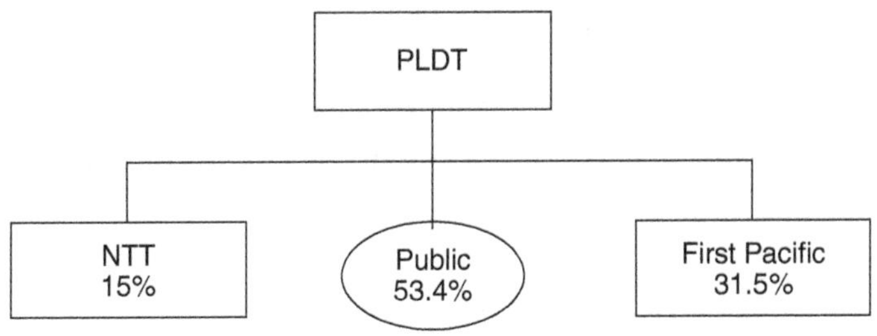

25. On 4 June 2002, John Gokongwei signed a memorandum of agreement with Anthony Salim, heir of Soedono Salim (Liem Sieo Liong), to buy from First Pacific part of its stake in PLDT and the Fort Bonifacio Land Corporation. The deal surprised the PLDT board, including First Pacific Executive Chairman and PLDT President and CEO Manuel V. Pangilinan, who was not informed that a sale was being negotiated. Pangilinan, in alliance with the Cojuangco family and the entire PLDT board, objected to the sale. The board's main objection was based on a board by-law, passed a couple of months before the event, that barred entry to the PLDT board by parties who were direct competitors of the company. Gokongwei was the majority stakeholder in PLDT competitor Digitel. Ironically, this was exactly what First Pacific did when it bought into PLDT in 1998, knowing full well that it held a majority stake in PLDT's then competitor, Smart Communications. Pangilinan, in alliance with Cojuangco and with the full support of the PLDT board, threatened to sue First Pacific, rallied the other stockholders including NTT and prepared a counter proposal to foil the sale. The memorandum of agreement for the sale between First Pacific and Gokongwei expired on 30 September 2002, thwarting Gokongwei's attempt to buy into PLDT.
26. In February 2003, it was reported that the Salims had begun talks with Telekom Malaysia to sell their shares in PLDT. Both Telekom Malaysia and First Pacific denied the report.[21] There were various reasons why the Salims may have wanted to unload their shares in PLDT and their Philippine investments, but one probable cause with respect to PLDT is that they found out that the shares for which they paid US$552 million could be taken over by the Philippine government. Reportedly, Anthoni Salim wanted to unload First Pacific's investments in the Philippines, where as of 2003, 70 per cent of its investments were located.[22]

NOTES

1. This table is constructed from information contained in the following two documents: (1) Internal Memorandum to President Joseph Estrada on "The Participation of Prime Holdings in the Philippine Long Distance Telephone Company (PLDT) and the Government's Right of Recovery", 26 October 1998; and (2) Confidential PCGG Document: *Summary Analysis of the PLDT Situation*, 11 August 1986. Both documents were found at the Philippine Center for Investigative Journalism library in Quezon City.
2. PCGG case no. 3, Alfonso Yuchengco versus Republic of the Philipppines, Presidential Commission on Good Government, Ferdinand Marcos, Imelda Marcos and Prime Holdings, Inc., 2 August 1988, pp. 9–10.
3. PLDT's Management's Discussion and Analysis of Financial Condition and Results of Operations for the Nine Months Ending 30 September 1998, p. 6.
4. All data for all three tables came from *Far Eastern Economic Review*, 11 February 1993.
5. *Far Eastern Economic Review*, 3 May 1993. According to Antonio Cojuangco: "I don't even know what happened — why Prime Holdings was formed, why the shares were turned back to the family. I don't know what the transactions were all about. I was just 16 years old when these occurred and even my mom doesn't know. The only thing we know is that we have them [PHI Stock certificates] in our folders."
6. G.R. 119292, Republic of the Philippines represented by the Presidential Commission on Good Government versus Sandiganbayan, Imelda Cojuangco, and the Estate of Ramon Cojuangco represented by Imelda Cojuangco and Prime Holdings Incorporated, 31 July 1998, p. 13.
7. G.R. 119292, pp. 7, 10.
8. G.R. 119292, pp. 13–14.
9. Ibid.
10. Ibid., pp. 14–15. Justice Panganiban wrote the Supreme Court's decision, with Justices Davide and Quisumbing concurring.
11. "First Pacific Acquires PLDT Stake for US$749 Million", First Pacific Press Release, 24 November 1998. See also PLDT's *Management's Discussion and Analysis of Financial Condition and Results of Operations for the Year Ended December 31, 1998*, p. 5.

12. *Asiaweek*, 20 November 1998.

13. *Philippine Daily Inquirer*, 23 August 2001.

14. *Philippine Daily Inquirer*, 5 and 9 December 1998.

15. *Philippine Daily Inquirer*, 26 May 2000 and 19 June 2002.

16. *Philippines Star*, 5 July 2003.

17. *Philippine Daily Inquirer*, 13 November 2002.

18. *Philippine Daily Inquirer*, 8 and 9 May 2002; *Manila Times*, 10 May 2002.

19. *Manila Times*, 11 June 2002.

20. See PLDT's corporate overview as of 2002 at http://www.pldt.com.ph.

21. See *Philippine Daily Inquirer*, 24 February 2003 and *Manila Times*, 25 February 2003.

22. *Philippine Daily Inquirer Inquirer*, 24 and 27 February 2003.

APPENDIX 3

Comparative Performance of the Telecommunications Sector in Selected ASEAN Countries

A3-1: Cellular Mobile Subscribers in Selected ASEAN Countries, 1990–2005

	Total No. of Subscribers (Thousands)						As % of Total No. of Telephone Subscribers			Fixed Line Teledensity		
Country	1990	1991	1995	1999	2001	2005	1999	2001	2005	1999	2001	2005
Malaysia	54.6	70.9	872.8	2,990	7,480	19,545	40.3	61.6	81.7	20.30	20.30	16.79
Philippines	9.8	33.8	492.7	2,750	11,700	34,779	48.5	77.9	91.2	3.88	4.24	4.00
Indonesia	21.0	24.5	218.6	2,220	6,520	46,910	26.8	47.5	78.6	2.91	3.45	5.73
Singapore	51.0	81.9	291.9	1,630	2,990	4,385	46.5	60.6	70.4	48.20	47.14	42.39
Thailand	66.3	210.0	1,087.5	2,340	7,550	27,379	31.0	55.5	80.1	8.57	9.87	10.59

Sources:
1. ITU, *Asia-Pacific Telecommunications Indicators* (Geneva: ITU, May 1993).
2. ITU, *World Telecommunication Development Report 1996/97* (Geneva: ITU, 1997).
3. ITU, *World Telecommunication Development Report 1999: Mobile Cellular* (Geneva: ITU, 1999).
4. ITU, *Asia-Pacific Telecommunications Indicators 2000* (Geneva: ITU, December 2000).
5. ITU, *Asia-Pacific Telecommunications Indicators 2002* (Geneva: ITU, November 2002).
6. ITU, *The Internet of Things*. Geneva: ITU, November 2005..
7. ITU, *Measuring the Information Society 2007: ICT Opportunity Index/World Telecommunication/ICT Indicators*. Geneva: ITU, January 2007.

Hereafter cited as "Various ITU Reports, 1993–2007".

A3-2: Cellular Telecommunications Cost in Selected ASEAN Countries, 1995–2006 (US$)

	Analog Tariff, 1995			Digital Tariff, 1995			1999				2001				2006			
									Per Min. Local Call			3-min. Local Call				Per Min. Local Call		
Country	CF	MS	3-min. Local Call	CF	MS	3-min. Local Call	CF	MS	Peak	Off-peak	CF	Peak	Off-peak	Cost of a Local SMS	CF	Peak	Off-peak	Cost of a Local SMS
Malaysia	19	22.9	0.34	—	—	—	121	15.29	0.08	0.04	47	0.43	0.31	0.04	—	0.10	0.10	0.01
Philippines	57	4.5	0.91	—	—	—	37	14.67	0.17	0.15	12	0.47	0.24	0.02	0.88	0.13	0.13	0.02
Indonesia	—	—	—	—	—	—	20	6.49	0.03	0.03	7	0.44	0.38	—	—	0.11	0.09	0.01
Singapore	7	31.7	0.42	7	31.7	0.42	6	23.90	0.12	0.12	—	0.31	0.31	0.03	17.78	0.14	0.14	0.03
Thailand	40	19.8	0.36	40	19.8	0.36	97	12.09	0.07	0.07	9	0.41	0.41	0.07	5.28	0.05	0.05	0.08

CF = Connection fee.
MS = Monthly subscription.
— = No data.
Sources: Various ITU Reports, 1993–2007.

A3-3: Fixed-line Telephone Waiting List in Selected ASEAN Countries, 1982–2005 (Thousands)

Country	No. of Applicants on Waiting List					Total Demand			Satisfied Demand (%)			Waiting Time (Years)		
	1982	1991	1995	1999	2005	1995	1999	2005	1995	1999	2005	1995	1999	2005
Malaysia	189,810	139,959	122.0	160.0	39.0	3,454.4	4,590.8	4,404.7	96.5	96.5	99.1	0.3	0.7	—
Philippines	132,026	705,045	900.2	900.2*	—	2,309.8	3,600.2*	—	61.0	75.0*	—	3.6	2.1*	—
Indonesia	182,477	307,414	117.5	—	—	3,408.3	—	—	96.6	—	—	0.2	—	—
Singapore	1,409	373	0.2	0	0	1,429.2	1,876.6	1,844.4	100.0	100	100.0	—	0	0
Thailand	386,820	1,299,439	1,083.5	419.5	351.8	4,565.5	5,635.2	7,386.5	76.3	92.6	95.2	1.9	1.2	2.2

Total demand = Satisfied demand + Waiting list.
* Data was for 1998.
— = No data.
Sources: Various ITU Reports, 1993–2007.

A3-4: Urban and Rural Fixed-line Teledensity in Selected ASEAN Countries, 1991–2001

	1991			1995			1999			2001		
Country	National	Urban	Rural	National	Urban	Rural	National	Urban	Rural	National	Urban	Rural
Malaysia	9.91	19.13	2.67	16.56	14.31	14.73	20.30	28.18	19.82	20.3	28.18	19.82
Philippines	1.03	—	—	2.09	9.58	0.95	3.88	14.55	1.61	4.24	26.47	0.75
Indonesia	0.68	1.32	0.39	1.69	—	—	2.91	16.33	2.27	3.45	26.11	2.31
Singapore	39.85	39.85	0	47.85	47.85	0	48.2	48.2	0	47.14	47.14	0
Thailand	2.73	8.09	1.16	5.86	33.94	2.94	8.57	37.13	4.88	9.87	45.23	5.23

— = No data.
Sources: Various ITU Reports, 1993–2002.

A3-5: Public Payphone Availability in Selected ASEAN Countries, 1991–2005 (Thousands)

	No. of Payphones				No. of Payphones per 1,000 Inhabitants			
Country	1991	1995	1999	2005	1991	1995	1999	2005
Malaysia	27.8	99.2	162.3	95.21	1.51	4.93	7.43	3.66
Philippines	3.1	8.0	11.1	—	0.05	0.12	0.15	—
Indonesia	25.3	108.3	269.2	402.87	0.13	0.56	1.29	1.90
Singapore	26.5	30.6	20.7	10.90	9.58	10.43	5.36	2.50
Thailand	26.1	49.4	139.3	363.89	0.46	0.83	2.29	5.67

Sources: Various ITU Reports, 1993–2007.

Table A3-6: Fixed-line Telecommunications Cost in Selected ASEAN Countries, 1991–2005 (US$)

Country	1991				1995				1999				2005			
	Residential		Business		Residential		Business		Residential		Business		Residential		Business	
	CF	MS	CF	MS	CF	MS	CF	MS	CF	MS	CF	MS	CF	MS	CF	MS
Malaysia	18	7.3	18	12.7	19	7.6	19	13.3	13	5.3	13	9.2	13	6.6	13	11.9
Philippines	11	8.6	13	23.6	13	10.1	16	23.1	8	9.0	7	20.5	36	11.9	64	22.9
Indonesia	169	4.1	154	4.1	311	9.1	400	13.8	38	2.9	57	5.0	30	3.4	46	5.9
Singapore	46	4.8	58	7.2	56	5.9	71	8.8	47	5.1	47	7.6	19	5.3	19	7.9
Thailand	145	3.9	145	3.9	133	4.0	133	4.0	89	2.6	89	2.6	83	2.5	83	2.5

CF = Connection Fee.
MS = Monthly Subscription.
Sources: Various ITU Reports, 1993–2007.

A3-7: Functions of the Regulatory Agencies

Does the Regulatory Agency Have Responsibility for:	Malaysia	Philippines
Granting licences?	No	Yes
Approving tariffs?	No	Yes
Setting conditions of interconnections?	Yes	Yes
Administering Universal Service Obligation (USO)?	Yes	No
Monitoring and enforcing compliance with regulations?	Yes	Yes
Undertaking conciliation and arbitration?	Yes	Yes

A3-8: Accountability, Autonomy, and Competency of the Regulatory Process

	Malaysia	Philippines
Who is the policy maker?	Minister of Energy, Communications, and Multimedia	Department of Transportation and Communications
Who is the regulator?	JTM, 1987–99 CMC, 1999 to the present	PSC, 1936–79 NTC, 1979-present
Who appoints the regulators?	Minister	President
How long do appointments last?	Minimum of two years, maximum of five years	No fixed tenure; depends on the President
Are regulators legally required to have certain qualifications?	No	Yes
Who can remove regulators?	Minister	President, Congress
Are there conflict of interest rules to prevent capture by operators?	Yes	Yes
What is the agency's source of financing?	Parliamentary allocation; collection from licence and other fees	Congressional allotment
Are draft documents published for comment?	Yes	Yes
Can regulatory decisions be appealed?	Yes	Yes
What are the agency's reporting obligations?	Public register of all matters	Website for public information

Tables A3-7 and A3-8 are adapted from Michael Kerf, Manuel Schiller, and Clemencia Torres, "Telecom Regulators: Converging Trends?" *Public Policy for the Private Sector*, Note no. 230 (World Bank Group Private Sector and Infrastructure Network, May 2001).

APPENDIX 4

JTM Regulatory Determinations List, 1995–99

Determinations	Title	Date of Issue
TRD 001/98	Telecommunications Regulatory Determination on Customer Access Arrangement — Implementation of Equal Access by Call-by-Call Selection	24 May 1998
TRD 002/98	Telecommunications Regulatory Determination – Technical Compliance of Customer Premises Equipment (CPE) — Standardization of CPE Alphanumeric Keypad	1 July 1998
TRD 003/98	Kenyataan Kawalseliaan Ketua Pengarah Telekomunikasi — Ketetapan Mengenai Penyaluran Trafik — Panggilan Terbitan dan Tamtan (Call Originations and Terminations)	8 June 1998
TRD 004/98	Kenyataan Kawalseliaan Telekomunikasi — Telefon Berbayar Sewaan	1 September 1998
TRD 005/98	Kenyataan Kawalseliaan Ketua Pengarah Telekomunikasi — Peruntukan Kod Ringkas Kepada Organisasi Bukan Pengusaha Rangkaian Telekomunikasi	15 June 1998
TRD 006/98	Kenyataan Kawalseliaan Telekomunikasi — Determination of Cost-Based Interconnect Prices and the Cost of Universal Service Obligation	15 June 1998
TRD 007/98	Telecommunications Regulatory Determination – Cellular Telecommunications Service — The Use of Subscriber Identity Module (SIM) Lock in the Cellular Network	1 January 1999
TRD 008/98	Telecommunications Regulatory Framework — Anti-Competitive Behavior in the Telecommunications Industry	16 November 1998
TRD 010/98	The Telecommunications Regulatory Framework — International Accounting Rate Settlement	1 January 1999
REG-T 002	Telecommunications Regulatory Framework for the 600 Services	15 May 1995, Revised, 1996
REG-T 004	Telecommunications Regulatory Framework for the 800 Services	22 June 1996

APPENDIX 4

JTM Regulatory Determinations List, 1995–99 (*continued*)

Determinations	Title	Date of Issue
REG-T 005	Telecommunications Regulatory Framework for the 1-800 Services	22 June 1996
REG-T 006 Version 1	Telecommunications Regulatory Framework for the Assignment Principles of the National Telecommunications Numbering Plan	15 June 1998
REG-T 007	Regulatory Framework for Telecommunications Network Boundaries	1 November 1997
REG-T 008	Regulatory Framework for the Provision of Internet Services	10 February 1998
REG-T 010 Version 1	Telecommunications Regulatory Framework for the Allocation of Short Codes within the National Telecommunication Numbering Plan	15 June 1998
REG-T 011	Regulatory Framework for the Provision of Personal Numbering Service	1 March 1999
REG-T 012	Regulatory Framework for the Provision of Caller Party Pays (CPP) Radio Paging Service	1 March 1999
REG-Q 001	Telecommunications Regulatory Framework for Service Quality in Mobile Cellular Telephone Service	
REG-Q 002	Telecommunications Regulatory Framework for Quality of Service in the Fixed Telephone Service	29 December 1997
REG-R 002	Regulatory Framework on the Sharing of Radiocommunications Infrastructure	24 May 1998
REG-R 003	General Regulatory Framework for Local Multi-Point Communications Service	
	Prosedur Permohonan Permit Impot	1 September 1998
	Approval of Telecommunication Equipment and Plant	13 October 1998
	General Framework for Wireless Local Loop	16 September 1997
	General Framework for Interconnection and Access	17 May 1996

APPENDIX 5

I. Agile's Ten-point Program to Restructure the Philippine Telecommunications Industry

1. The resolution of interconnection problems by drafting and promulgating clear interconnection implementation rules and regulations.
2. The establishment of a new pricing regulatory regime for wholesale services by providing guidelines for access pricing reform.
3. The establishment of a new pricing regulatory regime for retail services by providing guidelines for rate rebalancing.
4. The establishment of a new accounting regime by drawing up a uniform chart of accounts and cost allocation manual.
5. Increasing public knowledge on regulatory issues by setting up a website for the NTC and establishing an NTC-Industry Cooperation Council.
6. The promotion of convergence by commissioning a report on the regulatory reform need for convergence and drawing up a comprehensive guidelines for Internet/value-added service/ Internet service providers.
7. The promotion of universal access, service, and obligation by drawing up a clear policy on the issue, and exploring the possibility of establishing a universal service or access fund.
8. The establishment of technical standards.
9. Licensing.
10. The management of frequency resources.

II. Outcomes of Agile Support for the NTC, as of 2002

1. The issuance of Memorandum Circular 14-7-2000 on the Implementing Rules and Regulations for the Interconnection of Authorized Public Telecommunications Entities on 14 July 2000.
2. The release of the Consultative Document on Wholesale Charging Regime, Access, and Interconnect Arrangements in July 2000.
3. The establishment in November 2000 of the NTC website, through which consultative documents, memorandum circulars, procedures for costumer complaints, and other information are made accessible.

4. The release of a comprehensive study on the outcomes of the SAS in January 2001.
5. The issuance of Memorandum Circular 6-9-2001 on the Implementing Rules and Regulations for Retail Pricing.
6. The release of the draft Revised Rules of Practice and Procedure for the NTC on 9 July 2001.
7. The issuance of the Implementing Rules and Regulations for Competitive Wholesale Charging for Interconnection Services (Memorandum Circular 09-07-2002) on 31 July 2002.
8. The issuance of Memorandum Circular 05-05-2002 on the Rules and Regulations on the Provision of High Speed Networks and Communications to IT Hub Areas on 13 May 2002.
9. The issuance of a draft memorandum circular on rules and regulations that authorized entities other than public telecommunications entities to operate public calling stations, offices, and telecenters on 17 July 2002.

Bibliography

Abrenica, Ma. Joy, and Gilberto Llanto. "Services". In *Philippine Economy: Developments, Policies and Challenges*, edited by Arsenio Balisacan and Hal Hill, pp. 254–82. New York: Oxford University Press, 2003.

Adam, Christopher, and William Cavendish. "Background". In *Privatizing Malaysia: Rents, Rhetorics and Realities*, edited by Jomo K.S., pp. 11–42. Colorado: Westview Press, 1995.

Agoncillo, Teodoro, *Revolt of the Masses: The Story of Bonifacio and the Katipunan*. Quezon City: University of the Philippines, 1956.

Agoncillo, Teodoro, and Milagros Guerrero. *History of the Filipino People*. 7th edition. Quezon City: R.P. Garcia Publishing Company, 1987.

Alburo, Florian, Caesarina Rejante, and Charito Arriola. "Development Planning in the Philippines". In *Development Planning in Asia*, edited by Somsak Tambunlertchai and S.P. Gupta, pp. 193–221. Kuala Lumpur: Asia and Pacific Development Centre, 1993.

Ali Abul Hassan. "The Role of the Government in a Privatization Exercise: Pre- and Post-Privatization". Paper delivered at the National Conference on Privatization: The Challenges Ahead, Kuala Lumpur, 7 October 1993.

Ali Abul Hassan bin Sulaiman. "Privatization — A Malaysian Success". In *Malaysia Inc.*, pp. 22–27. Kuala Lumpur: Lim Kok Wing Integrated S.B., 1995.

Amsden, Alice. *Asia's Next Giant: South Korea and Late Industrialization*. New York: Oxford University Press, 1989.

NOTE: Interviews and primary documents are cited fully in the endnotes and are not repeated in the bibliography.

Andaya, Barbara Watson, and Leonard Andaya. *A History of Malaysia.* Hong Kong: Macmillan Press, 1982.

Anderson, Benedict. "Cacique Democracy in the Philippines". In *The Spectre of Comparisons: Nationalism, Southeast Asia and the World,* pp. 192–226. New York and London: Verso, 2000.

Aoki, Masahiko, Hyung-ki Kim, and Masahiro Okuno-Fujiwara, eds. *The Role of Government in East Asian Economic Development.* Oxford: Clarendon Press, 1997.

Aquino, Tomas. "Philippine Telecommunications in the Asia-Pacific Region". *The Columbia Journal of World Business* (Spring 1989): 73–82.

———. "The Philippines". In *Telecommunications in the Pacific Basin,* edited by Eli Noam, Seisuke Komatsuzaki, and Douglas Conn, pp. 177–200. Oxford: Oxford University Press, 1994.

Baldwin, Robert, Colin Scoot, and Christopher Hood, eds. *A Reader on Regulation.* New York: Oxford University Press, 1998.

Balisacan, Arsenio, and Hal Hill, eds. *The Philippine Economy: Development, Policies and Challenges.* New York: Oxford University Press, 2003.

Bardhan, Pranab. "The New Institutional Economics and Development Theory: A Brief Critical Assessment". *World Development* vol. 17, no. 9 (1989): 1389–95.

———. "Corruption and Development: A Review of Issues". *Journal of Economic Literature* 35 (September 1997): 1320–46.

Bates, Robert, and Anne Krueger. *Political and Economic Interactions in Economic Policy Reform.* Oxford: Blackwell Press, 1993.

Bauer, Johannes. "Market Power, Innovation and Efficiency in Telecommunications: Schumpeter Reconsidered". *Journal of Economic Issues,* June 1997.

———. "Competitive Processes in Network Industries: Toward an Evolutionary Perspective". Institute of Public Utilities Working Paper 97-02. http://www.bus.msu.edu/ipu/competit.htm (accessed 5 June 2000).

Beesley, Michael, and Stephen Littlechild. "Privatization: Principles, Problems and Priorities". *Lloyds Bank Review,* no. 149 (July 1983): 1–20.

Bhagwati, Jagdish. "Directly Unproductive, Profit-seeking (DUP) Activities". *Journal of Political Economy* 90, 5 (October 1982): 989–1002.

Bortolotti, Bernardo, Juliet D'Souza, Marcella Fantin, and William L. Megginson. "Privatization and the Sources of Performance Improvements in the Global Telecommunications Industry". *Telecommunications Policy* 26 (2002): 243–68.

Boss, Helen. *Theories of Surplus and Transfer: Parasites and Producers in Economic Thought.* London: Unwin Hyman, 1990.

Breyer, Sam. "Typical Justifications for Regulation". In *A Reader on Regulation,* edited by Robert Baldwin, Colin Scoot, and Christopher Hood. New York: Oxford University Press, 1998.

Broad, Robin. *Unequal Alliance: The World Bank, the International Monetary Fund and the Philippines.* Berkeley: University of California Press, 1988.

Buchanan, James, Robert Tollison, and Gordon Tullock, eds. *Toward a Theory of Rent-Seeking Society.* College Station: Texas A&M University Press, 1980.

Caroll, John. *The Filipino Manufacturing Entrepreneur: Agent and Product of Change.* Ithaca: Cornell University Press, 1965.

Case, William. "Semi-Democracy in Malaysia: Withstanding the Pressure for Regime Change". *Pacific Affairs* 66, no. 2 (Summer 1993): 183–205.

Castells, Manuel. *The Rise of the Network Society, the Information Age: Economy, Society and Culture,* vol. I. Cambridges, Mass.: Blackwell Publishers, 1996.

Chang Ha-Joon. *The Political Economy of Industrial Policy.* London: Macmillan Press, 1994.

Chaundry, Kiren Aziz. "The Myth of the Market and the Common History of Late Developers". *Politics and Society* 21, no. 3 (September 1993): 245–74.

———. "Economic Liberalization and the Lineages of the Rentier State". *Comparative Politics,* October 1994, pp. 1–25.

Cheong, Sally. *Bumiputera Controlled Companies in the KLSE.* 2nd edition. Kuala Lumpur: Corporate Research Services, S.B., 1993.

———. *Bumiputera Entreprenuers in the KLSE.* Vols. 1 and 2. Kuala Lumpur: Corporate Research Services, 1997.

Cook, Paul. *Competition and Its Regulation: Key Issues.* University of Manchester Centre on Regulation and Competition Working Paper Series Paper no. 2 (October 2001).

Cook, Paul, and Colin Kirkpatrick, eds. *Privatization Policy and Performance: nternational Perspectives.* Hertfordshire: Prentice-Hall, 1995.

Coronel, Shiela. *Pork and Other Perks: Corruption and Governance in the Philippines.* Manila: Philippine Center for Investigative Journalism, 1998.

Corpus, O.D. *The Philippines.* New Jersey: Prentice-Hall, 1965.

———. *The Roots of the Filipino Nation.* Quezon City: Aklahi Foundation, 1989.

Cremer, Helmuth, Farid Gasmi, Andrei Grimaud, and Jean-Jacques Laffont. *The Economics of Universal Service: Theory.* Washington: Economic Development Institute of the World Bank, 1998.

Crouch, Harold. "From Alliance to Barisan Nasional". In *Malaysian Politics and the 1978 Elections,* edited by Harold Crouch, Lee Kam Hing, and Michael Ong, pp. 1–11. Kuala Lumpur: Oxford University Press, 1980.

———. *Economic Change, Social Structure and the Political System in Southeast Asia: Philippine Development Compared with the Other ASEAN Countries.* Singapore: Institute of Southeast Asian Studies, 1985.

———. "Authoritarian Trends, the UMNO Split and the Limits to State Power". In *Fragmented Vision: Culture and Politics in Contemporary Malaysia,* edited by Francis Loh Kok Wah and Joel S. Kahn, pp. 21–43. St Leonards, NSW: Allen and Unwin, 1992.

———. "Malaysia: Neither Authoritarian nor Democratic". In *Southeast Asia in the 1990s*, edited by Kevin Hewison, Richard Robison, and Garry Rodan. St Leonards, NSW: Allen and Unwin, 1993.

———. *Government and Society in Malaysia*, St. Leonards, N.S.W.: Allen and Unwin, 1996.

Cullinane, Michael. "Playing the Game: The Rise of Sergio Osmena, 1898–1907". In *Philippine Colonial Democracy*, edited by Ruby Paredes, pp. 70–113. New Haven: Yale University Southeast Asia Studies, 1988.

Cutler, Terry. "The Development of Malaysia's Multimedia Super-Corridor". Paper read at the Third Biennial Australia-Malaysia Conference, Australian National University, Canberra, 25–27 March 2002.

Daroy, Petronilo Bn. "On the Eve of Dictatorship and Revolution". In *Dictatorship and Revolution: Roots of People Power*, edited by Aurora Javate-De Dios, Petronilo Bn. Daroy, and Lorna Kalaw-Tirol, pp. 70–131. Manila: Conspectus Foundation Incorporated, 1988.

Daud bin Isahak. "Meeting the Challenges of Privatization in Malaysia". In *Restructuring and Managing the Telecommunications Sector*, edited by Bjorn Wellenius, Peter Stern, Timothy Nulty, and Richard Stern, pp. 118–21. Washington: World Bank, 1989.

Dauvergne, Peter, ed. *Weak and Strong States in Asia-Pacific*. St Leonards NSW: Allen and Unwin, 1998.

De Dios, Emmanuel. "The Erosion of the Dictatorship". In *Dictatorship and Revolution: Roots of People Power*, edited by Aurora Javate-De Dios, Petronilo Bn. Daroy, and Lorna Kalaw-Tirol, pp. 1–25. Manila: Conspectus Foundation Incorporated, 1988.

———. "A Political Economy of Philippine Policy-Making". In *Economic Policy-Making in the Asia-Pacific Region*, edited by John Langford and K. Lorne Brownsey, pp. 109–48. Canada: Institute for Research on Public Policy, 1990.

———. "Philippine Economic Growth: Can It Last?" http://www.asiasociety.org/publications/philippines/economic.html (accessed 15 January 2000) and in *The Philippines: New Directions in Domestic Policy and Foreign Relations*, edited by David Timberman, pp. 49–84. Singapore: Institute of Southeast Asian Studies, 1998.

De Dios, Emmanuel, and Paul Hutchroft. "Political Economy". In *The Philippine Economy: Development, Policies and Challenges*, edited by Arsenio Balisacan and Hal Hill, pp. 45–75. New York: Oxford University Press, 2003.

De Guzman, Raul, and Mila Reforma. *Government and Politics in the Philippines*. Singapore: Oxford University Press, 1988.

Doherty, John. *A Preliminary Study of Interlocking Directorates: Among Financial, Commercial, Manufacturing and Service Enterprises in the Philippines*. Manila, no publisher, 1979.

———. "Who Controls the Philippine Economy: Some Need Not Try as Hard as Others". In *Cronies and Enemies: The Current Philippine Scene*, edited by Belinda Aquino, pp. 7–36. Honolulu: Philippine Studies Program, Center for Asian and Pacific Studies, University of Hawaii, August 1982.

Doner, Richard. "Approaches to the Politics of Economic Growth in Southeast Asia". *Journal of Asian Studies* 50, no. 4 (November 1991): 818–49.

Doronila, Amando. *The Fall of Estrada: The Inside Story.* Pasig City: Anvil Publishing and *Philippine Daily Inquirer*, 2001.

D'Souza, Juliet, and William Megginson. "The Financial and Operating Performance of Privatized Firms during the 1990s". *Journal of Finance* 54, no. 4 (August 1999): 1397–438.

Duch, Raymond. *Privatizing the Economy: Telecommunications Policy in Comparative Perspective.* Ann Harbor: University of Michigan Press, 1991.

Esfahani, Hadi Salehi. "The Political Economy of the Telecommunications Sector in the Philippines". In *Regulations, Institutions and Commitment*, edited by Brian Levy and Pablo Spiller, pp. 145–201. New York: Cambridge University Press, 1996.

Esman, Milton. *Administration and Development in Malaysia.* Ithaca: Cornell University Press, 1972.

Evans, Peter. "Predatory, Developmental, and Other Apparatuses: A Comparative Political Economy Perspective on the Third World State". *Sociological Forum* 4, no. 4 (December 1989): 561–87.

———. "The State as Problem and Solution: Predation, Embedded Autonomy and Structural Change". In *The Politics of Economic Adjustment*, edited by Stephan Haggard and Robert Kaufman, pp. 139–81. New Jersey: Princeton University Press, 1992.

———. *Embedded Autonomy: States and Industrial Transformations.* New Jersey: Princeton University Press, 1995.

Evans, Peter, and James E. Rausch. "Bureaucracy and Growth: A Cross-national Analysis of the Effects of 'Weberian' State Structures on Economic Growth". *American Sociological Review* 64, no. 5 (October 1999): 748–65.

Evans, Peter, Diethrich Rueschemeyer, and Theda Skocpol, eds. *Bringing the State Back In.* Cambridge: Cambridge University Press, 1985.

Fink, Carsten, Aaditya Mattoo, and Randeep Rathindran. "Liberalizing Basic Telecommunications: The Asian Experience", Development Research Group, World Bank, November 2001. http://econ.worldbank.org/files/2726_wps2718.pdf (accessed 14 May 2003).

Fong Chan Onn. "The Malaysian Telecommunications Services Industry: Development Perspective and Prospects". *Columbia Journal of World Business* Spring 1989, pp. 83–100.

———. "Malaysia". In *Telecommunications in the Pacific Basin,* edited by Eli Noam, Seisuke Komatsuzaki, and Douglas Conn, pp. 132–54. Oxford: Oxford University Press, 1994.

Foster, C.D. *Privatization, Public Ownership and the Regulation of Natural Monopoly.* Oxford: Blackwell, 1992.

Francis, John. *The Politics of Regulation: A Comparative Perspective.* Oxford: Blackwell Press, 1993.

Fujio, Hara. *Formation and Restructuring of Business Groups in Malaysia.* Tokyo: Institute of Developing Economies, 1993.

Funston, John. *Malay Politics in Malaysia: A Study of UMNO and PAS.* Kuala Lumpur: Heinemann Educational Books, 1980.

Funston, John, ed. *Government and Politics in Southeast Asia.* Singapore: Institute for Southeast Asian Studies, 2001.

Galal, Ahmed. *Public Enterprise Reform and the World Bank: Approach, Practice and Challenges.* Washington, Dc: World Bank, 1989.

Galal, Ahmed, and Bharat Nauriyal. *Regulating Telecommunications in Developing Countries: Outcomes, Incentives and Commitment.* Policy Research Working Paper 1520. Washington: World Bank, October 1995.

Galal, Ahmed, and Mary Shirley, eds. *Does Privatization Deliver? Highlights from a World Bank Conference.* Washington: International Bank for Reconstruction and Development, 1994.

Gavino, Jacinto. "A Critical Study of the Regulation of the Telephone Utility: Some Options for Policy Development". Ph.D. dissertation, College of Public Administration, University of the Philippines, 1992.

Gomez, Edmund Terence. *Politics in Business: UMNO's Corporate Investments.* Kuala Lumpur: Forum, 1990.

———. *Money Politics in the Barisan Nasional.* Kuala Lumpur: Forum Publications, 1991.

———. *Political Business: Corporate Involvement of Malaysian Political Parties.* Townsville: James Cook University, 1994

———. *Chinese Business in Malaysia: Accumulation, Accommodation and Ascendance.* Surrey: Curzon Press, 1999.

———. "Political Business in Malaysia: Party Factionalism, Corporate Development and Economic Crisis". In *Political Business in East Asia,* edited by Edmund Terence Gomez, pp. 82–114. London: Routledge, 2002.

Gomez, Edmund Terence, and K.S. Jomo, *Malaysia's Political Economy: Politics, Patronage and Profits.* 2nd edition. Cambridge: Cambridge University Press, 1999.

Government of Malaysia. *Second Malaysia Plan, 1971–1975.* Kuala Lumpur: Government Press, 1971.

———. *Fourth Malaysia Plan, 1981–1985.* Kuala Lumpur: Government Press, 1981.

———. *Mid-term Review of the Fourth Malaysia Plan 1981–1985.* Kuala Lumpur: Government Printers, 1984.

———. *Guidelines for Privatisation.* Kuala Lumpur: Economic Planning Unit, Prime Minister's Department, 1985.

———. *Privatisation Masterplan.* Kuala Lumpur: National Printing Department, 1991.

Gutierrez, Eric. *The Ties That Bind: A Guide to Family, Business and Other Interests in the Ninth House of Representatives.* Pasig, Metro Manila: Philippine Center for Investigative Journalism, 1994.

Gutierrez, Eric, Ildefonso Torrente, and Noli Narca. *All in the Family: A Study of Elites and Power Relations in the Philippines.* Quezon City: Institute for Popular Democracy, 1992.

Haggard, Stephan. "Interests, Institutions and Policy Reform". In *Economic Policy Reform: The Second Stage,* edited by Anne Krueger, pp. 21–61. Chicago: University of Chicago Press, 2000.

Haggard, Stephan, and Robert Kaufman, eds. *The Politics of Economic Adjustment.* Princeton: Princeton University Press, 1992.

Haggard, Stephan, and Mathew McCubbins. *Presidents, Parliaments and Policy.* Cambridge: Cambridge University Press, 2001.

Haggard, Stephan, and Steven Webb, eds. *Voting for Reform: Democracy, Political Liberalization and Economic Adjustment.* Washington, DC: World Bank; and New York: Oxford University Press, 1994.

Hariss John, Janet Hunter, and Colin Lewis, eds. *The New Institutional Economics and Third World Development.* London: Routledge, 1995.

Hariss-White, Barbara, and Gordon White. "Corruption, Liberalization and Democracy". *IDS Bulletin* 27, no. 2 (April 1996): 1–5.

Harper, T.N. *The End of Empire and the Making of Malaya.* Cambridge: Cambridge University Press, 1999.

Hashim, Rita Raj. "Privatisation of Telecommunications in Malaysia". Master in Public Administration (MPA). thesis, University of Malaya, 1986.

Hawes, Gary. *The Philippine State and the Marcos Regime: The Politics of Export.* Ithaca: Cornell University Press, 1987.

Held, David. "Privatization: Policies, Methods and Procedures". In *Privatization: Policies, Methods and Procedures.* Manila: Asian Development Bank, 1985.

Hills, Jill. "A Global Industrial Policy, US Hegemony and GATT: The Liberalization of Telecommunications". *Review of International Political Economy,* Summer 1994, pp. 257–79.

Hirschman, Charles. "Development and Inequality in Malaysia: From Puthucheary to Mehmet". *Pacific Review* 62, no. 1 (1989).

Hollnsteiner, Mary. *The Dynamics of Power in a Philippine Municipality*. Quezon City: Community Development Research Council, University of the Philippines, 1963.

Hooker, Virginia. *Writing a New Society: Social Change through the Novel in Malay*. Sydney: Allen and Unwin, 2000.

Horowitz, Donald. *Ethnic Groups in Conflict*. Berkeley: University of California Press, 1985.

Horwitz, Robert Britt. *The Irony of Regulatory Reform: The Deregulation of American Telecommunications*. New York: Oxford University Press, 1989.

Hua Wu Yin. *Class and Communalism in Malaysia*. London: Zed Books, 1983.

Hughes, Helen, ed. *Achieving Industrialization in East Asia*. Cambridge: Cambridge University Press, 1988.

Hutchcroft, Paul. "Oligarchs and Cronies in the Philippine State: The Politics of Patrimonial Plunder". *World Politics* 43, no. 3 (April 1991): 414–50.

———. "Booty Capitalism: Business-Government Relations in the Philippines". In *Business and Government in Industrializing Asia*, edited by Andrew MacIntyre. Ithaca: Cornell University Press, 1994.

———. *Booty Capitalism: The Politics of Banking in the Philippines*. Ithaca: Cornell University Press, 1998.

Hwang, In-Won. Changing Conflict Configurations and Regime Maintenance in Malaysian Politics: From Consociational Bargaining to Mahathir's Dominance. Ph.D. dissertation, Australian National University, 2001.

Ileto, Reynaldo. *Pasyon and Revolution: Popular Movements in the Philippines, 1840–1910*. Quezon City: Ateneo de Manila University Press, 1979.

International Telecommunications Union (ITU). *Asia-Pacific Telecommunications Indicators*. Geneva: ITU, May 1993.

———. *World Telecommunication Development Report 1996/97*. Geneva: ITU, 1997.

———. *General Trends in Telecommunication Reform 1998: World. Vol. I*. Geneva: ITU, 1998.

———. *Trend in Telecommunication Reform 1998: Asia-Pacific. Vol. V.* Geneva: ITU, 1998.

———. *World Telecommunication Development Report 1999: Mobile Cellular*. Geneva: ITU, 1999.

———. *Trend in Telecommunication Reform: Convergence and Regulation 1999: Executive Summary*. Geneva: ITU, 1999.

———. *Asia-Pacific Telecommunications Indicators 2000*. Geneva: ITU, 2000.

———. *World Telecommunication Indicators*. Geneva: ITU, March 2001.

———. *Asia-Pacific Telecommunications Indicators 2002*. Geneva: ITU, November 2002.

———. *World Telecommunications Development Report 2002: Reinventing Telecoms.* Geneva: ITU, 2002.

———. *The Internet of Things.* Geneva: ITU, November 2005.

———. *Measuring the Information Society 2007: ICT Opportunity Index/World Telecommunication/ICT Indicators.* Geneva: ITU, January 2007.

Islam, Iyanatul. "Political Economy and East Asian Economic Development". *Asian-Pacific Economic Literature* 6, no. 2 (November 1992): 69–101.

Ismail Muhd. Salleh and H. Osman-Rani. *The Growth of the Public Sector in Malaysia.* Kuala Lumpur: Institute for Strategic and International Studies, 1991.

Jabatan Telekomunikasi Malaysia. *Annual Reports,* 1979, 1981, 1982, 1984–86, 1993, 1994–96.

Jawan, Jayum. *The Iban Factor in Sarawak Politics.* Shah Alam: Malindo Printers Sdn. Bhd., 1993.

Jesudason, James. *Ethnicity and the Economy: The State, Chinese Business and Multinationals in Malaysia.* Singapore: Oxford University Press, 1989.

Johnson, Chalmers. *MITI and the Japanese Miracle: The Growth of Industrial Policy 1925–1975.* Stanford: Stanford University Press, 1982.

Jomo, Kwame Sundaram, ed. *The Sun Also Sets: Lessons in Looking East.* 2nd edition. Kuala Lumpur: INSAN, 1985.

Jomo, Kwame Sundaram, ed. *Privatising Malaysia: Rents, Rhetorics and Realities.* Colorado: Westview Press, 1995.

———. "Malaysia's Privatisation Experience". In *Privatisation Policy and Performance: International Perspectives,* edited by Paul Cook and Colin Kirkpatrick. Hertfordshire: Prentice-Hall, 1995.

Jomo, K.S., Christopher Adam, and William Cavendish. "Policy". In *Privatizing Malaysia: Rents, Rhetorics and Realities,* edited by K.S. Jomo, pp. 81–97. Colorado: Westview Press, 1995.

Jomo, K.S., and E.T. Gomez. "Rents and Development in Multiethnic Malaysia". In *The Role of Government in East Asian Economic Development,* edited by Aoki, Masahiko, Hyung-ki Kim, and Masahiro Okuno-Fujiwara, pp. 342–72. Oxford: Clarendon Press, 1997.

———. "The Malaysian Development Dilemma". In *Rents, Rent-Seeking and Economic Development: Theory and Evidence in Asia,* edited by Mustaq Khan and K.S. Jomo, pp. 274–303. Singapore: Cambridge University Press, 2000.

Jomo, K.S., and R.J.G. Wells, eds. *The Fourth Malaysia Plan: Economic Perspectives.* Kuala Lumpur: Malaysian Economic Association, 1983.

Jones, Leroy P. *Performance Evaluation for Public Enterprises.* Washington, DC: World Bank, 1991.

———. "Malaysia". In *Does Privatization Deliver? Highlights from a World Bank Conference*, edited by Ahmed Galal and Mary Shirley. Washington, DC: Bank, 1994.

Kahn, Alfred. *The Economics of Regulation: Principles and Institutions*, vol. I. New York: John Wiley and Sons, 1970.

Kang, David. "Transaction Costs and Crony Capitalism in East Asia". *Comparative Politics* 35, no. 4 (July 2003): 439–58.

Kennedy, Laurel Beatrice. "Privatization and Its Policy Antecedents in Malaysian Telecommunications". Ph.D. dissertation, Ohio University, November 1990.

———. "Telecommunications". In *Privatizing Malaysia: Rents, Rhetoric and Realities*, edited by K.S. Jomo, pp. 219–35. Boulder, Colorado: Westview Press, 1995.

Kerf, Michael, Manuel Schiller, and Clemencia Torres. "Telecom Regulators: Converging Trends?" *Public Policy for the Private Sector*. Note no. 230. World Bank Group Private Sector and Infrastructure Network, May 2001.

Kerkvliet, Benedict. "Towards a More Comprehensive Analysis of Philippine Politics: Beyond Patron-Client, Factional Framework". *Journal of Southeast Asian Studies* 26, no. 2 (September 1995): 401–19.

———. "Land Regimes and State Strengths and Weakenesses in the Philippines and Vietnam". In *Weak and Strong States in Asia-Pacific*, edited by Peter Dauvergne, pp. 158–74. St Leonards, NSW: Allen and Unwin, 1998.

Khan, Mustaq. "A Typology of Corrupt Transactions in Developing Countries". *IDS Bulletin* 27, no. 2 (1996): 12–21.

Khan, Mustaq, and K.S. Jomo, eds. *Rents, Rent-Seeking and Economic Development: Theory and Evidence in Asia*. Singapore: Cambridge University Press, 2000.

Khasnor Johan. *The Emergence of the Modern Malay Administrative Elite*. Singapore: Oxford University Press, 1984.

Khong Kim Hoong. "The Early Political Movements Before Independence". In *Government and Politics of Malaysia*, edited by Zakaria Haji Ahmad, pp. 11–39. Singapore: Oxford Universirty Press, 1987.

Khoo Boo Teik. *Paradoxes of Mahathirism: An Intellectual Biography of Mahathir Mohamad*. Kuala Lumpur: Oxford University Press, 1995.

Khoo Khay Jin. "The Grand Vision: Mahathir and Modernization". In *Fragmented Vision: Culture and Politics in Contemporary Malaysia*, edited by Francis Loh Kok Wah and Joel S. Kahn, pp. 44–76. Sydney: Allen and Unwin, 1992.

Kohli, Atul et al. "The Role of Theory in Comparative Politics: A Symposium". *World Politics* 48, no. 1 (October 1995): 1–49.

Koike, Kenji. "Dismantling Crony Capitalism under the Aquino Government".In *National Development and the Business Sector in the Philippines*, edited by Aiichiro Ishii et al., pp. 206–59. Tokyo: Institute of Developing Economies, 1988.

Krasner, Stephen. "Approaches to the State: Alternative Conceptions and Historical Dynamics". *Comparative Politics*, January pp. 223–46.

Krueger, Anne. "The Political Economy of Rent-Seeking Society". *American Economic Review* 64, no. 3 (June 1974): 291–303.

———. "Government Failures in Development". *Journal of Economic Perspectives* 4, no. 3 (Summer 1990): 9–23.

———. *Political Economy of Policy Reform in Developing Countries.* Cambridge: MIT Press, 1993.

Kunio, Yoshihara. *The Rise of Ersatz Capitalism in Southeast Asia.* Singapore: Oxford University Press, 1988.

Laffont, Jean Jacques, and Jean Tirole. *Competition in Telecommunications.* Cambridge, Mass.: MIT Press, 2000.

Lallana, Emmanuel, Rodolfo Quimbo, and Lorraine C. Salazar. *Business@philippines.com: E-Commerce Policy Issues in the Philippines.* Manila: Carlos P. Romulo Foundation for Peace and Development, 1999.

Lai Yew Wah and Tan Siew Ee. "Towards Effective Planning in Malaysia: Some Strategic Issues". In *Development Planning in Mixed Economies,* edited by Miguel Urrutia and Setsuko Yukawa, pp. 107–55. Tokyo: United Nation University, 1988.

Lande, Carl. *Leaders, Factions and Parties: The Structure of Philippine Politics.* Monograph Series no. 6. Michigan: Southeast Asia Studies Yale University, 1965.

Lee, Cassey. Telecommunications Reform in Malaysia. Working Paper no. 20, Centre on Regulation and Competition, Institute for Development Policy and Management, University of Manchester.

———. "Institutional and Policy Framework for Competition and Regulation in Malaysia". Paper read at Regulation, Competition and Development: Setting an Agenda, CRC International Workshop, University of Manchester, 4–6 September 2002.

Lent, John. "Telematics in Malaysia: Room at the Top for a Selected Few". In *Transnational Communications: Wiring the Third World,* edited by Gerald Sussman and John Lent, pp. 165–99. California: Sage Publications, 1991.

Leong Choon Heng. "Late Industrialization along with Democratic Politics in Malaysia". Ph.D. dissertation, Harvard University, 1991.

Levy, Brian, and Pablo T. Spiller. *Regulations, Institutions and Commitment: Comparative Studies of Telecommunications.* Cambridge: Cambridge University Press, 1996.

Lim Kit Siang. *Samy Vellu and Maika Scandal.* Kuala Lumpur: Democratic Action Party, 1992.

Lipton, Michael. "Market, Redistributive, and Proto-Reform: Can Liberalization Help the Poor?" *Asian Development Review* 13, no. 1 (1995): 1–35.

Loh Kok Wah, Francis, and Joel S. Kahn, eds. *Fragmented Vision: Culture and Politics in Contemporary Malaysia.* Sydney: Allen and Unwin, 1992.

Lowe, Vincent. "Malaysia and Indonesia: Telecommunications Restructuring". In *Telecommunications in the Pacific Basin: An Evolutionary Approach,* edited by

Eli Noam, Seisuke Komatsuzaki, and Douglas Conn, pp. 118–31. New York: Oxford University Press, 1994.

Machado, Kit. "Changing Aspects of Factionalism in Philippine Local Politics". *Asian Survey* 11 (December 1971): 1182–99.

MacIntyre, Andrew, ed. *Business and Government in Industrializing Asia.* Ithaca: Cornell University Press, 1994.

Mackie, Jamie. "Economic Growth in the ASEAN Region: The Political Underpinnings. In *Achieving Industrialization in East Asia,* edited by Helen Hughes. Cambridge: Cambridge University Press, 1988.

Mahathir Mohamad. *The Malay Dilemma.* Singapore: Times Books International, 1981.

———. "'New Government Policies', Memo to Senior Government Officials, 28 June 1983". In *The Sun Also Sets: Lessons in Looking East,* edited by K.S. Jomo. 2nd edition. Kuala Lumpur: INSAN, 1985.

———. "Malaysia: The Way Forward (Vision 2020)". Speech at the Malaysian Business Council, Kuala Lumpur, 28 February 1991.

Magno, Alexander R. "The Philippines in 1995: Completing the Market Transition". In *Southeast Asian Affairs 1996,* pp. 285–99. Singapore: Institute for Southeast Asian Studies, 1996.

———. "Between Populism and Reform: Facing the Test of May 1998". In *Southeast Asian Affairs 1998,* pp. 199–212. Singapore: Institute for Southeast Asian Studies, 1998.

Manapat, Ricardo. *Some Are Smarter Than Others: The History of Marcos' Crony Capitalism.* New York: Aletheia Publications, 1991.

———. *Wrong Number: The PLDT Monopoly.* The Animal Farm Series. No place and date of publication.

Mansell, Robin. *The New Telecommunications: A Political Economy of Network Evolution.* London: Sage Publications, 1993.

March, James, and Johan Olsen. *Rediscovering Institutions: The Organizational Basis of Politics.* New York: The Free Press, 1989.

Marcos, Ferdinand. *Revolution from the Center: How the Philippines Is Using Martial Law to Build a New Society.* Hong Kong: Roya Books, 1978.

Marx, Karl. *The Eighteenth Brumaire of Louis Bonaparte.* New York: International Publishers, 1935.

Mauzy, Diane. "Malaysia in 1986: The Ups and Downs of Stock Market Politics". *Asian Survey* 27, no. 2 (February 1987): 231–41.

———. "Malaysia: The Shaping of Economic Policy in a Multi-ethnic Environment: The Malaysian Experience". In *Economic Policy-making in the Asia-Pacific Region,* edited by John Langford and K. Lorne Brownsey, pp. 273–98. Canada: Institute for Research on Public Policy, 1990.

———. "Malay Political Hegemony and 'Coercive Consociationalism'". In *The Politics of Ethnic Conflict Regulation*, edited by John McGarry and Brendan O'Leary. London: Routledge, 1993.

Maxfield, Sylvia, and Ben Ross Schneider, eds. *Business and the State in Developing Countries.* Ithaca: Cornell University Press, 1997.

May, Glenn. "Civic Ritual and Political Reality: Municipal Elections in the Late Nineteenth Century". In *Philippine Colonial Democracy*, edited by Ruby Paredes, pp. 13–40. New Haven: Yale University Southeast Asia Studies, 1988.

May, Ron. "State, Society and Governance: Reflections on a Philippines–Papua New Guinea Comparison". In *Weak and Strong States in Asia-Pacific*, edited by Peter Dauvergne, pp. 60–76. St Leonards, NSW: Allen and Unwin, 1998.

McChesney, Fred. "Rent Extraction and Rent Creation in the Economic Theory of Regulation". *Journal of Legal Studies* 16 (January 1987): 101–18.

McCoy, Alfred, ed. *An Anarchy of Families: State and Family in the Philippines.* Wisconsin: Center for Southeast Asian Studies, University of Wisconsin, 1993.

McGarry, John, and Brendan O'Leary, eds. *The Politics of Ethnic Conflict Regulation.* London: Routledge, 1993.

McVey, Ruth, ed. *Southeast Asian Capitalists.* Ithaca: Cornell University, 1992.

Means, Gordon. *Malaysian Politics.* London: University of London Press, 1970.

———. *Malaysian Politics: The Second Generation.* Singapore: Oxford University Press, 1991.

———. "Soft Authoritarianism in Malaysia and Singapore". *Journal of Democracy* 7, no. 4 (1996): 103–17.

Megginson, William, and Jeffrey Netter. "From State to Market: A Survey of Empirical Studies on Privatization". *Journal of Economic Literature* 39 (June 2001): 321–89.

Mehmet, Ozay. *Development in Malaysia: Poverty, Wealth and Trusteeship.* Kent: Croom Helm, 1986.

Melody, William. "Telecom Reform: Progress and Prospect". *Telecommunications Policy* 23 (1999): 7–34.

Migdal, Joel. *Strong Societies and Weak States: State-Society Relations and State Capabilities in the Third World.* Princeton, New Jersey: Princeton University Press, 1988.

Migdal, Joel, Atul Kohli and Vivienne Shue, eds. *State Power and Social Forces: Domination and Transformation in the Third World.* Cambridge: Cambridge University Press, 1994.

Milne, R.S. *Government and Politics in Malaysia.* Boston: Houghton Mifflin Company, 1967.

Milne, R.S., and Diane K. Mauzy. *Politics and Government in Malaysia.* Singapore: Federal Publications, 1978.

———. *Malaysian Politics under Mahathir*. London: Routledge, 1999.

Miralao, Leoncio Jr. "Philippine National Development Planning". In *National Development and the Business Sector in the Philippines*, edited by Aiichiro Ishii, et al., pp. 78–127. Tokyo: Institute of Developing Economies, 1988.

Mitchell, Timothy. "The Limits of the State: Beyond Statist Approaches and Their Critics". *American Political Science Review* 85, no. 1 (March 1991): 77–96.

Mohamed Said Mohamed Ali. "Privatization — Lessons to Be Learnt from Local Experience". Paper presented at the National Conference on Privatization — The Challenges Ahead, 7–8 October 1993, Kuala Lumpur.

Monge, ricardo and Francisco Galema. *The Philippines Reform on Telecommunications: The Impact of USAID's Support*. Arlington, Virginai: Carana Corporation, 1999.

Montes, Manuel. "The Business Sector and Development Policy". In *National Development and the Business Sector in the Philippines*, edited by Aichiro Ishii, et al., pp. 23–77. Tokyo: Institute of Developing Economies, 1988.

———. *Financing Development: The Political Economy of Fiscal Policy in the Philippines*. Manila: Philippine Institute for Development Studies, 1991.

Munro-Kua, Anne. *Authoritarian Populism in Malaysia*. New York: St Martin's Press, 1996.

Murphy, Kevin, Andrei Shleifer, and Robert Vishny. "Why Is Rent-seeking So Costly to Growth?" *American Economic Review* 83, no. 2 (May 1993): 409–14.

Nabli, Mustapha, and Jeffrey Nugent. "The New Institutional Economics and Its Applicability to Development". *World Development* 17, no. 9 (1989): 1333–47.

Naidu, G., and Cassey Lee. "The Transition to Privatization: Malaysia". In *Infrastructure Strategies in East Asia: The Untold Story*, edited by Ashoka Mody, pp. 27–49. Washington, DC: World Bank, 1997.

Nelson, Joan, ed. *Economic Crisis and Policy Choice: The Politics of Adjustment in the Third World*. Princeton, New Jersey: Princeton University Press, 1990.

Nelson, Joan et al. *Fragile Coalitions: The Politics of Economic Adjustment*. Washington, DC: Overseas Development Council, 1989.

Nemenzo, Francisco. "From Autocracy to Elite Democracy". In *Dictatorship and Revolution: Roots of People Power*, edited by Aurora Javate-De Dios, Petronilo Bn. Daroy, and Lorna Kalaw-Tirol, pp. 221–68. Manila: Conspectus Foundation, 1988.

Newbery, David. *Privatization, Restructuring and Regulation of Network Utilities*. Cambridge, Mass.: MIT Press, 1999.

Ng Chee Yuen and Toh Kin Woon. "Privatization in the Asian-Pacific Region". *Asian-Pacific Economic Literature* 6, no. 2 (November 1992): 42–68.

Noll, Roger. Telecommunications Reform in Developing Countries. AEI-Brookings Joint Center for Regulatory Studies Working Paper 99–10, November 1999.

———. "Telecommunications Reform in Developing Countries". In *Economic Policy Reform: The Second Stage*, edited by Anne Krueger, pp. 183–242. Chicago: University of Chicago Press, 2000.

North, Douglass. "Institutions and Economic Growth: An Historical Introduction". *World Development* 17, no. 9 (1989): 1319–32.

———. *Institutions, Institutional Change and Economic Performance.* Cambridge: Cambridge University Press, 1990.

Nowak, Thomas, and Kay Snyder. "Clientelist Politics in the Philippines: Integration or Stability". *American Political Science Review* 68 (1974): 1147–70.

Olson, Mancur. *The Logic of Collective Action.* Cambridge: Cambridge University Press, 1965.

Ong, Michael. "Government and Opposition in Parliament: The Rules of the Game". In *Government and Politics of Malaysia*, edited by Zakaria Haji Ahmad, pp. 40–55. Singapore: Oxford University Press, 1987.

Patalinghug, Epictetus. *Philippine Privatization: Experience, Issues and Lessons.* Quezon City: University of the Philippines Center for Integrative and Developmental Studies, 1996.

Paredes, Ruby, ed. *Philippine Colonial Democracy.* New Haven: Yale University Southeast Asia Studies, 1988.

Parreñas, Julius Caesar. "Transition and Continuity in the Philippines, 1992". In *Southeast Asian Affairs 1992,* pp. 269–81. Singapore: Institute for Southeast Asian Studies, 1992.

———. "The Philippines: Reaping the Fruits of Reform and Stability". In *Southeast Asian Affairs 1997,* pp. 242–45. Singapore: Institute for Southeast Asian Studies, 1997.

Peltzman, Sam. "The Economic Theory of Regulation after a Decade of Deregulation". *Brooking Papers on Microeconomics*, 1989.

Petrazzini, Ben. *The Political Economy of Telecommunications Reform in Developing Countries: Privatization and Liberalization in Comparative Perspective.* Westport, Connecticut: Praeger, 1995.

Pinches, Michael. "The Philippines' New Rich: Capitalist Transformation Amidst Economic Gloom". In *The New Rich in Asia: Mobile phones, McDonalds and Middle-class Revolution*, edited by Richard Robison and David S.G. Goodman, pp. 105–36. London and New York: Routledge, 1996.

Polanyi, Karl. *The Great Transformation: The Political and Economic Origins of Our Time.* Boston: Beacon Press, 1957.

Posner, Richard A. "The Social Costs of Monopoly and Regulation". *Journal of Political Economy* 83, no. 4 (August 1975): 807–28.

Puthucheary, Mavis. *The Politics of Administration: The Malaysian Experience.* Kuala Lumpur: Oxford University Press, 1978.

———. "Privatisation — Proceed, But with Caution". In *The Sun Also Sets: Lessons in Looking East*,edited by K.S. Jomo, 360–69. 2nd edition. Kuala Lumpur: INSAN, 1985.

———. "The Shaping of Economic Policy in a Multi-Ethnic Environment: The Malaysian Experience". In *Economic Policy-making in the Asia-Pacific Region*, edited by John Langford and K. Lorne Brownsey, pp. 273–98. Canada: Institute for Research on Public Policy, 1990.

Putzel, James. *A Captive Land: The Politics of Agrarian Reform in the Philippines.* London: Catholic Institute for International Relations, 1992.

Rabushka, Alvin, and Kenneth Shepsle. *Politics in Plural Societies: A Theory of Democratic Instability.* Ohio: Charles E. Merrill Publishing Company, 1972.

Radin Soenarno Al Haj and Zainal Aznam Yusof. "The Experience of Malaysia". In *Privatization: Policies, Methods and Procedures.* Manila: Asian Development Bank, 1985.

Ramamurti, Ravi, "Why Haven't Developing Countries Privatized Deeper and Faster?" *World Development* 27, no. 1 (January 1999): 137–55.

Ramamurti, Ravi and Raymond Vernon. *Privatization and Control of State-owned Enterprises.* Washington, DC: Economic Development Institute of the World Bank, 1991.

Ramos, Fidel. *Developing as a Democracy: Reform and Recovery in the Philippines, 1992–1998.* Hong Kong: Macmillan Publishers, 1998.

Ranis, Gustav. "The Role of Institutions in Transition Growth: The East Asian Newly Industrializing Countries". *World Development* 17, no. 9 (1989): 1443–53.

Ranis, Gustav, and Syed Akhtar Mahmood. *The Political Economy of Development Policy Change.* Cambridge: Blackwell Press, 1992.

Ratnam, K.J. *Communalism and the Political Process in Malaya.* Kuala Lumpur: University of Malaya Press, 1965.

Ravenhill, John. "State and Market". *Asian Studies Review* 16, no. 3 (April 1993): 111–21.

Rivera, Temario. *Landlords and Capitalists: Class, Family and State in Philippine Manufacturing.* Quezon City: University of the Philippines Center for Integrative and Developmental Studies and the University of the Philippines Press, 1994.

Robison, Richard, Mark Beeson, Kanishka Jayasuriya and Hyuk-Rae Kim. *Politics and Markets in the Wake of the Asian Crisis.* London: Routledge, 2000.

Rodrik, Dani. "Understanding Economic Policy Reform". *Journal of Economic Literature* 34, no. 1 (March 1996): 9–41.

Ros, Agustin. "Does Ownership or Competition Matter? The Effects of Telecommunications Reform on Network Expansion and Efficiency". *Journal of Regulatory Economics* 15 (1999): 65–92.

Sadka, Emily. "Malaysia: The Political Background". In *The Political Economy of Independent Malaya: A Case-study in Development,* edited by T.H. Silcock and E.K. Fisk, pp. 28–58. Berkeley: University of California Press, 1963.

Salazar, Lorraine Carlos. "Privatisation, Patronage and Enterprise Development: Liberalising Telecommunications in Malaysia". In *The State of Malaysia: Ethnicity, Equity and Reform,* edited by Edmund Terence Gomez. London: RoutledgeCurzon, 2004.

———. "First Come First Served: Privatization under Mahathir". In *Reflections: The Mahathir Years,* edited by Bridget Welsh. Washington D.C.: School of Advanced and International Studies, Johns Hopkins University, 2004.

Samsudin Hitam. "Development Planning in Malaysia". In *Development Planning in Asia,* edited by Somsak Tambunlertchai and S.P. Gupta, pp. 167–92. Kuala Lumpur: Asia and Pacific Development Centre, 1993.

Sandholtz, Wayne. "Institutions and Collective Action: The New Telecommunications in Western Europe". *World Politics* 45, no. 2 (1993).

Sapolsky, Harvey M., Rhoda J. Crane, W. Russell Neuman, and Eli Noam. *The Telecommunications Revolution: Past, Present and Future.* New York: Routledge, 1992.

Saunders, Robert, Jeremy Warford, and Bjorn Wellenius. *Telecommunications and Economic Development.* Baltimore: Johns Hopkins University Press, 1983.

———. *Telecommunications and Economic Development,* 2nd edition. Baltimore: Johns Hopkins University Press, 1994.

Schamis, Hector. "Distributional Coalitions and the Politics of Economic Reform in Latin America". *World Politics* 51 (January 1999): 236–68.

Schneider, Ben Ross. *Politics within the State: Elite Bureaucrats and Industrial Policy in Authoritarian Brazil.* Pittsburgh, Pennsylvania: University of Pittsburgh Press, 1991.

Scott, James C. "Patron-Client Politics and Political Change in Southeast Asia". *American Political Science Review* 66, no. 1 (March 1972): 91–113.

Searle, Peter. *Politics in Sarawak 1970–1976: Iban Perspective.* Singapore: Oxford University Press, 1983.

———. *The Riddle of Malaysian Capitalism: Rent-seekers or Real Capitalists?* St Leonards, NSW: Allen and Unwin, 1999.

Serafica, Ramonette. "Tests of Efficient Industry Structure in Telephone Service Provision: Revealing the Past and Potential Direction of Philippine Telecommunications". Ph.D. dissertation, University of Hawaii, December 1996.

Shapiro, Helen, and Lance Taylor. "The State and Industrial Policy". *World Development* 18, No. 6 (1990): 861–78.

Shirley, Mary. "Bureaucrats in Business: The Roles of Privatization versus Corporatization in SOE Reform". *World Development* 27, no. 1 (1999): 115–36.

Shirley, Mary, et al. *Bureuacrats in Business: The Economics and Politics of Government Ownership*. Washington, DC: World Bank, 1995.

Shleifer, Andrei, and Robert W. Vishny. "Corruption". *Quarterly Journal of Economics* 108, no. 3 (August 1993): 599–617.

Shy, Oz. *The Economics of Network Industries*. Cambridge: Cambridge University Press, 2001.

Sidel, John. *Capital, Coercion and Crime: Bossism in Philippines*. California: Stanford University Press, 1999.

Simbulan, Dante. *A Study of the Socio-economic Elite in the Philippine Politics and Government, 1946–1963*. Ph.D. dissertation, Australian National University, 1965.

Smith, Peter, and Gregory Staple. *Telecommunications Sector Reform in Asia: Towards a New Paradigm*. Washington, DC: World Bank, 1994.

Smith, Peter, and Bjorn Wellenius. "Mitigating Regulatory Risk in Telecommunications". *Public Policy for the Private Sector*. Note no. 189, July 1999.

Smith, Peter, and Bjorn Wellenius. "Strategies for Successful Telecommunications Reform in Weak Governannce Environments". Draft paper, 31 March 1999. http://www1.worldbank.org/wbiep/trade/papers_2000/BP2telcm.pdf (accessed 19 October 2000).

Snyder, Richard. "After Neoliberalism: The Politics of Reregulation in Mexico". *World Politics* 51 (January 1999): 173–204.

Spiller, Pablo, and Carlo Cardilli. "The Frontier of Telecommunications Deregulation: Small Countries Leading the Pack". *Journal of Economic Perspectives* 11, no. 4 (Autumn 1997): 127–38.

Stauffer, Robert. "Philippine Authoritarianism: Framework for Peripheral Development". *Pacific Affairs* (June 1977): 365–86.

Stigler, George. "The Theory of Economic Regulation". *Bell Journal of Economics and Management Science* 2 (Spring 1971): 3–21.

Stiglitz, Joseph. "The State and Development: Some New Thinking", International Roundtable on the Capable State, Berlin, Germany, 8 October 1997. http://www.worldbank.org/html/extdr/extme/jssp100897.htm.

———. "Redefining the Role of the State: What Should It Do? How Should It Do It? And How Should These Decisions Be Made?" Speech presented at the 10th Anniversary of MITI Research Institute, Tokyo, Japan, 17 March 1998.

———. "Reflections on the Theory and Practice of Reform". In *Economic Policy Reform: The Second Stage*, edited by Anne Krueger, pp. 551–84. Chicago: University of Chicago Press, 2000.

———. "More Instruments and Broader Goals: Moving toward the Post-Washington Consensus", The WIDER Annual Lecture, Helsinki, January 1998. In *Joseph Stiglitz and the World Bank: The Rebel Within*, edited by Ha-Joon Chang. London: Anthem Press, 2001.

Sussman, Gerald. "Telecommunications Transfers: Transnational Corporations, the Philippines and Structures of Domination". Third World Studies Center, University of the Philippines Dependency Series no. 35, June 1981.

———. "Banking on Telecommunications: The World Bank in the Philippines". *Journal of Communications* (Spring 1987).

———. "Telecommunications for Transnational Integration: The World Bank in the Philippines". In *Transnational Communications: Wiring the Third World*, edited by Gerald Sussman and John Lents, pp. 42–65. California: Sage Publications, 1991.

———. "The Transnationalization of Philippine Telecommunications: Postcolonial Continuities". In *Transnational Communications: Wiring the Third World*, edited by Gerald Sussman and John Lent, pp. 125–49. California: Sage Publications, 1991.

Syed Hussein Mohamed. "The Asian Telecommunications Market — What are the Options?" Paper presented at the Mobile Communications 1992 Conference, 25–26 August 1992, Kuala Lumpur.

———. "Corporatization and Partial Privatization of Telecommunications in Malaysia". In *Implementing Reforms in the Telecommunications Sector*, edited by Bjorn Wellenius and Peter Stern. Washington, DC: World Bank, 1994.

———. "New Regulatory Framework for Communications and Multimedia Sector: Subsidiary Legislation". Communications and Multimedia Commission notes. No date.

Tan, Samuel K. *The Filipino Muslim Armed Struggle, 1910–1972*. Manila: Filipinas Foundation, Inc., 1977.

———. *A History of the Philippines*. Quezon City: Department of History, University of the Philippines, 1987.

Tarjanne, Pekka. "Preparing for the Next Revolution in Telecommunications: Implementing the WTO Agreement." *Telecommunications Policy* 23 (1999): 51–63.

Task Force on Human Settlements. *Infrastructure/Utilities Development in the Philippines Technical Report Part IV: Telecommunication*. Quezon City: Development Academy of the Philippines, March 1975.

Thompson, Mark. "Off the Endangered List: Philippine Democratization in Comparative Perspective". *Comparative Politics* 28, no. 2 (January 1996): 179–205.

Tiglao, Rigoberto. "The Consolidation of the Dictatorship". In *Dictatorship and Revolution: Roots of People Power*, edited by Aurora Javate-De Dios, Petronilo Bn. Daroy, and Lorna Kalaw-Tirol, pp. 26–69. Manila: Conspectus Foundation Incorporated, 1988.

Timberman, David. *A Changeless Land: Continuity and Change in Philippine Politics*. Singapore: Institute of Southeast Asian Studies, 1991.

Toh Kin Woon. "Privatization in Malaysia: Restructuring or Efficiency?" *ASEAN Economic Bulletin* 5, no. 3 (March 1989): 242–58.

Tullock, Gordon. "The Welfare Cost of Tariffs, Monopolies and Theft". *Western Economic Journal* 5 (1967): 224–32.

Vasil, R.K. *Politics in a Plural Society*. Kuala Lumpur: Oxford University Press, 1971.

Vickers, John, and George Yarrow. "Economic Perspectives on Privatization". *Journal of Economic Perspectives* 5, no. 2 (Spring 1991): 111–32.

Vogel, Steven. *Freer Markets, More Rules: Regulatory Reform in Advanced Industrial Countries*. Ithaca: Cornell University Press, 1996.

Von Vorys, Karl. *Democracy Without Consensus: Communalism and Political Stability in Malaysia*. New Jersey: Princeton University Press, 1975.

Wade, Robert. *Governing the Market: Economic Theory and the Role of Government in East Asian Industrialization*. New Jersey: Princeton University Press, 1990.

———. "East Asia's Economic Success: Conflicting Perspectives, Partial Insights, Shaky Evidence". *World Politics* 44 (January 1992): 270–320.

Wallsten, Scott. "Telecommunications Privatization in Developing Countries: The Real Effects of Exclusivity Periods". Draft paper, 12 May 2000.

———. "Does Sequencing Matter? Regulation and Privatization in Telecommunications Reforms". *World Bank Policy Research Working Paper*, February 2002.

Waterbury, John. "The Heart of the Matter: Public Enterprises and the Adjustment Process". In *The Politics of Economic Adjustment*, edited by Stephan Haggard and Robert Kaufman, pp. 182–220. New Jersey: Princeton University Press, 1992.

Wellenius, Bjorn. "Telecommunications Reform — How to Succeed". *Public Policy for the Private Sector*. Note no. 130. The World Bank Group, October 1997.

Wellenius, Bjorn, and Gregory Staple. "Beyond Privatization: The Second Wave of Telecommunications Reforms in Mexico". *World Bank Discussion Paper No. 341*. Washington, DC: World Bank, 1996.

Williamson, John, ed. *The Political Economy of Policy Reform*. Washington, DC: Institute for International Economics, 1994.

———. "On Markets and Regulations". Paper delivered at the University of California at Sta. Cruz Conference, 20–21 November 1998.

———. "What Should the World Bank Think About the Washington Consensus?" *World Bank Observer* 15, no. 2 (August 2000): 251–64.

Woo, Jung-en. *Race to the Swift: State and Finance in Korean Industrialization*. New York: Columbia University Press, 1991.

Woo-Cumings, Meredith, ed. *The Developmental State*. Ithaca: Cornell University Press, 1999.

Wurfel, David. *Filipino Politics: Development and Decay*. Ithaca: Cornell University Press, 1988.

Yarrow, George. "A Theory of Privatization". *World Development* 27, no. 1 (1999): 157–68.

Zainuddin Hj. Abdul Rahman. *Malaysia's Privatisation Policy: The Rationale, Policy and Process of Privatisation in Malaysia.* Kuala Lumpur: Economic Planning Unit of the Prime Minister's Department, 1997.

Zakaria Haji Ahmad, ed. *Government and Politics of Malaysia.* Singapore: Oxford University Press, 1987.

———. "Malaysia: Quasi-Democracy in a Divided Society". In *Democracy in Developing Countries,* Vol. 3, edited by Larry Diamond, Juan Linz, and Seymour Lipset. Boulder: Lynne Rienner, 1989.

Zhou Huizhong. "Rent-Seeking and Market Competition". *Public Choice* 82, nos. 3–4 (1995): 225–41.

Newspapers and Magazines

Aliran Monthly
Asian Wall Street Journal
Asiaweek
Business Times (Malaysia)
Business World (Philippines)
Computimes (Malaysia)
Daily Globe (Philippines)
Far Eastern Economic Review
Forbes Magazine
ID (Malaysia)
Malaya (Philippines)
Malaysiakini
Malaysian Business
Manila Bulletin
Manila Chronicle
Manila Standard
Manila Times
Newsday (Philippines)
New Straits Times
Philippine Daily Inquirer
Philippine Graphics
Philippine Star
Singapore Business Times
Singapore Straits Times
The Edge
The Globe Spectrum
The Star
The Sun
Time Magazine

Index

About the Author

Born in Rosales and raised in Villasis, Pangasinan in Northern Philippines, **Lorraine Carlos Salazar** is a Visiting Research Fellow at the Institute of Southeast Asian Studies, Singapore and an Assistant Professor of Political Science at the University of the Philippines, Diliman, Quezon City. She earned her Ph.D. from the Australian National University in 2004 and a B.A.-M.A. Political Science (Honours) degree from the University of the Philippines in 1996. Her research and teaching interest are on comparative political economy issues in Southeast Asia, focusing on the politics of market reform, political dynamics, and contemporary developments in the region. She has also done work on telecommunications, e-commerce, and information and communications technology policies and issues.

www.ingramcontent.com/pod-product-compliance
Lightning Source LLC
Chambersburg PA
CBHW021544260326
41914CB00001B/165
9789812303820